I0605695

CIRCLING ROUND EXPLICITNESS

Also by Raymond Tallis and published by Agenda

Freedom: An Impossible Reality

Logos: The Mystery of How We Make Sense of the World

Of Time and Lamentation: Reflections on Transience

Seeing Ourselves: Reclaiming Humanity from God and Science

Circling Round Explicitness

THE HEART OF THE MYSTERY OF HUMAN BEING

Raymond Tallis

Dedicated to Bob Doede and Jens Zimmermann as a token of my gratitude for unforgettable visits to Trinity Western University and Regent College and their wonderful friendship, kindness and intellectual stimulation.

First published in 2025 by Agenda Publishing

Agenda Publishing Limited
PO Box 185
Newcastle upon Tyne
NE20 2DH
www.agendapub.com

ISBN 978-1-78821-790-3

British Library Cataloguing-in-Publication Data
A catalogue record for this book is available from the British Library

Typeset by JS Typesetting Ltd, Porthcawl, Mid Glamorgan
Printed and bound in the UK by CPI Group (UK) Ltd, Croydon, CR0 4YY

EU GPSR authorised representative:
Logos Europe, 9 rue Nicolas Poussin, 17000 La Rochelle, France
contact@logoseurope.eu

Contents

Part IV Circling round what-is

"Uniquely within us nature opens her eyes and sees that she exists."
– Friedrich Schelling

Preface

For nearly 60 years, I have been haunted by a thought, an intuition, lying just beyond the edge of anything I am able to think. Notwithstanding the numerous published and yet more numerous unpublished pages where I have circled that thought from different distances, it has eluded me. It relates to something for which I have settled on the unsatisfactory name of "explicitness". The present volume is my most sustained effort to make landfall.

Explicitness is one of the three fundamental mysteries in which our lives are wrapped, though most of the time they are hidden under what we take for granted. The first is that there is Something rather than Nothing. The second is that that Something has an order which, according to the standard story, ultimately gives rise to and sustains life. The third is that that Something (which I shall call "what-is") is made explicit, courtesy of entities that also make themselves explicit. For conscious subjects, what-is presents itself as *that*-it-is or (more ponderously) "that-it-is-the-case". The most arresting aspect of this mystery is the "I" itself, the blush of "who" awoken amidst the deserts of "what".

Concerning the first mystery there is nothing to be said for those who, like myself, lack religious belief and who do not share the faith that the authority of science extends to metaphysics. The "creative accounting" of physicists who imagine they can explain the origin of Something out of Nothing by (for example) appealing to instability in the so-called quantum vacuum expressed in a bubbling brew of virtual particles popping in and out of existence, allowing Nothing to wobble into Something[1] is no more persuasive than the Creation stories offered by sacred texts.

As for the second mystery, the anthropic argument that we humans would not exist to puzzle over the order of the universe if it were entirely chaotic rather than ordered, seems only to highlight rather than address the problem. It does not deal with the fact that the kind of order exemplified by living organisms seems highly improbable, given that there are many more ways of being disordered than ordered – as we are reminded by the relentless rise of entropy in an untended universe. And any kind of complexity, above that of an individual hydrogen atom requires fine-tuning of the relevant constants to unimaginable orders of precision. According to the physicist Luke Barnes the chances of the universe being "biophilic" are 1 in 10^{135}.[2]

The present book focuses on the third mystery: that of explicitness; that what-is has been revealed (though how completely no-one can say) to beings like you and me who have also been revealed (again incompletely) to themselves. In what follows I do not presume to lessen this mystery but rather to highlight it, even to get closer to it. The mystery of explicitness is one from which many traditional lines of ontological, metaphysical and epistemological inquiry take their rise, though this lineage has been insufficiently recognized. The assimilation of that-it-is to what-is lies behind many wrong directions taken in philosophy, most importantly in the embrace of scientism.

The challenge I have faced in writing this book (and, less directly, in writing its predecessors) is that of making a journey to a place where I have already arrived. While this is true of much philosophy, it is more clearly evident in respect of explicitness. The attempt to grasp explicitness, to make it explicit, is analogous to the endeavour to land on oneself, or in a place already occupied by oneself.

Thought cannot get hold of itself, even less stand for the sum of articulate awareness, of all thoughts, any more than a hand can grasp the totality of all grips. We are especially successful at evading our own grasp. The psychiatrist R. D. Laing captured one aspect of this elusiveness when he said that "The truth I am trying to grasp is the grasp that is trying to grasp it".[3] Any endeavour to make explicitness explicit is an attempt to reach out to that which comprises one's act of reaching.

There is another barrier to making explicitness explicit and indeed to bringing any fundamental philosophical inquiry to a point of genuine arrival. It is the dynamism of consciousness. The ease with which we talk to ourselves propels us past anything we are thinking. While sentences may seem to be still when they are sitting on the page of a written work, they are elusive rivulets of fading meaning when we are thinking them. Even when we are least distracted, there is a cognitive momentum propelling us past any place on which we are trying to alight. We slither away like skaters needing to keep moving to remain upright and this prevents us from arriving and staying arrived. It is easier to *host* thoughts – in the sense of inwardly reciting them as we gallop forward – than to immerse ourselves in them or to be had by them. We fly past the wakefulness buried in the most commonplace thought. The challenge is to be (as Yeats characterized Caesar at a crucial moment of decision-making) "Like a long-legged fly upon the stream / His mind moves upon silence".[4] There have been moments when the idea of explicitness is, suddenly and unexpectedly, revelatory. But this does not last. Consequently, this book – its frequent engagement with the standard menu of philosophical topics, its necessary sideways glances at the thoughts of others, its occasional descent into "Gotcha!" polemic, and its apparatus of footnotes – sometimes seems a doomed attempt to return to, or reconstruct, flashes of lightning out of the long rumbles of thunder they have prompted.

Among the grounds for apologizing for what follows, perhaps the most obvious is the name of my philosophical protagonist: "explicitness". It sounds clumsy, even dowdy, lacking the glamour of a visible classical ancestry; at any rate, it does not wear its Latin origin – from "*explicare*", past participle of "to unfold" – on its sleeve. And it has not enjoyed the distinction of being treated as an equal in the company of terms such as "Being", "Matter", "Mind", "Necessity" and other members of the aristocracy of philosophical concepts. Nor does it have the kudos of a technical term or carry the suggestion of novelty promised by a neologism.

Its marginalization in philosophical discourse is a consequence of its very nature which results in its being overlooked – just as looking fails to see itself – though a truly reflective critical philosophy should not take it for granted. What follows is, among other things, affirmative action for something that is no less real for lacking the obtrusiveness, ubiquity and implacability of "matter", the glamour and ethereality of "mind", for being neither a stuff nor a force.

Enough already. I shall proceed in the faith that a concept that eludes satisfactory definition will define the territory that belongs to it by being put to work. My approach to explicitness shall, for much of the time, be one of indirection, by engaging with some of the traditional preoccupations of philosophy whose current treatment often marginalizes explicitness. I hope this will cast an unfamiliar light on familiar philosophical problems. While I have explored these topics in previous books, my revisiting them is not just a repetition or summary of earlier work. If the outcome of the inquiry is less solution or explanation than a description or even celebration, this is because explicitness is a mystery by which we are engulfed, in which we are dissolved, rather than a problem or a cluster of problems "out there".[5]

Before I bring this Preface to a close, I want to distinguish the thought (or pre-thought) that I hope will burn at the heart of this book from a sentiment expressed by Martin Heidegger: "Man alone of all beings, when addressed by the voice of Being, experiences the marvel of all marvels: that what-is is".[6] If this is interpreted as astonishment that there is something rather than nothing – entirely justified though it is – it is not the same as the equally appropriate astonishment and delight at the transition from what-is to that-it-is. I am prompted to highlight this distinction as a result of the experience of giving a daylong seminar on this topic when it was thought that my concern was the Heideggerian one: astonishment that things exist; that there is something rather than nothing; or that things are the way they are. All of these are appropriate grounds for reflection – and there are certain Heideggerian overtones to my inquiry – but they are not the theme of this volume.

Ultimately, what follows is another attempt to reciprocate the gift of life with the gift of astonished attention. Because, in circling round explicitness, we are circling round "Man, the Explicit Animal",[7] around ourselves.

I am indebted to Professor Bob Doede and Professor Jens Zimmermann for their careful reading of the manuscript of this volume and the very helpful suggestions they made. And I am enormously grateful to Steven Gerrard for his continuing generous support for my writing and the care he has taken in seeing this volume through the press. Truly the publisher from heaven.

Overture: making explicitness explicit

(1)

The purpose of the present volume can be summarized in four words: to make explicitness explicit. A five word summary of its more realistic ambition would be: to make explicitness *more* explicit. It is motivated by my conviction that we do not sufficiently appreciate our extraordinary capacity to make nature, or parts of it, explicit and, what is more, to make ourselves, and our relationships with other entities, which we encounter as "our surroundings", explicit. We consequently fail to see a fundamental aspect of the kind of beings we are and to engage in the right way with many of the preoccupations of philosophy.

In what follows, I shall most often characterize explicitness as a *transition*: from *what-is* to *that-it-is* or from *what-is* to *what-it-is-that-is*. This simplification impoverishes what is or should be a rich concept. There are many other ways of characterizing the transition that is explicitness or making-explicit. Here are some, in no particular order (notwithstanding my numerous attempts to order them):

1. From what-is as stuff (in the form of objects, processes, events) that simply is to what-is-the-case.
2. From whatever-is to whatever-is-the-case.
3. From is to is-the-case.
4. From Is to "Is".
5. From the "phenomenalization" of what-is in experience to the articulation of what-is in discourse, especially where this involves signs whose arbitrariness makes explicit their role in making explicitness explicit.

6. From what-is to signs, signals, information, that are about, or refer to, what-is.
7. From what-is to the objects presented to subjects, as the *world* inhabited by those subjects.
8. From stuff to sense, to intelligibility – where "sense" and "intelligibility" encompass experience, knowledge, purpose, significance.
9. From Being to the presence, appearance, manifestation, or revelation of Being or beings.
10. From Being to the articulation of presence and absence.
11. From what-is to an environment, a location transformed into a habitat, and ultimately a world that is someone's world faced by that someone.
12. From Being to beings as the Given.
13. From what-is to the possibility of possibility.
14. From basic stuff designated by the term "matter" to mattering.
15. From Being to ex-istence – which may be variously construed as "that which *ex*-ists", or "stands forth, comes out, emerges, appears, becomes visible, comes to light";[1] is uncovered, disclosed.
16. From Being to being-there.[2]
17. From Being to facts, truth, reality.
18. From what-is to what-is-lived.

All of these transitions are (perhaps) gathered up in "*Tattva*" – a Sanskrit word meaning "thatness", "reality", or "truth".

I cannot rank the closeness to their prey of these attempts to grasp explicitness, not the least because they reflect different aspects of this irreducibly multifaceted, complex, and frankly tatty, concept I have mobilized in the endeavour to refer to what is, after all, the necessary condition of all concepts. Circling, indeed. In the text that follows I have most often used the transition from "what-is" to "that-it-is" or to "what-it-is-that-is" to capture explicitness, hoping that the reader will appreciate that this is shorthand for a notion with a rich hinterland, hinted at in 1–18 above.

To make explicitness entirely explicit, fully to capture the dimensions of what is opened up by it, it would anyway be necessary to perform something impossible: to specify, identify, articulate a pre-explicit realm as a starting point for the transition. The problem is that both ends of the transition from "what-is" to "that-it-is", or from "what-is" to "what-it-is-that-it-is", are necessarily made present in words that are a late, sophisticated expression of explicitness.

It might be argued that making explicitness explicit is no more difficult than something that linguists and grammarians do all the time when they talk about language. Do they not use words such as "word", and write sentences about sentences, using phrases such as "parts of speech" and "subject of sentence", and technical terms such as "case", "modality" and "tense"? It remains, however, uncertain how far linguists and grammarians really do get outside of, or rise above, or dig beneath, language to see what it is and to observe its limits, even to take the measure of the refractive index of the many lenses it interposes between us and what-is. It is reasonable to suspect that a viewpoint necessary to see language for what it is would not be able to express this; it would fall silent.

The limitations of a head-on approach to making explicitness explicit are betrayed in my preferred choice of terms in the text that follows to characterize explicitness: the transition from "what-is" to "*that*-it-is" or "what-it-is-*that*-is". "That" does scant justice to (say) the explicitness of sense perception – as when for example (according to the accepted version of events) electromagnetic radiation is transformed into light that is both luminous and illuminating, making what-is (including itself) a field of vision, part of a revealed world. "That" does not capture the difference between the photosensitivity of a leaf and the experience of a conscious subject seeing that she is surrounded. It does equally scant justice to the many *discourses* that present what-is as a nexus of facts and propositions, assertions and denials, and that may ascend to the point where the totality of things is called "the universe" and everything is, as it were, placed in inverted commas.

And "what" as in "*what*-is" does not seem a sufficiently primordial point of departure, if only because pre-explicit being seems "pre-what" in the sense of being prior to classification, to being identified as an entity, a what, of a certain kind. Explicitness, moreover, does not deliver an array of discrete bits: the experienced world is not a mosaic of separate entities, though such entities may be highlighted. Any "what-is" is, at least implicitly, part of a larger whole, of an individual or collective world, reflected in discourses that weave a vast fabric out of what is sensed, spoken, written, thought, and engaged with in action.

Paradoxically, "that-it-is" or "That-it-is-what-it-is" as a destination of the transition designated by explicitness seem not only rather bare but also pleonastic: "that", "it", and "is" seem to try to reach out to the same object. They nevertheless still fall short, separately or together, doing little justice to a hinge moment in the history of Being (if we can imagine Being having a history in the absence of conscious subjects gathering its successive phases together) when what-is becomes something *that* is – typically *for* a conscious subject.

A leaner definition, describing explicitness as "the transition from Is to 'Is'", (4) above, is still unsatisfactory because language remains obtrusively present at both ends of the transition as a consequence of which the profundity of the transition is insufficiently registered by the importation of inverted commas.

Besides, explicitness is not always a matter of linguistic or even pre-linguistic articulation. It goes deeper than this, as when – to return to our previous example – electromagnetic radiation becomes visible light, and entities bathed in it are made present to conscious subjects harvesting that light. If the inability to get outside of language to make explicitness explicit therefore seems important, it is because explicitness begins outside of, before, language and what is said. It begins, perhaps, somewhere between sensation and perception; or even between sensation and shared or otherwise acknowledged perception. I add this because there is a justified sense that, while there is something intrinsically social about full-blown explicitness, explicitness goes deeper than the social realm, though the latter is a necessary condition of many of its manifestations.[3]

"That", if it is taken propositionally rather than simply marking the revelation of what-is, seems to begin too far down the track. After all, non-propositional toothache makes a tooth explicit, sunbathing is not reducible to propositional "that-bathing", and my gaze makes light explicit and, through this, a scene that it lights up. Explicitness has its seed in the first hint of aboutness in sentience from which it flowers into a world that is created, maintained, expanded through individual minds enriched by the boundless community of other minds in which they participate and, indeed, find much of their being.

The reference to "aboutness" is a cue for us to open the door to something that has been knocking for a while and is rather more familiar: consciousness, which is that in virtue of which entities – including the conscious subject – become explicit. As we shall discuss in Chapter 3, neural theories of consciousness endeavour to make a particular variety of what-is (nervous systems) into the brewery of explicitness, so that parts of the universe, assumed to be a vast acreage of insentient stuff, become worlds anchored on subjects. Invoking consciousness is also a reminder of the many kinds and layers of explicitness, encompassing perception, memory, thought, emotion, and individual and collective awareness of what-is, creating ever more complex modes of that-it-is. Consciousness, or mind, is that in virtue of which what-is is made explicit. If I have preferred "explicitness" to "mind", it is because there is less temptation to think of explicitness as a stuff or a substance, even less one interacting with a parallel stuff or substance called "matter".

The reference a few paragraphs back to "the history of Being" and the example of the transformation of electromagnetic radiation into visible light is a reminder of the kind of thinking that philosophical inquiries centrally illuminated by the notion of explicitness may wish to avoid. Behind it is the assumption that what-is is something to be identified with matter, understood as the primordial stuff of a physical world whose most authentic portrait is that provided by physics. This is evident in (14) above, the characterization of explicitness as the transition from basic stuff designated by matter to "mattering".

According to the standard view, matter arrived 13.8 billion years ago and mattering some considerable time later – somewhere between 10 billion (with the emergence of single-celled organisms) and 200,000 (with the arrival of *Homo sapiens*) years later.[4] It would be in the spirit of what follows to try to avoid the notion that what-is is essentially insentient matter and that-it-is is the *product* of sentient matter emerging from insentient matter when it acquires a certain form. Nevertheless, this is a default position that has a gravitational attraction even at times to the present writer, brought up in the twentieth century and continuing to live in the twenty-first. But it must be resisted, given the problems with the idea (discussed in Chapter 3) of the brain as "the brewery of explicitness" generating that-it-is out of the what-is of neural activity.

One final preliminary. While explicitness does not begin with propositional that-it-is, there is no ending there, either. Making explicit is an unfinished journey. And it is probably unfinishable; for "what-is" does not dissolve without remainder in explicitness.[5] There is always more to be made explicit.

(2)

As it is made explicit, so what-is is, in a special sense, taken outside of itself. It acquires an existence (or "ex"-istence) alien to itself in the presence it has to a conscious subject – a presence that is not reciprocated, unless that which is made explicit is another conscious subject. The etymology of "existence" – derived from the present participle of Latin *existere* to "stand forth, come out, emerge; appear, be visible, come to light" – underlines this idea. It is a central insight of Heidegger's philosophy, where it refers both to the capacity of human being-there to stand outside, take a stance on, itself, and its capacity to create an openness in Being.[6]

Let me illustrate the sense in which that which is made explicit, comes to ex-ist, to stand outside of, and be reunited with itself, with the example of a humble material object. A log is not, of itself, a "log" – for it is not intrinsically a member of a class; not, at least, until it is taken out of itself in order to be asserted as having a token identity, as an instance of a class. After all, it can be assigned to many classes. Even when it is pointed out mutely, employing the carnal indexicality of the index finger, its "thisness" goes beyond the this it is as it is nailed down to itself.

One of the most profound and, from the standpoint of human history, most portentous modes of separation and reattachment is the extraction of quantity from an entity and its reattachment to that entity. When an entity is ascribed its mass, its length, or some other physical property – as when I say that the table is 2 feet long – it is alienated from itself and aligned with an indefinite class of entities of equal length. The extraction of quantities from what-is opens a path to a world-picture which eventually leads to a vision of what-is reduced to quantities. Such a world, emptied of qualities, will be homogeneous.

Identity as "the condition of being the same with something described or asserted"[7] may seem rather commonplace. Nothing could be further from the truth. The separation of an entity from itself, permits it to be subsequently *id-entified* with itself as the such-and-such that it is, and thus to be reattached to itself as that (or the same) thing. The idea that something is "the same thing" over time explicitly connects its temporal slices with each other.

The separation and reattachment is most explicit and elaborate in personal identity. When we are baptized, introduced by another to a third party, or answer the question "Who are you?", we are fastened, or fasten ourselves, to ourselves by our proper name. Even though that name may be shared with many others as is usually the case, "I am x", picks out a token identity. "John Smith" for example does not stand for a subset of persons who belong to a class that is systematically contrasted with the subset of persons who are called Bill Smith. The declaration that "I am Raymond Tallis" or acceptance of the assertion that "You are Raymond Tallis" is a hub that reaches out to all the many ways in which I stand outside of myself to encounter myself as that which I am.

Not all philosophers have overlooked the numinous fact that, courtesy of conscious subjects, entities can be separated from and reattached to themselves. That in virtue of which Being is separated from and connected with itself, securing the transition from what-is to that-it-is, is a not-too-distant cousin of what Jean-Paul Sartre characterizes as a "Nothingness" opened up in the plenum of Being.[8] The present investigation of explicitness will, however, take place at some distance from Sartre and from Heidegger – notwithstanding the reference in (16) above to "being-there".

The sense in which an entity is distanced from and reattached to itself by being identified as a such-and-such of a certain kind is also evident in the fact, already noted, that any entity may be classified in different ways. An instance of what-is may be identified, or engaged with, as "a cat", "a mammal", "that damned animal", "a furry thing", "a possession", "our pet", "another drain on our resources", etc. No entity is entirely gathered up in the class or classes to which it is assigned, any more than it is dissolved in the sense we make of it. The fantasy that what-is dissolves without remainder in explicitness is the other side of the belief in the "paninformationalist" vision of the universe.

The fact that entities can be classified and reclassified is expressed less formally in our sense that something is "like" something else or what something is "like". Or indeed unlike. It hardly needs to be spelled out why likeness and unlikeness are relational rather than intrinsic properties of what is – and that to judge of something that it is "like" something else, being like some other individual or earning its place in some general category of being, is to separate it from and reattach it to itself.

The manner in which entities made explicit can be detached from and reattached to themselves is also evident when we separate them from what we regard as their properties and then reattach those properties, as predicates, to them; most obviously, when the entities have been made the subject of a sentence – as in "The cat is black". What-is is teased out into elements that may be detached from one another as predicates of objects themselves presented as the subjects of sentences. When we say of an individual cat that it is black, furry, greedy, etc., we are separating it virtually – in time in the case of speech, in space in the case of a written text – from the properties that in reality cannot be separated from it, prior to reattaching those properties in the realm of explicitness.

Here or hereabouts is the breeding ground of "ness" monsters that litter philosophical endeavours to grasp the nature of things. ("Explicitness" itself must surely be supreme among such monsters – an Ur Ness Monster – closely followed by Sartre's "Nothingness". Explicitness is Ness-ness.) In Western philosophy, the most fertile womb of ness monsters has been Platonist thought where the classes to which things are assigned are taken not merely to be entities in themselves – as Forms, or Ideas – but to be the only entities that are fundamentally real, because they are revealed by the intelligence rather than sense experience. Contrary to Platonic understanding, the classes to which objects are assigned fall short of full-blown being. It is the classes rather than the objects that lack reality. An entity can be reduced to its membership of a particular class only by a fantasy of shaving off other predicates that can be correctly, indeed are necessarily, ascribed to it.[9]

This seems an appropriate juncture to glance at one of the most interesting (though seemingly commonplace) technical terms in philosophy: the existential quantifier. It is mobilized in propositions that assert the existence of an entity (a subject) in order to ascribe certain features to it (typically expressed as predicates):

> There is an x, such that for some or all x's such-and-such is the case

As Kant pointed out in response to Anselm's ontological argument for the existence of God: existence is not a predicate.[10] We could express this by saying that existence is a pre-predicate: it is that in virtue of which what-is is separated from itself in order that properties – *its* properties – can be ascribed to it. To put it another way: the existence of an entity is asserted – that-it-is – in order to provide a locus for the things that are said about it, to transform it into a predicate-bearer. So, we (or those of us who are philosophically inclined) say "There is an x such that x is a cat and x is black". In some respects, "There is an x" is a more conventional way of capturing what lies at the heart of the transition from what-is to that-it-is. This is overlooked when the existential quantifier is taken too much for granted and the work to be done in relation to logic – and, indeed, to thought – is seen to begin only beyond the existential quantifier, which is merely the ground on which the properties stand. What makes the existential quantifier extraordinary is that it opens the space – or marks the already-opened space – in which the "that" of "that-it-is" or "that-it-is-the-case" can be detached from what-is and is then available to be reconnected with it, to be ascribed to it.

The existential quantifier points to the most fundamental and universal manifestation of explicitness – the perception of, or the assertion of something, that it *is*. There is a kind of inner displacement that renders it open to being united with itself in an identity. It is this making-explicit that is overlooked in the anti-psychologistic philosophy of thought and logic.

(3)

To propose, of what-is, *that*-it-is is to build on any primordial revelation, by articulating and sharing it. It is a reminder that making explicit is as much an activity as it is passive reception. This is particularly obvious when explicitness takes propositional form: propositions, after all, are occurrents that exist only insofar as they are *proposed* – etymologically "put forward" or "set forth". A token proposition is an act not a thing, though it may assume thing-like status when it is written down and occupies space. (Or material event-like status when it is spoken out loud.) The fact that something can be made visible on the page, or audible in speech, or neither audible nor visible when it is being thought, indicates how it is not tied to anything that qualifies as a material event.

The sharing of the that-it-is in propositions and assertions "sediments" into the parallel realm of "thatter" (*vide infra*) whose most substantial manifestations may be gathered up into the idea of the semiosphere – the realm of signs which supplements and in some cases marginalizes the biosphere, the realm of living things to which all organisms belong.

The attempt to apportion the contributions of passive receptivity and the spontaneous activity of the mind, the relative contributions of what-is and the subject, in delivering a portrait of what-is as what-it-is-that-is, has been an enduring preoccupation of philosophy. The propositional faculty of the mind, capable of asserting of what-is *that*-it-is, of proposing what-is, is an obvious pushback against "pure" receptivity.

An even more obvious pushback is the capacity to say of something that does not exist, *that* it does not exist. Non-existences are pure explicitness; instances of that-it-is in the absence of a corresponding what-is. They are most fully developed within language, though they may also be present in an unsuccessful search for something we have in mind.

Propositions that propose of a what-is that it *may* exist, or that it is unlikely to exist, or that it ought to exist, also illustrate what we might call (relatively) untethered explicitness corresponding to the fact that a putative what-is can be entertained as a possibility, which can in turn be denied, or discovered to be untrue – or indeed true. Negative propositions, questions, expressions of the possibility of certain existents, are particularly striking manifestations of the spontaneous or generative activity of the mind. It is the realm of the "free-floating explicitness" of thought. By contrast, what-is lacks modality.

The contrast between the receptivity and the spontaneous activity of the mind is most influentially associated with Kant and *The Critique of Pure Reason* for whom receptivity delivers only sensations.[11] The organization of sensory experience into a world of entities set out in space and time and causally connected with one another is the product of the spontaneous, or synthetic, activity of the mind. One need not, of course, be as radical as Kant in distinguishing the receptive and the spontaneous functions of the mind in order to do justice to the complexity of the transition from what-is to that-it-is. What we must, however, acknowledge is that the transition does not have a pure, innocent, beginning. In part this is because cognition is *in medias res*: revelation is informed by a history of cognition – our own history and that of the various communities of minds to which we belong.

What-is, therefore, is not merely given to a passively receptive subject who experiences it as it is in itself before it is made explicit. The "what" of what-is is at the very least contaminated with (general) sense: there is no sharp separation between sensing an entity and making sense of it – as we shall discuss presently. Indeed, one might be tempted to argue along with many thinkers that "pure" what-is, existing prior to its being made explicit is mythical, though this would require one to question the mainstream assumption that (conscious) beings that make what-is explicit arrived late in the multi-billion year history of the universe. To doubt that there was stuff long before there were any creatures making sense of it is bold, verging on the foolish.[12]

(4)

We may think of the human individual emerging as a maker of explicitness during early post-uterine life, in the months in which a neonate becomes an infant increasingly aware that there are things "out there", circling her embodied proto-self, things which can be shared with others, initially made explicit through a conjoined attention often highlighted in pointing.[13] The pointing will, in the case of the parent usually be accompanied by words – "Where's mummy?", "Look at that doggy!" – that soon lift objects from the mere singular thisness of something that is, and is present, to the generalized whatness of an entity that is named and classified.

After the earliest phases of development, the passage from sensing to sense-making, from what-is to that-it-is and what-it-is-that-it-is is iterative, cumulative: the circle of sense widens, and the realm of explicitness, curated by a widening community of minds, grows in scope and complexity. Perception is shaped by concepts and concepts feed on perceptions. Sentience and discursive meaning nourish each other, building complex edifices of what-it-is-that-is-out-there, ultimately piecing together something that could reasonably be called a world picture.

As just noted, there seems to be a being of what-is prior to any "what-kind-ness" revealed through "that-it-is", an Order of Being preceding an Order of Knowing.[14] Nevertheless, as members of a community of language speakers, we are often first introduced to instances of what-is through terms that name the types to which they are assigned: "Look at the doggy!". We meet the "thisness" of what-is before and around us via the mediation of general terms that order things under types and in doing so make a kind of sense of them. Cognition, that is to say, comes steeped in *re*cognition. When I perceive what is out there, I do not (usually) perceive first a dollop of stuff which is then made sense of. This is a reminder that the "what" in "what-is" is neither particular nor general, neither a token nor a type, but the ancestor of both tokenization and classification.

The division into pieces of what-is, moreover, depends on the direction and grain of my attention, itself guided by my previous experience and my present interests, and of those around me; by the interests, that is, of a subject or subjects for whom it is an object. To take a simple example, whether the what-is in question is a fallen leaf, a tree, a wood, a wooded landscape, or a wider territory, is not determined independently of the process of making explicit. The direction and scale of interest may be involuntary – as when a pain-causing event gatecrashes the consciousness of a subject – or discretionary, as when a subject scans her surroundings with a view to shaping a plan of action or simply to enjoy it.

In the extreme case of Kantian transcendental idealism (or, in more recent iterations, such as the evolutionary anti-realism of contemporary writers such as Donald Hoffman),[15] the contribution of the object (or of what-is imagined in the absence of a conscious subject) to that-it-is is highly problematic. If there is intrinsically nothing located in space or time and of itself engaging in causal interactions at least in part dictated by a spatial-temporal location, it is difficult to understand why a particular embodied subject should experience itself as located in a given parish of what-is defined at least in part by spatio-temporal locations and causal connections; indeed, as being conscious of one world rather than another.[16]

I mention this to highlight the complexity of the territory opened up by the transition from what-is to that-it-is. Whether or not one accepts the cognitive pessimism according to which what-is will be forever hidden behind the processes necessarily associated with the transition to that-it-is, one thing is undeniable. It is that there is no pure reception of what-is: there is no giving of the given without an active, many-leveled process of taking. What-is does not have built into it all the wherewithal necessary for givenness. The universe, typically understood as insentient matter, is not intrinsically addressed to the conscious subjects who construct the realm of explicitness. Nor is it self-revealing in conscious subjects understood as being composed of the same material as the rest of the universe. While the idea of a pure, uncontaminated Given is a "myth" as Wilfrid Sellars proclaimed,[17] givenness of any kind – mediated or unmediated, direct, or indirect – deserves more fathomlessly puzzled attention than it will receive in this work. At the very least, for Sellars, as Eric Watkins expresses it, "what is immediately given lacks the kind of structure that would be required to justify propositional knowledge".[18]

There is a "taking", therefore, necessary to turn what-is into the Given, such that the latter is "given to be an instantiation of a certain [general] property"[19] – in short something gathered under a concept. It is, consequently, a fundamental mistake to imagine that the contents of consciousness are a kind of transcript written by causal interaction between the conscious subject (or her body or her brain) and the object or event or state of affairs of which she is conscious.

Indeed, if we believe that dumb, numb matter is the primordial form of what-is – as materialist neurophilosophers typically do – the emergence of what-is as the Given, as a world populated by present entities, is inexplicable. More has to be brought to the party and, while there may be arguments as to how much of this more is imported by the conscious subject and how it does this, that it *is* needed cannot be denied.

The various modes of the transition from what-is to that-it-is and to what-it-is-that-it-is are interwoven with our status as persons, and a largely implicit "we" with whom we cognitively cohabit in a community of minds. We need to acknowledge the special nature of the transition from what-is to that-it-is when the "what" is a who – a being who makes his or her own existence explicit, as in the lifelong "that-I-am" of first-person being, of conscious subjects who make explicit both themselves and the entities that surround them and bring about the transition from what-is to that-it-is. A conscious subject, a person, makes her existence explicit in the lived corporeity of the body, in the condition of being an embodied subject.[20]

The way we are presented to ourselves as a third- or no-person body in which we develop as first-person beings, in which we are "ambodied", is an extraordinary mode of explicitness. One of the most striking aspects of it is the way our own bodies become the primordial examples of entities to which is to be ascribed an existence that transcends what is experienced of them – of objects in the weighty sense; of things that exist independently of our awareness of them.

(5)

I have already referred to different levels at which, or modes in which, items are made explicit. The conventional story begins with sensation, proceeds to object perception, and ascends to higher levels of explicitness, most notably the natural sciences that attempt to grasp the physical world and the arts and social sciences which aim to arrive at a vantage point on the human world.[21] The most striking platform for this ascent is the frozen explicitness in our densely signposted human world. In addition to the natural offerings or "affordances" of the environment, and the tapestry of discourse, the human world is packed wall-to-wall with artefacts. Since they have been constructed to perform certain functions, they wear their meaning, their what-it-is-that-they-are, on their sleeves, like the bottle in *Alice in Wonderland* that says "Drink me". The labelling of what-is as what-it-is-that-it-is adds another layer of explicitness – as in the case of artefacts which are obliged to broadcast the category to which they belong (a bottle of sauce labelled "tomato sauce") or their individual identity (a street wearing its own name at both ends.) They take their place in contexts of practical action and local meaning and in some cases

– illustrated by a judge's gavel – are assimilated into the fabric of institutions where discourse is woven into the roles we may assume and functions we may perform, that are integrated with those of other human subjects, and the rules, regulations, laws and customs that shape and usually facilitate our daily existence.

Reflecting on these riches makes "explicitness" seem a woefully inadequate, barren term to encompass the openness, uncoveredness, infinitely various revelation of waking consciousness expressed in so many ways. It seems a thin and po-faced way of capturing how we humans who, as a consequence of the torrent of transitions from what-is to that-it-is, have become such complex entities; how we face a rich world that faces us, a world into which we peer, inquire and investigate; a realm, in which we live our lives, relating to what is "out there", that looks back at us.

In search of something a little less unsatisfactory than "explicitness" I have elsewhere talked about "thatter" (as opposed to matter) as the made-explicitness of the material universe and all that goes on in it.[22] Unfortunately, the choice of this word and its echo may make it seem as if the transition to that-it-is generates a kind of stuff, parallel to matter. Consequently, it risks reviving the ghost of dualism, though the rootedness of explicitness in that which is made explicit should exorcise the ghost somewhat. Describing the act of making explicit as "thattering" may be a little less unsatisfactory because there is even less a hint of ghostly substances, and it acknowledges the active role or spontaneity of the conscious subject. I have also written about the "thatosphere" as that part of what-is which is made explicit, a realm (not a stuff) additional to both the biosphere and the domain of inanimate matter, created and sustained by the community of human subjects living out their shared and separate lives.[23] Any substance attributable to that realm is, of course, borrowed from material objects and embodied conscious subjects.

Neologisms such as "thatter", "thattering" and the "thatosphere", notwithstanding their ugliness, do endeavour to capture the difference between a putative physical universe and a world in which conscious subjects, whose spaces and times are at least in part defined by dimensions other than those of physical spacetime, pass their lives. This may seem to entrain the assumption that what-is, the starting platform of the journey towards explicitness, is matter. We should be cautious, given that "matter" is looking increasingly like something gathered under a concept that has been shaped out of that evolving form of collective explicitness we call physics – and what is more, is looking ever more problematic.[24] We perhaps need to be equally hesitant about characterizing explicitness as that in virtue of which matter comes to matter (in the sense of being of concern to a conscious subject), though I have found this difficult to resist, as item (14) in the list several pages back, attests. After all, much of what is made explicit is the mere background, hinterland, or context, of what matters to us.

If we do accept (insentient) matter as a basic or primordial what-is, then mattering – of things to conscious beings who matter to themselves and to each other – becomes inexplicable. This is something we can live with – indeed we have no choice – so long as it is acknowledged that the deployment of terms such as "emergence" or "complexity" does not remediate this, does not close the gap between matter and mattering. Invoking transitional stages in which, for example, there is emergence of biological entities for which physical events become "stimuli", such that other events causally related to them can be designated as "responses", and parts of what-is become an environment or *umwelt*, as a stepping stone towards a *world* faced by a conscious subject, seems closer to tendentious (re-)description than explanation.

The appeal to "emergence" offers nothing to diminish the mystery of the transition from what-is to that-it-is, such that the former becomes the Given. Or indeed account for the distance opened between local stimuli that trigger responses and (for example) *facts* – which do not have a location – that guide behaviour. Facts are not stimuli – or not, at least, in the way that events in the natural environment are. Nor are conscious memories, wherein that which was once explicit is revived, relived, rehearsed, at another time. Memory seems to have a double dose of explicitness: the passage from what-is to perception of what-is; and the passage from perception to recall of perception. In the form of memory, that-it-is is curated separately from what-is by a subject whose continuity over time is underwritten at least in part by the continuous, enduring basic viewpoint of a living body that is embraced by itself in the complex manner we shall discuss as "ambodiment", and as selfhood.

Agency operates in a space of possibility woven out of that which is and has been made explicit – including aspects of the agent herself. This space marks the distance between the immediate responses of an organism, which are conventionally represented as being causally related to stimuli, and actions where the agent genuinely *enacts* what she does in virtue of knowing what she is doing: knowing the reason for her actions and the (at present invisible) goal or goals that she has – an explicit but as yet unrealized future that is fulfilled by the actions that serve her own future. This is an important aspect of the mode in which time in conscious subjects is made explicit – as tensed time, as measured or timed time, and, ultimately, as the post-tensed time of the clock and calendar.[25] The transition from what is taken to be physical time to post-tensed time involves an extraordinary mode of making time explicit: the set of actions and technologies that go under the heading of "timing".

But we are getting ahead of ourselves. It is enough that we have affirmed the riches hidden inside the po-faced term "explicitness". Let us continue exploring these riches.

(6)

A philosophy that gives prominence to explicitness may not seem to lack precedents or – to put it less politely – may not be all that original. Consider the opening of *Tractatus Logico-Philosophicus*

> 1. The world is all that is the case
>   1.1 The world is the totality of facts not of things.[26]

It is possible to read this as Wittgenstein identifying a fundamental distinction between the universe as a vast aggregation of basic stuff such as matter (of "things") and the human world (of "facts") and read the former as the totality of what-is and the world (the realm in which humans live and have their being) as the totality of that-it-is.

Wittgenstein's elaboration of this opening shows, however, the need for caution. He asserts that "The world divides into facts"[27] which sounds fine. But he then construes this as follows: "What is the case – a fact – is the existence of states of affairs. A state of affairs (a state of things) is a combination of objects (things)".[28] This suggests that combinations of objects or things would be sufficient to generate "facts". To accept this is to overlook the transition between what-is and that-it-is or what-it-is-that-is. We are entitled to support this conclusion only if "the world" is identified with the natural, material world, rather than being the product of the shared consciousness of subjects, of communities of minds. Most importantly, "is the case" seems to be slipped in free of charge. There are, however, no "natural" facts generated by what-is, not the least because this would not allow space for facts to be *of* or *about* what-is – to keep their distance from what-is. Arrangements of objects are not facts-in-waiting, waiting to be elevated to the status of being "the case"; nor would they be, even if they were self-arranging. States of affairs do not of themselves "ex-ist" in the sense we discussed earlier, and objects do not add up to combinations without the assistance of the organizational faculty of consciousness. Seeing facts as inherent in nature is a first step on the slippery slope towards the paninformationalism that lies at the heart of so much confusion in contemporary philosophy.[29]

Wittgenstein did not need to be told this. After all, he famously identified the limits of the world – the sum of that which is the case – with the limits of language: "The limits of my language mean the limits of my world".[30] It implies that, after all, the world that he said was made of facts, not of things, was not the natural world, of stars, lakes and mountains, even less the material world of physical stuff, or the universe that preceded conscious subjects by billions of years.

This position was reinforced in *Philosophical Investigations* where the language defining the limits of the world was even more limiting because it was shaped by conventions, constituted out of language games. In his Introduction to the *Tractatus* – that Wittgenstein hated – Bertrand Russell stated that "the essential business of language is to assert or deny facts".[31] Wittgenstein would come to oppose this on the grounds that language has many other functions, as is evident in greetings, commands, questions, jokes, etc. But it is not necessary to make this point to see that to characterize "the business of language" in the way that Russell did is to overlook the meta-fact that there are no facts simply out there waiting to be asserted or denied. Nor is there a ground floor of language which is up close to facts that are "on the ground". By the time we have arrived at even the most elementary facts, we have drifted a long way from innocently mirroring "things", "stuff out there", replicating their structures with the structures of linguistic elements such as sentences and propositions. Facts are not tethered to "a combination of things"; even less *are* they a combination of things.

Most importantly for our present concerns, there is no "the case", except where "states of affairs" crystallize out of what-is in the seas of individual, intersubjective and collective consciousness. It is this that marks the distance between what-is (in-itself) and a countless multitude of possible worlds available to be unpicked and shared as facts.

A few more words about facts – if only to dispose of the idea that there is anything matter of fact about them. It is conscious subjects who unpack what-is into facts: facts are, as the etymology tells us, "made" – *facta*. While they are not "made up", facts are not entirely brute. That is why it requires something more than a brute to make them out of what-is. Contrary to what John Searle said, when he contrasted "brute" (physical) and "social" facts, *all* facts are social.[32] While, in the absence of a conscious subject, the table does not abandon its position next to the chair, it is not, however, in and of itself "next to the chair" or "in the room next door" or "240,000 miles from the moon". We should not doubt that the stuff of Mont Blanc and Mount Everest exists in the absence of conscious subjects making it explicit. Nevertheless, we can reasonably deny that, in the absence of such consciousness, neither mountain has a discrete, bounded existence that separates it from the rest of the bumpy surface of the earth. The boundaries of a mountain are drawn by convention. Nor does Mont Blanc, of itself, have the relation of being "so many miles from Mount Everest", or of "being smaller than Everest" or, indeed, bigger than my thumb.

That the relations between entities exist only insofar as they are made explicit by conscious subjects is evident from the fact that there is no limit to the number of such relations – not even the law of contradiction. That is why, *pace* Plato, it is not a problem, even less a contradiction, that an entity can be both "bigger than" and "smaller than" other objects. Of course, objects and their intrinsic nature place constraints on the facts about them when they enter the world of a conscious observer. The bedside table correctly reported as being "240,000 miles from the moon" cannot at the same time be "250,000 miles from the moon". To repeat, facts are "made" but they are not "made up".

What-is, therefore, is no more an aggregation of facts than it is an aggregation of the referents of propositions or assertions. Trees whisper in our ears but one thing they do not whisper is "tree"; nor do they whisper "trees whispering". In sum, Searle's idea of lower-level (brute) facts that exist independently of human beings, their consciousness shared in discourse, and their institutions, does not stand up. Facts are the product of the process we discussed earlier in which entities are separated from themselves, from their properties, which enables those properties to be (re-)attached to them as predicates, and the entities to become the explicit bearers of their properties.

It may seem unnecessary to make this point were it not that there are increasing numbers of thinkers who believe that facts, or more generally "information", are inherent in the world. The physicist John Wheeler famously asserted that "Each elementary quantum is an elementary act of 'fact creation'".[33] All physical processes, we are asked to believe, are information processes. For Wheeler, the very stuff of the universe is information: this is the paninformationalism that he also expressed as "IT = BIT" or "IT from BIT".

The strange idea that facts and information can exist without being entertained by conscious subjects is easier to understand when we remember that our encounter with facts is usually mediated by speech and language. In the form of utterances, they seem like physical events and, frozen as written propositions, they look like physical things. Hence the origin of the almost willful overlooking of explicitness in the use of the word "information" to apply to goings on that have nothing to do with the exchange of knowledge or understanding shared between conscious subjects, in short with explicitness. It is the most egregious attempt to collapse the distance between what-is and that-it-is. If there is one thing that justifies the present book, it is that paninformationalism is becoming more widespread in the discourse of philosophers, biologists, psychologists, and even physicists.

Paninformationalism is a particularly striking example of the overlooking of explicitness and the failure to see the extraordinary nature of the transformation of what-is into that-it-is. When I say "extraordinary" I do not exaggerate; for explicitness is that in virtue of which we humans have been able to weave an extra-natural world, haunted by explicit possibilities that draw on an absent past and point to an imagined future. Paninformationalism, that gathers explicitness into the physical world, hides this. It also makes it easier to accommodate explicitness in a naturalistic world picture according to which objective, quantitative science, notably fundamental physics and biology, will ultimately provide us with a way of understanding this aspect of ourselves.

(7)

The realm of "facts" is most intimately associated with that of thought. In this realm possibilities are entertained as to how things are, or are not, or might be. Truths and falsehoods, factuals and counterfactuals, are established and connected with other truths and falsehoods. Reported or recorded observations are hoovered up in, or are invoked to justify, general statements. Logical connections are established between possibilities. The instantiation of those logical connections in spoken or written tokens, and their being set out in a temporal sequence – as when the tokens of initial axioms (protases) precede the tokens of conclusion (apodoses) – confers on states of what-is the status of beginnings and endings.

We are close here to something we shall cover in our discussion of thought: the false path of anti-psychologist (indeed "de-psychologizing") approaches to thought and indeed other mental entities. At its heart is a belief expressed by Gottlob Frege in his classic – and hugely influential – paper "The Thought". Here he argues that "A property of thought will be called inessential which consists in, or follows from the fact that, it is apprehended by a thinker".[34] And he adds that "Thoughts can be true without being apprehended by a thinker and are not wholly unreal even then".[35] There could be no clearer expression of the tendency in analytical philosophy – and the scientism to which analytical philosophers influenced by Frege allied themselves – to overlook explicitness; to fail to see that it is prior to all the gymnastics of "p" and "not-p". This false path takes us to a place where thoughts can exist without being thought and truth can exist without being asserted and modalities such as possibility, probability and actuality, and even necessity, can emerge without the assistance of a conscious subject – something that has become central to some interpretations of quantum physics.

Modality in the absence of conscious subjects may seem plausible in the case of necessity. The necessity built into nature does not require such subjects who are, after all, late arrivals in the order of things. Or so it would seem. However, the interpretation of rigorously uniform patterns of the unfolding of what-is into a material or *de re* necessity is mind-dependent. As Hume pointed out, it is conscious subjects who transform the way things reliably happen in the overwhelming majority of cases into being *fated* to happen, with preceding events prescribing their successors – *causing* them. It is this that gives us the ground floor of the normative sense – of how things ought to be based on how they usually are; of a *de re* necessity that will ensure that thunder *must* succeed lightning rooted in justified expectation.

Whatever one thinks of the idea of *de re* or causal necessitation – and most philosophers after Hume, Kant, and David Lewis think little of it[36] – it is true that this could not be entertained or ascribed to the world out there in the absence of complex and sophisticated explicitness. *De re* necessity is something *felt* rather than a relationship between material objects or events, not the least because objects and events have to be identified and separated prior to being connected. While the appeal to *de re* necessity is a way of registering the uniformity of nature as non-negotiable, it is misleading.

As for *de dicto* necessity – the connection of necessity between assertions, propositions, or facts, such that, for example, p necessarily implies not-(not-p) – this is even more clearly downstream of explicitness.

(8)

As we have already noted, making-explicit is not item by item: explicit objects, states of affairs and events, are inseparable from *worlds* to which they belong. Worlds are lit up situation by situation, unfolding scene by unfolding scene, rather than flash by flash, object by object. There are innumerable modes of connectedness between aspects of that-it-is. They reflect a sense of evolving reality that is, to a variable degree, coherent at any given time and coherent over time.

With respect to such worlds, there is expectation, prediction, surprise, error, and true or false assertions. Conscious subjects making what-is explicit generate *possibilities* and retrospectively confer on what-is the status of fulfilled or realized possibility. Among the many reasons why this is important is that *agency* operates within a space of possibility woven out of that which is made explicit: actions are the realization of envisaged possibilities. This is central to the nature of free will and to the case against determinism. The animal that is most explicit – *Homo sapiens* – also has the most developed freedom.

There is another, connected, reason for focusing on possibilities in the pages that follow. The metaphysical confusions besetting much of quantum mechanics – in particular the Schrodinger Wave Function which is understood by some to portray the fundamental reality of the physical world – arise out of thinking of possibilities (and, more specifically probabilities) as being part of what-is. In fact, they are no more part of nature than the mode of explicitness that is science part of nature. This latter observation is central to any understanding of the relationship between the entertainment of possibility and the possibility of agency and how the science of physics, far from providing theoretical evidence of our lack of freedom, in practice enhances our ability to manipulate, and act as free agents in, the physical world. Physical science cannot accommodate itself; there is nothing natural about natural science.

Possibilities, then, are entities that exist only insofar as they are made explicit – in being entertained by conscious subjects, and believed, asserted, or denied, to correspond to anything that did, does, or will exist. This is particularly obvious in the case of possibilities that have non-existent entities as their intentional objects: they are the true-born children of explicitness. The exercise of specifying what-is-not in order to deny that it exists is a reminder of the inappropriateness of seeking that-it-is in what-is.

In perhaps the most influential argument in Western philosophy (set out in his very short poem *On Nature*), Parmenides begins with the seemingly incontrovertible assertion that what-is-not is not.[37] In a relatively few lines he travels from this starting point to the extraordinary claim that the sum total of things is an unchanging, homogeneous, sphere. We need not engage with his argument at present except insofar as to note that the very idea of what-is-not – entities that paradoxically have a specificity despite lacking existence[38] – underlines the irreducible and real distance between what-is and that-it-is or that-something-is. That Parmenides' assertion that "what-is-not is not" should have paved the way to extraordinary conclusions that have had such huge influence in Western thought is an indirect reminder of the depths hidden in the very process of making what-is explicit.

(9)

I mentioned at the outset that explicitness is overlooked because it is presupposed in any discourse. But there is another reason, already touched on in this Overture: the currently predominant materialistic, scientistic mode of thought, embraced by many philosophers, cannot accommodate it. "Explicitness" is not usually rejected by name; but that in virtue of which what-is becomes that-it-is is buried in stories that reduce consciousness to physical events, usually in the brain. That-it-is, so we are reassured, can be assimilated to, found in, what-is – usually understood as certain forms of organic matter.

Anyone who resists the collapsing of that-it-is into what-is is accused of being a dualist, for whom consciousness, or the mind, is a kind of substance parallel to the matter that makes up the overwhelming majority of what-is. The mind, except insofar as it is reduced to brain activity, is dismissed as a pre-scientific ghost in the machinery of the body. As we have already noted, one can accept the reality of explicitness, and its being other than what-is, without embracing the notion that it is a kind of substance or stuff. If we think of "mind" as a term that gathers up the many modes of explicitness, we shall, perhaps, not be tempted to think of it as a substance or even (as in the case of "property dualism") as a property.

Even those who reluctantly accept that there is a problem presented to materialism by consciousness, sometimes signal their reluctance by suggesting that phenomenal consciousness is an illusion constructed by the brain for reasons that are typically connected with evolutionary forces and survival advantage. By this means, the transition between what-is and that-it-is can be ignored, or at least downplayed.

Connected with this is the idea of fundamental reality, revealed at the end of scientific inquiry, as what-is somehow present and yet viewed from nowhere by no-one. The very idea of "the view from nowhere" is as contradictory as it is meant to sound.[39] Equally contradictory, though less obviously so, is the idea of what-is as being in itself "real". What-is does not have the property of being real – or indeed unreal. It becomes real when it is encountered, made explicit, as a result of which it is proposed to be of a certain nature.

Since we have mentioned dualism, it is worth noting that it does not do any more justice to explicitness than does materialist monism. The kind of localized engagement that makes what-is explicit is as problematic for mind-stuff as for matter-stuff, as is evident from the difficulty, highlighted by Descartes, of locating mind-stuff in a body that defines the locality in which engagement takes place.

Equally unhelpful is idealism, according to which all is mind. It offers no account of the starting point of the transition from what-is to that-it-is; nor of the difference between that which is to be made explicit and the explicitness that captures it. Idealism seems to make everything explicit, even if the mind in question is a collective or absolute mind that might store explicitness in a way that is unavailable to an individual mind – a view whose most recent iteration is the cosmopsychism born out of the embarrassments encountered by panpsychism. At any rate, the gap between what-is and its being the case is overlooked as completely in idealism as it is in materialism.

And so, in what follows, I shall endeavour to avoid idealism and dualism as carefully as I shall endeavour to avoid materialist monism. The world we live in is composed of neither matter-dissolving mind nor of mind-swallowing matter or both co-existing rather uncomfortably. The absence of an ontology emerging out of the discussion in the following pages is not, I believe, a failure. The reduction of the All to one or two kinds of stuff is, for my taste, too homogenizing and will erect insurmountable barriers to explaining, or recovering, the irreducible variousness of the universe.

(10)

The present volume, in which I endeavour to make explicitness explicit, is motivated by the conviction that overlooking explicitness, or taking it for granted, will result in a point-missing approach to many philosophical preoccupations. Anyone who is looking for examples of this philosophical failing is spoiled for choice. I list some of the most egregious examples at the end of this Overture. Paninformationalism and panpsychism, which have sections to themselves, are perhaps the most spectacular, indeed barefaced, instances; but many others are addressed in pages that follow. Here, however, is one example to whet the reader's appetite and convince her that, unlike Victor Borges' legendary uncle, I am not offering a cure for which there is no disease.

In his recent *Helgoland,* physicist and philosopher Carlo Rovelli argues that: "[*A*]*ny* interaction between two physical objects can be seen as an observation. We must be able to treat any object as an 'observer' when we consider the manifestation of other objects to it. Quantum theory describes the manifestation of objects to one another".[40] *Any* interaction? Interactions in the universe before there were conscious subjects making what-is explicit? The force of "physical" here is entirely obscure. Likewise, the basis of the difference between a physical object and an observation. And in what sense are objects *manifested* to other objects? Are they manifested to each other? Do they acknowledge each other? And what are we to make of the fact that we normally confer the status of "observer" on relatively rare, highly complex entities – organisms that show signs of awareness. Rovelli, in short, seems to be getting explicitness as part and parcel of the physical world so that the interaction between parts of what-is can generate that-it-is.

There is a particular irony in this passage from a physicist because physics began by opening up the widest distance between what-is and that-it-is. And yet it ends up by making that distance invisible: what-is is confused with the that-it-is of observation; or (more specifically in physical science) the "how-much-it-is" of quantitative observations, displaces what-is. In the realm of pure what-is there are no questions, no answers, no truths, no falsehoods, no problems and their solutions – and no observations.

Beyond its polemic purpose, the present volume is an attempt to arrive at a beginning from which I should have set out many years ago; and to land on a place that I have been circling above or around at various distances in previous books and, indeed, much of my writing life. It is part of a project to reflect on the nature of an astonishing creature whose flesh gave birth to words and who, among many uses to which she put those words, employed them to give Being a name – many names – and even to place the universe itself in inverted commas, gathering it up in segmented puffs of air.

If the result proves to be a reminder that philosophy is, to borrow what Borges said of art, "the perpetual immanence of a revelation that never comes",[41] so be it. And if the problems that are addressed in the chapters to come are not bathed in the mystery of explicitness as I would hope, they will at least be side-lit by it.

And if, finally, this Overture has proved at times to be more of a plod than a dance, I apologize to the reader. I hope that the pages to come are less hard going and the reader will find some enjoyment in this correction of a fundamental cognitive wrong, in which cognition overlooks itself, as I circle round explicitness in different orbits.

AFTERWORD: SOME SYMPTOMS OF A TENDENCY IN PHILOSOPHY TO OVERLOOK EXPLICITNESS

My commitment to making explicitness explicit is in large part motivated by the belief that overlooking explicitness lies at the root of many wrong turns in philosophy – fundamental ones such as the embrace of materialism and more technical ones such as the mind-body-identity theory and paninformationalism. They have in common a collapsing of the distance between conscious subjects and their worlds, between that-it-is and what-is.

Here are some examples:

1. The appeal to "emergence" as an explanation of the origin of conscious subjects in a material universe that has nothing in it to account for such subjects (Chapter 2).
2. The belief that there are "easy" problems of consciousness (Chapter 3).

3. The claim that there can be physicalist/causal accounts of the transformation of impingements on an organism into sensations, perceptions, appearances, signals. The notion that the transition from what-is to that-it-is can be generated by causal interactions between portions of what-is (Chapter 3).
4. Functionalist/behaviorist accounts of the mind (Chapter 3).
5. The notion that consciousness or information is everywhere in the universe; that what-is is that-it-is; that what-is is "about" itself (Chapter 3).
6. A universal scientism that closes the gap between conscious subjects and that of which they are conscious, most obviously manifest in a materialism that ultimately reduces what-is to a mathematical structure (Chapter 4).
7. A tendency for science – which opens the greatest distance between what-is and what-it-is-that-it-is – to overlook not only itself but also that in virtue of which it is possible (Chapter 4 and *passim*).
8. The assumption that metaphysics has been replaced, or made redundant, by natural science – in particular physics (Chapter 4).
9. The belief that the body (or the brain) accounts for first-person being, self and the "am" of the person is the "is" of the body (Chapters 3 and 5).
10. Extreme enactivism that threatens to collapse the distance between the embodied subject and her world (Chapter 6.)
11. The failure to do justice to the distinctive nature of selfhood and first-person being, to the point of denying the reality of the self and of personal identity (Chapter 6).
12. Overlooking the fundamental distinction between mere happenings and those doings that are manifestations of agency (Chapter 7).
13. Failing fully to acknowledge the nature of thought as free-floating explicitness (Chapter 8).
14. Naturalizing possibility and probability (Chapter 9).
15. Not acknowledging the complex, ambivalent nature of "reality" that is downstream of explicitness (Chapter 10).
16. Deflationary or redundancy accounts of the nature of truth (Chapter 10)
17. An overall anti-psychologistic tendency in philosophy symptomized in the reduction of mental entities to functional connections, thoughts to their types linked by logic, emptying the notion of truth, and the grammatical deflation of first-person being (*Passim*).

It is worth mentioning that there is an opposite tendence to that of overlooking explicitness: elevating it to a kind of stuff. The most striking example is the "Nothingness" that Jean-Paul Sartre puts to so much work in *Being and Nothingness.* But that is a story for another time.[42]

PART I

The unholy trinity

CHAPTER 1

Looking in the mirror: a vision of the unholy trinity

1.1 INTRODUCING THE UNHOLY TRINITY

Philosophy has many things going for it. Not the least is that it doesn't require expensive equipment. Even the armchair – that iconic piece of furniture invoked by those who denigrate the discipline – is dispensable. Philosophers worth their salt can philosophize while standing up or walking the dog. This said, the rewards for mobilizing even a modest amount of equipment are disproportionate to the outlay. Consider what can be revealed of our nature when we look into a humble mirror and reflect on what is reflected back to us. And for this purpose, the outlay on equipment can be even more sparing: the mirror can be imaginary.

Indeed, to focus on our thoughts, we need to switch our attention from an actual mirror to the idea of one – not the least so we shall not be distracted by a spot on our nose or a crumb in a beard. The occupant of the mirror who also occupies this page is not a particular individual at a particular moment in space and time; not a bald-headed, grey-bearded individual; rather *the idea* of an individual looking at himself, realized in the self-directed gaze in a mirror. Up close and (im)personal.

So, there we are: staring at ourselves in an imaginary mirror, endeavouring to make ourselves present to ourselves, to bathe in one of the most literal dimensions of explicitness, being given to ourselves in a paradigm of revelation. Our gaze into the mirror sees our gaze out of the mirror, though we know that the returning gaze is not a gaze at all but a blind image of a seeing gaze.

We might characterize what we see, apparently as far behind the mirror as we are in front of it, as "The Unholy Trinity": the It, the I, and, mediating between them, the living creature in virtue of which the "It" becomes an "I"; the Thing, the Organism, and the Person; or the Blob, the Beast, and (in my case) the Bloke. Usually, I see the entity in the mirror as the Bloke, rather than a blob or a beast. What I see staring out at me staring at the Bloke is myself, the person I happen to be, an entity whose state, standing, and welfare is of more intense, and more continuous, interest to me than is that of any other entity in the universe.

I shall come to the Bloke in due course. Before this I shall look past him and his status as a person or even as a living creature and unpeel the material object that seems to underpin the other items that I see.

1.2 THE BLOB

I start here because "Tallis-as-It" seems the most straightforward member of the trilogy and hence a good place to start. Things-as-material-objects are relatively unpuzzling compared with beasts and blokes. They occupy a certain quantity of space at a particular location at any given time. And they have a (relatively) stable existence over time – something that distinguishes them from events and processes, properties, and qualities, which seem more elusive.

On closer inspection, the qualifications for being an object become less well-defined.[1] "Spatial location" seems more applicable to a pebble than to, say, a fog. What's more, the kind of attention we pay to an item may determine whether it counts as an object, an event, or a process. We may think of a mountain as an event in, or a process taking place over, geological time. Contrariwise, we may think of a puff of smoke as a short-lived, unstable object, expanding into extinction, though referring to it by means of a concrete noun makes it seem a little more thing-like than it does when it is encountered without the mediation of words.

These uncertainties do not seem to apply to the body in the mirror. It has clearly defined edges and any changes within it are either hidden away (as in the case of the circulation of the blood, the incessant biochemical processes of the beast or the thoughts of the person) or so gradual as not to be registered. Of course, it is hardly surprising that the skinful that is Raymond Tallis' body seems to be a spatially well-defined object in good standing, because it is Raymond Tallis – and his attention scaled to his sense of space and time – who is registering it. To anticipate something to which we shall return (in Chapter 5), the embodied (or ambodied) "I" who grants his body the status of a material object, who senses that "I am this" where "this" is a thing, is the archetypal or primordial material object on which our sense of the fundamental nature of other objects – trees, sticks, and the bodies of our fellows – is founded.

There are many reasons for judging the body in the mirror to be classified as a material object alongside other entities such as the table also visible in the mirror. As already noted, RT's body – like the table – occupies a discrete portion of space, and it endures in time. The body and the table share a common space and have reciprocal spatial relations: I am behind the table and the table is in front of me. We are separated by an equal distance and that distance between us can be altered either by my moving myself or by the table being moved. Like my body, the table blocks the view of some of the things behind it. The visibility of the two items seems to depend on similar circumstances: opening the curtains to let in the daylight or switching on a lamp in a dark room makes both the table and my body visible. And of course, they both cast shadows whose shape maps their shape and whose sharpness reflects the locations and intensity of the source of the light that falls upon them. They are occupants of the same view.

There are other important similarities. The table and my body are divisible into parts that occupy regions of the space occupied by the whole object. They are both solid, relatively impenetrable. They are subject to the same physical laws. If I press on the table, it presses back on me: our actions and reactions are equal and opposite. My feet leave an imprint on the carpet just as do the table's feet. Besides, there must be something fundamental that RT's body and the table have in common if the former can sit down to the latter and eat a meal. And the same applies to his clothes, given that they can be donned and doffed, and in between donning and doffing they conform loosely or tightly to his form.

Further evidence of the thinginess of body in the mirror includes pressure marks imprinted on chairs and cushions and other items to which he has recourse when he wishes "to take the weight off his legs". Such pressure marks may also be seen when one part of his body presses on another, as when he crosses his legs. His participation in the gravitational field is evident to those who carry him to a place of safety after they have found him in a state of collapse. Like the objects around him, his body has a measurable weight: his body and the table can be quantified in kilograms. When he walks to the shops or squeezes a dishcloth, Newton's third law of motion ("Action and reaction are equal and opposite"), is applied to both parties – legs and pavement, hands and cloth – without fear or favour. The interchange of heat is equally democratic. When he deposits his hindquarters on a cool surface, they donate some of their warmth to the said surface, strictly in accordance with the second law of thermodynamics; and the donation goes the other way when he sits on a hot radiator. The stuff of his body is very similar, for a while, to that of his future corpse, vacated of organic life and personhood.

So, there we have it: my body, furniture, and the walls of the room, are "out there" as fundamentally similar contents of the world. Time to move on to the next member of the trilogy.

Not so fast. Being like a table does not make the body-as-thing entirely straightforward. The way we experience so-called physical objects – how they manifest to us – has been discredited, so we are to believe, by developments in physics, especially over the last century or so. And what physics says is difficult to dismiss. Its authority – its claim to have the last say on what-is – seems to be upheld by success in extending our powers. There is its capacity to predict with an unimaginable accuracy how things in the physical world under certain carefully controlled conditions will unfold. Even more impressive is the extent to which physics enhances our capacity to act on the world through technology. Thanks to recent developments, I can talk to a particular individual in Australia without raising my voice. When physics speaks, therefore, we have reason to listen. (Just how uncritically we should listen is a matter for Chapter 4.)

If, however, we accept that physics has the last word on the overwhelming majority of what-is, the stuff of the world, the stuff that surrounds us, the stuff we interact with and depend on, including (to bring us back to the body in the mirror) the stuff of which we are made, then we are in trouble. There is a vast difference between objects as they are present to us when we engage with them and objects as they are seen through the lens of fundamental science. That gap – characterized by Wilfrid Sellars as being between the "manifest" and the "scientific" image[2] – was famously popularized by the great physicist Arthur Eddington in his discussion of "The Two Tables".

He begins his Introduction to his Gifford Lectures as follows:

> I have settled down to the task of writing these lectures and have drawn up my two chairs to my two tables. Two tables! Yes there are duplicates of every object about me – two tables, two chairs, two pens ... One of them has been familiar to me from earliest years. It is a commonplace object of that environment which I call the world. How shall I describe it? It has extension; it is comparatively permanent; it is coloured; above all it is *substantial*. By "substantial" I do not merely mean that it does not collapse when I lean upon it. I mean that it is constituted of "substance" and by that word I am trying to convey some conception of its intrinsic nature ... Table No 2 is my scientific

> table. My scientific table is mostly emptiness. Sparsely scattered in that emptiness are numerous electric charges rushing about with great speed but their combined bulk amounts to less than a billionth of the bulk of the table itself .... There is nothing *substantial* about my second table. It is nearly all empty space – space pervaded, it is true, by fields of force but these are assigned to the category of "influences", not of "things".[3]

And, if we believe that physics is the true portrait of what-is, the second table is the only one which is really there.

It is worth commenting on Eddington's claim that there is nothing substantial about the table at which he is writing, notwithstanding its appearance of solidity. For it should undermine our confidence that the first, lowliest, member of the unholy trinity – The Thing – should not cause puzzlement; or, at least, cause less than the Beast and the Bloke.

First, there is the philosopher Susan Stebbing's famous criticism that Eddington confused the world with the physical world and the physical world with the world of the physicists, the latter being constructed out of theoretical models.[4]

Secondly, as we shall discuss in Chapter 4, in the century since Eddington was writing his Gifford Lectures, the metaphysical authority of physics has been undermined by many layers of uncertainty as to how its fundamental theories should be interpreted. The table of the physicist has become even less substantial, indeed ethereal. So-called classical objects such as tables and chairs and mirrors and human bodies reflected in them are an embarrassment to an account of the world that sees the fundamental elements of which such entities are composed as (perhaps) not having a definite existence when they are not being observed or, worse, as being mere mathematical structures.[5]

The first possibility is particularly associated with Neils Bohr, though it has been a dominant view among quantum physicists for nearly a century. The Nobel prize-winning physicist Eugene Wigner expressed this particularly clearly. We should understand the appearance of an objective world, he says, "not as a primary reality, but as an emergent aspect of a primary process that can be described as 'experiencing' ... [T]he 'paradoxes' of quantum physics are an invitation to abandon the metaphysical assumption of the primacy of the objective dimension of reality".[6]

The second, also associated with Bohr, is that "words like electron, photon, or atom should be regarded ... as useful models that consolidate in our imagination what is actually only a set of mathematical relations connecting observations".[7]

We are, it seems, obliged to believe that the fundamental stuff of the world is insubstantial. This makes it even more difficult to explain the origin of solid, macroscopic objects, not least the macroscopic microscopes – and the scientists who use them – that make the microscopic realm visible to the macroscopic microscopist if reality is to be found at the microscopic level.

For the present, it is sufficient to note that what science tells us about material bodies makes it more than a little awkward when we are looking in the mirror and the focus of our attention is not any old object such as a table but the body of the person who seems to be looking back at themself – when Eddington's double vision gets up close and personal. Are we to accept that the head we see in the mirror is "mostly emptiness" concocted out of electrical charges whose "combined bulk amounts to less than a billionth of the bulk" of the head that sees itself in the mirror? Can we really accept that there is "nothing substantial" in the body we are looking at – not even the head that is looking at it, and seeing that it is looking at it – because it is emptiness pervaded by "fields of force". Such fields do not seem to add up to the "thinginess" of the item that supports itself on its hands and closes its eyes to minimize distraction from the puzzle that, courtesy of his fellow heads, he has set himself.

Is this not the conclusion we must draw when the findings of physics are applied to the physical objects that are human bodies – which will, of course, include those of physicists? It seems that the things physicists leave behind in the theorizing that makes what-is explicit are things they take for granted in the work they do to generate and test those theories.

So much, then, for the easy bit: the Blob! The egregiously shaped body I see in the mirror is not a substantial thing nor is the mirror that makes it visible. The thought "I am (largely) a void" cuts no ice with me, however often I articulate it. If I recite it, I do it without really thinking it.

Perhaps it shouldn't cut any ice. After all, the criterion for solidity is not an unbroken continuity that is evident even at the nanoscopic levels. Solidity is interactive: my hand and the table are set to match each other as each presses against each. Between the hand and the table there is a relationship of equals, with both parties being resistant to deformation or penetration. My body is more like a block of wood than a splash of liquid or a puff of air. I cannot push my hand through a table because hand and table are equally distant from splashes of liquid and puffs of gas.

Even so, it is difficult to shrug off the fact that most of the space beneath – and indeed within – my skin is in some sense empty. That beneath the interactions between my head, the mirror, and the light that creates a portrait of a solid head, a head that looks as solid as it feels in my hand, or as the roof-beam on which I accidentally bang it, there is a meeting of entities that are 99.99999999 per cent void. That that void is in fact filled with electron clouds – with indeterminate boundaries – does not seem to fill it.

And so, we arrive at the first item of unfinished business – where we might have least expected it – if we take material objects such as human bodies, tables, and mirrors, as straightforward, least enigmatic, stuff, as the place where business begins. How did what is "(mostly) emptiness" come to encounter itself as a solid body – both directly and indirectly courtesy of its image in the mirror – surrounded by other solid bodies including the mirror itself and the other contents reflected, bathed in the same light as bathes the body puzzled by its image in the mirror? If the body is (mostly) a void, it is a special kind of near-void that can make itself, and the items it interacts with, into seemingly continuous substances. And then, joining with other voids with which it can come together in what we might call a community of minds, or its subset the community of physicists, expose the apparent substances as their original (near) void. It seems that, if the world were as physics portrays it, then physics would not be possible.

The mystery will become either more or less compelling – it is not clear which – when, as in Chapter 4, we look critically at the scientific mirror in which we look at the material world, including our own bodies and the mirrors in which we see ourselves. The apparent evaporation of substance, of solid stuff, may be a consequence of placing quantification at the centre of things; of the reduction of the world to what is revealed by measurement; of what-is to how-much-it-is. This process, en route to the (almost-complete) void, burns off the colour of my eyes and skin as secondary qualities, the warmth of my hands as emergent properties, and the shine of my bald head and, indeed, the visibility of the Blob, as the transformation of that which is not in itself luminous into that in virtue of which the mirror reveals me to myself.

Steady. We are getting ahead of ourselves. We shall return to these issues later. For the present, we notice how quickly we run into difficulties when we reflect on that which we make explicit; and that is before we even address the question – the process, the miracle – of our capacity to make things explicit. Already the ground beneath our feet, the solid starting point of the material world, is turning to quicksand.

1.3 THE BEAST

Next up, the Beast.

The body that experiences itself as a body "out there", in the present case courtesy of a mirror, is not after all any old thing, entirely on a par with the chairs on which it sits, the clothes into which it fits itself, or the mirrors whose reflecting surfaces it exploits in order to encounter itself as something out there. We must, that is, not overstate the ontological democracy suggested in the previous section. The body is more than a thing: it is an organism seeing an organism.

The organism came into being by organic means, growing *in utero* where it helplessly assembled itself into an entity whose toes were the far end of its world. It was named "Raymond Tallis" on its howling emergence into the world, a name that it came to identify as, and with, "I".

The human organism shares many features with other organisms, as set out beautifully by Thomas Suddendorf:

> Like all living *organisms*, humans metabolize and reproduce. Your genome uses the same dictionary as a tulip and overlaps considerably with the genetic makeup of a banana. You are an *animal.* Like all animals, you have to eat other organisms – whether plant, fungus, or animal – for sustenance. You tend to approach things you want to eat while avoiding things that want to eat you, just as spiders do. You are a *vertebrate.* Like all vertebrates, your body has a spinal cord that leads up to the brain. Your skeleton is based on the same blueprint – four limbs and five digits – as that of a crocodile. You are a *mammal.* Like all placental mammals, you grew inside your mother and after birth received her milk (or someone else's). You are a *primate.* Like other primates, you have an immensely useful opposable thumb … You are a *hominid.* Like all hominids you have shoulders that allow your arms to fully rotate. Your closest living animal relative is a chimpanzee.[8]

The Beast has properties not seen in the overwhelmingly inanimate universe. The most arresting is that, by interacting with the part of nature it appropriates as its environment, it has, from its beginning as a single diploid cell, assembled itself from the material served up in the maternal womb. To this extraordinary achievement Humberto Maturana and Francisco Varela gave the term "autopoiesis".[9] It is captured poignantly by A. E. Housman:

> From far, from eve and morning
> And yon twelve-winded sky,
> The stuff of life to knit me
> Blew hither: here am I.[10]

During the nine months of the Beast in the mirror's mindless self-assembling, self-structuration, and differentiation, in which his accumulating tissues played an increasing role in creating the environment within which his genes would be expressed, he was of course at best dimly aware of the outside world. It would be many years before he learnt that his early development took place roughly at the time Stalin declared the beginning of the Cold War, Trans Australia Airlines made their first international flight, Clement Atlee revealed his plan for Indian Independence, Pope Pius XII announced the appointments of 12 new cardinals, and the first general purpose computer began operation. An unimaginable world was awaiting him, as "the stuff of life to knit me" was harvested by his growing body through the pipeline of the umbilical cord.

The embryologist Lewis Wolpert has argued that, in the miraculous journey beginning with single-celled zygote born of the fusion of a sperm and an oocyte and resulting in an entity that would ultimately have sufficient wit to catch a bus, run a business, bring up a child, simulate icy politeness, take a position on the Oxford comma, or puzzle over his nature when looking at his image in a mirror, the most portentous step was gastrulation. It marked the emergence of an embryonic architecture, in which an inner layer is differentiated from an outer layer: a more important landmark, Wolpert said, than birth, marriage and death.[11] Indeed, it is probably the most significant evolutionary innovation in the animal kingdom. Herein lies the root of all insides and outsides, of surfaces opposed to depths, and ultimately of the embodied subject who distinguishes the "I" in here from the "it" out there.

Thus, the first and most important phase in the cooking of Raymond Tallis (and you, gentle reader) into the rawest of raw youths. And then there was the process of birth – a transition from an interior, a womb without a view, to a boundless exterior.

It hardly needs saying, even less repeating, that the processes by which beasts put themselves together and, often with the help of other beasts, subsequently curate what has been assembled, are extraordinarily complex in any organism. In specimens of exotic megafauna such as Tallis, however, they are even more mind-boggling. They operate at a multitude of levels: the cell, the organ, the system, and the whole body. In addition to the visible activities – such as eating and defecating, drinking and urinating, shivering, and sweating, self-propulsion towards this and away from that – there are innumerable goings on, most of which are unknown to the Beast or even to the Bloke. His beating heart, gurgling gut, breathing chest, and creaking joints offer a highly censored story of the events beneath and within his skin. Countless micro-agents – enzymes, first- and second-messengers, hormones, to name but a few – regulate every aspect of his bodily chemistry in the battle to maintain the precisely coordinated functions exercised through a variety of structures established to resist the universal tendency towards the ultimate disorder of thermodynamic equilibrium.

The chemical processes of life seem to be transcended by any organism that picks out a parish of what-is as its environment, transforms its physical surroundings into its habitat, most spectacularly in the case of a human organism that *faces* a world, and even, as we are imaging at present, faces its own face.

It seems that I am drifting towards talking about the Bloke, an entity belonging to a species who invented puns, bollards, cantatas, libel suites, power stations, art galleries, and the rules of war. Time to move on.

1.4 THE BLOKE

Looking at my body in the mirror, I am confronted by my gaze looking back at me. I am inclined to say that my gaze enters my gaze – like a finger entering a fingerstall or that my eye is poking my eye out – but this would not be accurate. What I am seeing are my eyes (something sufficiently remarkable in itself) and I only *infer* that they are looking. I cannot, after all, see their looking, see that in virtue of which my face, my head, my body, and the room behind me, are visible. They are opaque, capable of bearing reflections of the windows of the room in which the mirror is situated. It is a paradigm case of how in looking we overlook the process of looking; of how the gaze which reveals what-is looks past itself, does not reveal itself.

The physical appearance of the eyes seems to be of something far too small to be the source of the revelation of a *mise-en-scène* in which I stand, framed by a mirror, as its protagonist. But their importance cannot be exaggerated. The claim by biologist Andrew Parker that "the invention of the eye [542 million years ago] was *the* decisive event in the Cambrian era"[12] gains plausibility when we contemplate the differences between sight and its two most versatile sensory rivals – hearing and touch. We can see moving objects and see *that* they are moving. And when it is a question of an object moving towards us, as threat (predator) or promise (prey or mate), sight gives us forewarning. This is true also of terrain and weather which we can see before we engage with, or are engulfed by, it.

Seeing is the paradigm of perceptual explicitness because there is the greatest, and certainly the most explicit, distance between the perceiving subject and the perceived object – hence sight as foresight. While we may "fore-hear" the predator in the shrubbery, the river that might drown us, hearing does not, at least in humans, deliver a coherent canvas as does sight. The object of visual perception is situated in a continuum of visibilia. It is related to other objects also located in a common space, most significantly to the body of the seeing subject. By contrast, sounds and smells are discrete deliverances.

Like so much of the organism of which they are a part, the eyes conceal how they do what they do. I cannot see what is going on in order that they should be visible as part of a mirror image looking out at me, when the light bounces off from my actual cornea, lands on the mirror, is returned to my eyes where it is harvested, and transformed into an image of the eye in question.

Without the aid of textbooks, I would not know that the light was refracted through a lens, and after crossing aqueous and vitreous humour, prompted a photo-electric reaction in my retina, and this in turn causes electrochemical discharges in my visual pathways travelling to my visual cortex and beyond. Even knowing all this, I still have no idea of how what happens in my eyes and their cerebral connections transforms an interaction between them and the light bouncing off surfaces into a revelation of a scene, of a surrounding world.

There is nothing, that is, in the what-is of the human organism that accounts for its capacity to transform the what-is of (some of) the world (including parts of the organism itself) to that-it-is; and its further capacity to transform perceptual that-it-is into objective knowledge and subsequently an understanding that connects what is known with what else is known. The mystery is placed in italics when we think of our ignorance of the activity in our brain – activity that is commonly believed to be the basis of our gaze and, indeed, of everything we feel, know and think.

In the light (or darkness) of what we cannot experience of our bodies, and the distance between what may be known of them and what is felt, and between what is known and what remains unknown, it is scarcely surprising that there is such a wide divergence between any account of what happens in my body and the story of my life.

That divergence is flagged up on the visible surfaces of the organism RT. Any hairiness he has on the face that is the centre of his attention (such as the beard that makes him look a generic old man) is elective. His clothes were bought rather than secreted *in situ* like fur. He does not therefore look particularly beastly. And this is appropriate because his life and that of his conspecifics is remote even from that of the species closest to him – the higher primates.

I have dwelt on this in other books. I shall not spell it out again but draw on it when we go beyond the mirror.[13] Suffice, perhaps, to respond to Augustine's sardonic observation – intended to pull us down a peg rather than as guidance to midwives – that, like other beasts, Inter faeces et urinam nascimur ("we are born between faeces and urine"[14]) by pointing out that we are the only creatures that brood on this fact – *and in Latin.*

The most mysterious characteristic of the Bloke is the consciousness that has the exquisitely regulated soupy scaffolding and scaffolded soup of living tissue as its necessary condition, though it falls far short of an adequate explanation. Somewhere between the simplest organism and the example of the exotic megafauna that is Raymond Tallis, what-is becomes that-it-is and the world or the sum total of things eventually acquires a name – "the world" or "the universe". What is most striking about the Bloke is the extent to which he is aware of the environment from which the Beast forged itself, of the world beyond that environment, and of himself. If there is nothing in the self-constructed organism that accounts for even elementary sentience – and there isn't, as the failures of coming chapters will remind us – there even less to account for the extraordinary multi-layered consciousness and self-consciousness of the Thing that is the Beast that is the Bloke.

What, then, is a person? What do all persons have in common?[15] Regarding the specimen in the mirror, we have established that he is a material object while admitting that he is not just a material object and acknowledging that the very idea of a material object is beset by theoretical difficulties. In addition to being a material object, he is an organism but not just an organism. What, in addition to a material object and an organism looking out of the mirror? The answer to that question is not short and it is in danger of breaking up into a list that lacks the radiance of the truth to which I am endeavouring to bear witness. Let me, therefore, simply pick out one or two features of the third member of the unholy three-in-one that seem of particular note.

As already noted, RT is conscious – but then so are other organisms. More to the point, he is self-conscious in a way, and to a degree, that does not seem to be the case with organisms, not even his nearest primate kin. This self-consciousness has many dimensions.

First, there is a relationship to what he and others call "his" body. He is related to it in countless ways. Most obviously when he looks in the mirror and, seeing his body, sees "himself". The gaze that is looking out at him is his own gaze, making his gaze "mine" and the body on which the gaze is anchored "me". Those eyes looking at "me" are his own eyes doing the looking. Less explicitly mediated identification with his body takes many forms, as we shall discuss in Chapter 5. They include (a) the body schema by which I am informed, for example, that my head and torso and limbs are, gathered up into a platform for explicit actions; (b) a sense of how I may look – as a material object and in the eyes of others judging my appearance and my behaviour; and (c), in between, as the primary source of my experiences and the primary agent of my agency. This body is a multi-sensory window on a multitude of worlds (locations, situations, predicaments) in which I locate myself and on the numerous theatres in which I enact my many roles, from itch-scratcher, to stroller, conversationalist, officeholder, activist, holidaymaker, fulfiller of short-, medium-, and long-term plans.

Some other non-human organisms seem to be able to acknowledge their own bodies in the mirror as something between theirs and themselves. This was most famously revealed in George Gallop's classic experiments in which chimpanzees who had had lipstick placed on their foreheads while under anaesthetic would try to wipe it off when placed in front of a mirror. This seemed to indicate that they had an inchoate sense that they *were* their bodies.[16] Such "this-is-me" awareness, however, falls far short of the kind of bodily self-awareness, associated with the appearance they might have in the eyes of others, and the curating of that appearance, that is normal for human beings. Or so it would seem from the relative underdevelopment of the community of minds of our nearest primate kin.

Seeing one's appearance through the real or imagined eyes of others – individual others or the generalized idea of others – is a mode of self-distancing and self-embrace that reveals the complexity of the idea that I am my body. My identity with my body is highlighted when I am aware of being perceived by another. Indeed, the other's gaze and the judgement I may read in it, may under certain circumstances nail me to my incarnation. Usually, however, a gap remains. Not even the most direct experience of my body in pain or pleasure or neutral proprioception entirely closes the distance between my body as an object of perception and its being me. That gap is widened by the memories, thoughts, preoccupations, responsibilities which distance me from the present state of my body, from the "what" the body is and the "who" I am.

Something of this gap is captured in the possessive when I think or speak of "*my* body" – the most intimate and omnipresent of our possessions, where "having" wobbles in and out of "being" or, as I shall call it, "amming". Looking in the mirror, I see a gaze that is my gaze, a smile that is my smile, and when I move the hand I see in the mirror, I move my hand – which is why I can understand the instruction "Move your hand". My hand is special and yet not special, as when I obey the command "All salute!" as part of a group. We often treat our hand as examples, entities that do whatever the other hands do.

Identity and possession are just two dimensions of the relationship between the person in the mirror and its unholy trinity siblings. And then there is judgement. I look in the mirror and see an elderly, white, male with certain other characteristics – bald, thin, etc. These words lift the body into a realm whence I see it as an instance of a general type or category. There is no limit to the distance that can be opened by my judgement on what I see of myself, of the irruption of "me" into "it": I may see someone looking serious, or trying to, stooped, badly dressed. My judgement can come from others whose views have been internalized: this is how I imagine I might be seen by others I imagine seeing and possibly judging me. There are many sources of what we think of as our "appearance", as how, in different circumstances, we might appear "in the eyes of others".

The others in question may be particular or general, real or imaginary, encircling eyes or a social media community. They provide numerous metaphorical mirrors in which I can see or define myself or feel myself defined. They furnish frameworks for the many stories of my lives, acted out in cognate realms that are wall to wall with possessions, tools and countless other artefacts – ranging from screws to buildings to bars of soap to cars – to assist the enactment of those stories.

The eye that looks at me therefore when I look at myself in the mirror is not the innocent eye of the organism, but a gaze informed by the collective – or by a multitude of internalized others who may belong to real or imaginary collectives. I stare at myself with eyes other than my own, a gaze woven as much out of words as of eyes. And I of course return the complement by adding to the mirrors in which others see themselves.

We are our bodies, we possess them (though, equally, we are possessed by them), we judge them directly or through the real or imagined gaze of others. In addition, we *use* our bodies. The thing, the organism, that I see in the mirror is the primary and universal agent of my agency. It is that in virtue of which I perform, or attempt to perform, actions in order to bring about or head off futures that I envisage, which I prescribe for myself or are prescribed for me. Shoving, pushing, running, and walking and keeping still, moving from here to over there in order to see, touch, hear, taste, smell, eat, fight, and mate, are aspects of agency we share with other organisms. But our agency is transformed by tools and they in turn are inspired by Aristotle's "tool of tools", namely the hand.

Of all the parts of the body, the hand is the one we most consciously *use* – exploring, communicating and manipulating our surroundings.[17] This primary "tool of tools" is assisted by a multiplicity of secondary tools, and many modes of cooperation with others sedimented into institutions, legislatures and constitutions. Among those higher-order tools the most versatile is the language that enables us to be co-present in a limitless variety of ways, with individuals and groups of others, drawing on the collective knowledge, expertise, understanding, of our conspecifics.

The varieties of the Bloke's consciousness – and the extent to which they are unique to the species to which he belongs – is a topic suitable for a book even fatter than this one.[18] For the present, however, it is worth reminding ourselves that the third member of the unholy trinity is blessed (or cursed) with many modes of perception of his body and the world around it; with several types of memory (habit, factual, episodic and autobiographical); and with different kinds of knowledge (of facts, of concepts, of himself) and of expertise. His consciousness is interwoven with that of a multiplicity of ill-defined but vast constituencies of others, communities of minds, to which he belongs as a paid-up member of a culture or cultures. He is an inheritor of shared pasts that shape the senses he makes of the events in the interval between womb and tomb, an anticipator of a multitude of collective futures, and a native of proliferating landscapes of technologies.

The Bloke then, is a long way from the Thing and the Beast. Hence the difficulty of seeing how the physical world – which includes the human body as a material object and as an organism – could give rise to people and the role they play in delivering the many-stranded transition from what-is to that-it-is.

To the catalogue of distances already spelled out, I want to add another, central to our nature as human beings: tensed time that is not seen elsewhere in the physical world. As I look into the mirror, I cannot see any direct presentation of the past and future: they are invisible to any kind of gaze that we might share with animals. We shall discuss the place of tensed time in the creation and curation of the sense of self in Chapter 6 and in underpinning agency in Chapter 7.

There is something else happening in this head, also not represented in the mirror image. I am referring to thinking. The huge distances that separate the Person from the Organism, Bloke from Beast, are most obviously instantiated in our thoughts to which Chapter 8 is devoted. Even if my facial expression seems to suggest that I am thinking, *what* I am thinking could only be guessed at (almost certainly inaccurately) from the image in the mirror. After all, I could not determine the thoughts I was having from a photograph taken of a mirror image of my face.

The hiddenness of thought is of a different kind from the hiddenness of the organic processes beneath, and indeed within, my skin. No inspection, however close, of my image in the mirror or even images of the interior of my body, could reveal that I was thinking or what I was thinking about. In part this is because my thoughts have intentionality and intentional objects: they are *about* states of affairs, actual or possible, particular and general, which are necessarily separate from the thought. Any "here" they have is hollowed out by an elsewhere. Irrespective of whether what is proposed in my thoughts is true – for example that Paris is the capital of France – or false – that there are flying horses – they do not exist in the world in any way comparable to their mode of existence in my thoughts.[19] Thoughts are not connected with their referents in the way that things in the world revealed in the mirror are connected: my eyes with my face, my hand with the table it is resting on. That is why I could not tell, had I not have privileged access to my thoughts in virtue of currently thinking them, that the RT who is looking in the mirror is thinking about his dinner, his professional duties, his upcoming holidays, or his endeavour to make sense of himself. As a thinker, he is, and he is not, in front of, and reflected in, the mirror.

1.5 FURTHER REFLECTIONS ON BEING A REFLECTION

Thus, some of the ways in which the Bloke looking at his body – a material object that is an organism – is different from the body he is looking at. Before we drag ourselves away from the mirror, it is worth reflecting again on the fact that what we are looking at is not ourselves but a mirror image. Nevertheless, a mirror image is an important, perhaps primordial, form of self-encounter, available to humans (and, as we have seen, perhaps other animals) so long as there are reflecting surfaces in the surrounding world. We can be present to ourselves in this way only through vision. Narcissus is confronted by himself in a way that Echo isn't. We see the eyes that see the world, the face that faces the world. This is different from our immediate acquaintance with our bodies, either through proprioception, enteroception, kinesthetic experience, the body schema, the beating heart, the inflating and deflating chest, gurgling gut, aching legs; or through engaging with the surrounding world pushing and shoving against things that push and shove back, walking uphill and down, gripping and kicking and so on.

The intimate relationship between the light of the world and the light of the mind, between vision and explicitness, is reflected in this arresting passage from the medieval philosopher and polymath Robert Grosseteste: "[Light] is more exalted and of a nobler and more excellent essence than all corporeal things. It has, moreover, greater similarity than all bodies to the forms that exist apart from matter, namely, the intelligences".[20] Of all the senses, vision is closest to knowledge: it is the most epistemic of the senses. Hence, we say "I *see* what you *mean*". And we look with words as well as eyes: from our earliest days we see the world through word-tinted, word-shaped spectacles. For the tidy-minded philosopher this may be a source of despair as well as wonder.

Our mirror image develops our sense of ourselves as material bodies on all fours with sticks and stones and as organisms with features overlapping with that of other living creatures. It contributes to the intuition – served up to us in showers of connected and disconnected, large and small inklings – that we are located in the natural world. We make ourselves explicit as things and as organisms and so set the terms of the puzzle of the unholy trinity.

One other important feature of the person not shared with the material object, or the organism deserves more attention than we have given it so far: our extraordinary capacity to combine our awareness and self-awareness with that of other persons. This capacity extends from sharing the present in gestures (notably pointing), language, and other symbolic systems used to communicate, to the creation of a community of minds, and spheres corresponding to it of artefacts, institutions, and (most relevant to the present inquiry) a vast realm of knowledge. My gaze is imbued with some of that knowledge when I look at the man in the mirror seeming looking back at me; most obviously when I see, or try to see, or pretend to see, myself as a material object or an organism.

Which brings us back to where we began: RT as a Thing seen through the lens of fundamental physics and his endeavour to engage with the thought that, contrary to his own feelings about the matter, his carnal being is largely empty space. The Bloke doesn't sincerely believe this of the Thing. But, being a philosopher, he enjoys hovering in the vicinity of vertiginous thoughts.

Let us suppose that the widely separate atoms of the material entities – tables and living human bodies – are gathered up by the synthetic activity of his consciousness, by his sense of *being* this Thing. How shall we explain this extraordinary capacity of his consciousness?

We think of consciousness as a late arrival in the order of things – requiring a macro-organism with a nervous system as its substrate. Even more recent is the self. What is more, "amming" requires a certain density of "being" that seems a far cry from the largely empty cloud of elementary particles which, according to the most powerful theories in physics, is what the body truly is. We require, it seems, substantive things to house the consciousness that itself transforms the thin cloud of particles into a substantive unity. If it were the gaze of the self-conscious subject turned in on itself that transformed the near emptiness into a substantive organism, into solid items in good standing, whence the origin of that gaze? And how did that gaze combine with other gazes of other organisms – fellow humans, fellow scientists – to create the lens that supposedly betrayed the fact that these organisms are largely emptiness?

The many-layered mystery of my seeing myself in the mirror, such that light bouncing off one object (my body), travelling to another (the mirror) and returning to the first object (the eyes), revealing the body to itself as itself headlines the mystery of explicitness. If we accept the standard story of our place in the order of things, we have to explain how, in a miniscule part of the universe (occupied by sentient creatures), over a small stretch of its longevity, light has become luminous – visible in itself and capable of rendering visible the objects that are its primary or secondary sources, including the very object – the Blob-Beast-Bloke – in virtue of which there is visibility. When electromagnetic radiation enters the living eye and is inexplicably ignited into dazzle and revelation, it ends an interval of billions of years between what-is and that-it-is, between Being and its being registered as existing.

While nothing could seem more straightforward than looking at oneself in the mirror, there could be no deeper mysteries than those deftly excavated by a gaze that unpeels the gazer as a material object, as a living organism, and as a person, that is to say as an entity looking at itself and thinking about looking at itself. The opacity of the head makes it potentially visible while its epistemic transparency transforms that potentiality into actual visibility. And while the Blob might seem easier to explain than the Beast or the Bloke, this overlooks the apparent fact that, if one embraces the physicists' account of the material world, the Bloke must be implicated in the construction of the macroscopic Blob – the transformation of an insubstantial buzz of nanoscopic probabilities into an item that can block and reflect light, sense it, and translate it into a revelation of a solid object. This problem arises before we encounter the mystery of the gaze, itself invisible, turning what-is into that-it-is, gathering up inanimate stuff into a world that is faced by, and lived in by, a conscious subject.

The Blob-to-Bloke story rests on the general acceptance that matter (or mass-energy) is universal and ancient. Animate matter, especially in the form of persons, by contrast, is rather localized and relatively recent. And conscious and self-conscious matter that identifies itself as "this thing here" and "that thing over there (in the mirror)" is very much a *parvenu* in the order of things and a rarity in the universe. If we believe that what-is unfolds according to universal laws, operating since the beginning of the universe preceding the Man in the Mirror by billions of years, questions arise: how are the members of the trinity related to each other; more particularly, how do the later members arise out of the earlier; how do beasts arise out of things, and persons out of beastly materials?

Many philosophers respond to this problem with hand-waving passing itself off as explanation. The hand-waver's favourite term is "emergence". Beasts, they say, "emerge" out of things because they have a certain *complex* structure; and persons emerge out of beasts because they have even more complex structures. We shall discuss this claim about the origin of explicitness, and the transition from what-is to that-it-is, in the next chapter.

CHAPTER 2

From things to persons: "emergence" as a non-explanation

2.1 GETTING FROM THE THING TO THE PERSON

When I contemplate what I see when I stand in front of a mirror, I can pay separate attention to the material object or The Thing, to the living Organism, and to the Person. It seems as if they are and yet are not the same entity.

Of course, they are the same entity. To point to one, is to point to all three. None can move without the others moving. They occupy the same space and interact with the same material objects: if the Blob trips over a stone, so do the Beast and the Bloke. The jacket I drape on the Thing encloses the organism and is worn by the Person. It is the same jacket maintaining the temperature of the Blob and delivering warmth to, and protecting the life of, the Organism; and is worn with a (very) modest degree of dash by the Person. When the Thing is destroyed, the Beast and the Person also disappear.

And yet this trinity, this three-in-one, is not entirely one. Its members can part company. A non-fatal but devastating head injury resulting in life-long coma removes the Person, while leaving the Organism intact: living, yes, but no longer me. An even more serious injury may result in a material body that is no longer a (living) Organism: a corpse that may, for a while, occupy roughly the same space and have many of the basic physical properties as the Organism. In short, the Organism can outlive the Person and the material body outlast the Organism.

The separation of the person from the organism, and of the organism from the material body, is of course one-way. The existence of the person is impossible without the organism maintaining itself in a narrowly defined range of states; and the life of the organism is likewise dependent on the material body also being in a narrow range of states.

While, therefore, the conscious person is in one important sense identical with the living organism and the latter with the material body, in other equally important senses the three are not identical: they are three, not one. Nevertheless, the thing is not separated from the person in the way that one thing is separated from another – as, for example, the body from the mirror in which it is reflected. And the person is not separated from the organism in the way that one person is separate from another.

A time-honoured way of capturing something of this kind of unity-and-separation is to compare it with that between (1) a block of marble and (2) the statue carved out of it. There may be further separations – as (3) between the statue and something valued as a work of art, an example of X's late style, or as a precious possession. And the inseparability is likewise captured. It is not possible to move, break, or hide behind (1) without moving, breaking, or hiding behind (2) and (3). When we accidentally drop and break a piece of marble, we also drop and break a statue, an example of X's late style, or a precious possession. The statue and the block of marble seem, therefore, to deliver a straightforward example of more than one object occupying the same space. They are made of the same sample of the same stuff, and they have the same weight, the same size and the same shape.[1] Whatever happens to the marble of which the statue is made also happens to the marble. If the marble is broken, chipped, moved from place to place, so, too, is the statue.

But this identity of constitution is not token identity. If, as we might reasonably assert, the block is defined by the sum total of its atomic constituents, then the removal of one atom would change the identity of the marble but leave the identity of the statue intact. It would still be a statue of David; indeed, *this* statue of David. The discussion of the multiple identities of a single piece of matter, of *constitutional* definitions of identity, remains unresolved. Instead of resolving the paradox of three-in-one, it only highlights it. The analogy with the block of marble and the statue does not deliver what is needed to cross the bridge from things to persons.

Whereas the person, as a conscious subject contemplating herself directly or in a mirror, makes the organism and the thing visible and distinct, it would not make sense to suggest that the statue makes the block of marble explicit as a block of marble – even less as a statue. Both the block of marble and the statue (not to speak of the items as a "precious possession") require the *viewpoint* provided by a conscious subject for them to be present and unified and consequently distinct. Likewise, when we consider the triune in the mirror, we must acknowledge that we are faced neither with more than one entity or with a single entity that paradoxically manages to fail to be identical with itself; rather we are looking at different aspects of a single entity, picked out by different modes of attention. That kind of attention is not available to be paid to themselves by pebbles or trees or most animals. The block of marble is not a statue for the leaf falling on to it, the moss growing over it, a slug crawling over its neck, or even for a chimpanzee catching sight of it. It is not even "a block of marble".

This matters. The example of the block of marble and the statue has been mobilized on many occasions to make the emergence of different kinds of entity from a single kind of stuff (typically matter) easier to understand and to help philosophers feel more relaxed about the fact that a single entity may be both (say) a piece of matter and something a bit more interesting. Unfortunately, unlike the piece of matter, which is available alike to humans and snails – and to moss and rain in the possibility of, and the constraints on, physical interactions – the form *qua* statue is available only to a conscious subject with a sensory system capable of harvesting the statue as a unity and, in addition, recognizing what kind of thing it is: its purpose, and its intentional object – in the case of our example, King David who lived many centuries ago.

The analogy between a portion of marble and a statue, therefore, does not take us across the gap between a material object and a person for whom that material object is their body, a body which they embrace, and by which they are embraced, as themselves, and as that in virtue of which they are located in a world that they inhabit as their world. What is missing in the block of marble–statue analogy is any account of *explicitness*.

This is so obvious that the seeming attractiveness of the analogy calls for explanation. The explanation is to be found in the terminology used to characterize the mystery as a problem of the "emergence" of conscious subjects from unconscious matter.

We shall come to "emergence" presently but first I want to examine another term used in this context: that of "realization" invoked to cross what Joseph Levine politely described as "the explanatory gap"[2] between the material world and phenomenal consciousness. As Tim Crane has described the explanatory gap, "we lack an adequate deductive explanation of all the truths about consciousness in physical terms".[3] That lack, of course, encompasses the fact that there are entities (human beings) who seek for and arrive at explanations, entities for whom there are truths (and falsehoods) – a twist that ties an interesting knot in the explanatory gap. It is the conscious subject who is asking the question about the transition from material objects to conscious subjects.

What, then, of the notion of "realization" which is often invoked in connection with the relationship between the block of marble and the statue; of the idea that the higher-order statue is "realized" in the basic stuff of marble, mere matter? "Realization" does a lot of work in philosophy – especially in the philosophy of mind. The neural theory of mind – the topic of the next chapter – is often parsed as the belief that mental phenomena such as perceptions, memories and thoughts are *realized* in neural activity. What is wrong with this?

A lot. The fundamental reason for rejecting the employment of realization in this context is that its home base is in human consciousness. Typically, we apply the concept of realization to something that exists as a possibility before it exists as an actuality – as when a statue is first envisaged and is subsequently realized in a piece of marble chosen by the sculptor. The transition from possibility to realization presupposes consciousness as the place where possibilities are envisaged – that is to say, generated (see Chapter 9 *et passim*). The concept of "realization" cannot therefore narrow the gap between consciousness and something as apparently remote from it as neural discharges. The unidirectionality of "realization" (block of marble to statue, ionic discharges to conscious experiences) further exposes its explanatory inadequacy. We do not say that the block of marble is "realized in the statue" or the ionic discharges "are realized in the conscious experiences". This asymmetry betrays how the explanatory gap between the thing and the person is not closed.

There is the additional problem of explaining how an entity can be two distinct things or processes – so that a single process can be both a neural discharge and a perception. The suggestion that these are two aspects of one thing relies on the assumption that one gets two separate aspects – which are typically the product of different modes of attention – out of one thing in the absence of consciousness.

I make this (perhaps too obvious) point in the service of a more general point: that the language used to "explain" the passage from material objects to conscious subjects, or to account for the origin of explicitness in a universe of wall-to-wall what-is, innocent that it is and of what kind of entity it is, typically involves terms that themselves presuppose explicitness.

So much for the language that bypasses the problem. What of the way the problem is formulated? Why, that is to say, do we think there is an uphill gradient of inexplicability as we pass from the Thing to the Person? After all, that there is Something rather than Nothing entirely lacks explanation, from which it would seem to follow that everything is equally unexplained. And what we take to be "basic what-is", namely matter, is becoming more enigmatic. As we discussed in the previous chapter, matter seems an increasingly unpromising explanatory jumping off place.

Why, then, does this way of framing the problem seem inescapable? The answer lies in the fact that the features of the Thing as "material object"

– such as solidity, being subject to gravitational forces, and obstructing (and reflecting) the light – are universal throughout the macroscopic world. In contrast, the features of the Organism – self-movement, homoeostasis, sentience – are found in only a small subset of material objects. They are (so we are to believe) of relatively recent origin – perhaps 500 million years ago in a 13.7 billion-year-old universe. As for the characteristics of the Person – self-consciousness, complex agency, thinking – these are restricted to an even smaller subset of entities of an even more recent vintage. It is this that justifies – or seems to many philosophers to justify – the idea of matter as the basic level of what-is, which every mode of being achieves. Hence its status (or lack of status) as the starting point for explanation, as the near side of the explanatory gap. Organisms and conscious subjects seem to require more explanation than rocks because they are relative rarities in the universe as it is at present and latecomers in its history. So, there is an additional story to be told about them, a story of how, in a world where matter was everywhere and its rule universal, they came into being.

Thus, the origin of the problem highlighted by the unholy trinity. If all things in the macroscopic realm – the realm of classical physics – are portions of matter with the universal properties of matter, how does a living organism, with its quite different properties – self-construction, self-maintenance, self-movement, and (possibly) sentience – arise? And there is a more compelling question: how does a small subset of living organisms acquire the characteristics of a person and become a self-conscious being whose successive states add up to a *curriculum vitae* at least in part directed, shaped, and embraced by itself *as* itself? What is it about the (material) human body that enables it to be conscious of, to look after, and think about itself? How can the light-mediated interaction between the body and the mirror make it possible for the body to see itself in the mirror as "I"? The overwhelming majority of items in the universe are entirely innocent of their own existence – neither knowing nor caring for themselves. For the greater part of the history of being, being had no history.

The mystery becomes deeper when we rephrase it as follows: How does the person arise in or out of a material object made of constituents whose history is defined by processes that, elsewhere in the universe, are not associated with sentience, even less personhood? How could unhaunted, apersonal matter generate a first-person being who in turn *faces* a third-person or no-person world and is conscious of herself doing so?

2.2 THE APPEAL TO EMERGENCE

Philosophers engaging with this mystery often try to make it manageable by framing it as a question of the *emergence* which will explain novel properties, states, dispositions, and powers out of constituents that do not possess them. They usually ascribe emergence to the assemblage of those constituents into certain patterns; into a structured whole that, in virtue of its structure and wholeness, has properties that can be fundamentally different from those of its parts. Persons, for example, emerge from bodies because of the many-layered ordering in those bodies – in intracellular structures, cells, organs, bodily systems – of their basic constituents and processes. This is how persons have properties and, more importantly, causal powers, that are dependent on constituent parts that do not have them. As is the way with discussions of emergence, the nature of that dependence is not always clear, encompassing causation, grounding, realization (*vide supra and infra!*), or a necessary connection (supervenience).

In thinking about emergence, it may be helpful to set aside the middle member of the unholy trinity, namely the Beast – at least for the present – if only because it is not clear where "matter" ends and "living matter" begins. The ambivalent status of viruses illustrates this. It is equally unclear where "organism" ends, and "person" begins – what counts as subjectivity of the kind we associate with "I". Let us, for the present, confine our investigation to the emergence of personhood in an apersonal, largely insentient material universe. (I shall return to the emergence of organisms briefly towards the end of this chapter.)

Behind the reassurance seemingly associated with the concept of emergence is that of an *explanatory gradient* and a continuity, resulting in a transition from what is considered to be a less puzzling kind of entity to what is considered to be a more puzzling one. Gradualness and continuity, it is suggested, make the transition less resistant to explanation than would be an abrupt jump from things to persons who are conscious subjects, such that pebbles started making sarcastic comments about adjacent pebbles. This gradualness is particularly impressive: billions of years seem like sufficient cover for what we might call "ontological creep" or "ontological sneak".[4] It seems that modes of being that "emerge" didn't just "pop up".

Emergence is typically thought of as being of two kinds: weak and strong.

"Weak emergence" refers to the claim that the distinctive properties of persons are dependent on those of the matter of which they are made but are sufficiently independent of those properties for the person to have causal powers (most strikingly expressed in agency) not seen elsewhere in the material world.

While, at the higher level (the person) there are genuinely novel features, we are assured that these can be fully explained by the processes that take place at the lower level (the thing). New powers arise from the interaction between constituents that do not individually have those powers. While, for example, mental properties are distinct from physical properties, having different causal powers from them, they are, nonetheless, properties of physical entities – typically identified as neural discharges. As Jessica Wilson has expressed it, weakly emergent features retain a subset of the causal powers of the stuff from which those features emerge, while strongly emergent features have a distinct causal power.[5]

This idea of emergence seems to offer a way of embracing materialism without reducing everything that happens in persons – conscious experiences, thoughts, voluntary activity – to what is seen more widely in the physical world: hence its description as "non-reductive physicalism". Physicalism affirms that "all natural phenomena are wholly constituted and completely metaphysically determined by fundamental physical phenomena, entailing that any fundamental-level physical effect has a purely fundamental physical cause".[6] Non-reductive physicalism accepts that mental entities are real, but they arise out of material events as a result of processes that are regulated by the laws that are seen elsewhere in the physical world. "Physicalism" implies a rejection of the suggestion that there is a distinctive substance called "mind" operating independently of the physical world. More radically, it claims to collapse the distance between entities that seem to be fundamentally different, starting from the position that they are fundamentally the same.

Such emergentism seems reassuringly consistent with the scientist's – notably the physicist's and biologist's – account of the evolution of the natural world. Mental phenomena are properties of physical objects, most obviously of brains. Scientism is honoured. If, however, it seems to the reader that weak emergence is more like a description, or a "one-fine-day" promise of explanation, than an explanation, then join the club.

One element of the physicalist non-reduction of weak emergence is something we have discussed in the previous section: the appeal to the idea of "realization"; more specifically, "multiple realization". The handiest analogy for understanding this in the case of consciousness is that of the many ways in which software programmes may be realized in a multitude of circuits in a hardware platform. While the hardware is to be understood as a collection of physical entities, the function delivered by the software is not. There is, therefore, a dissociation between the physical constituents of the hardware and the function it serves.

The analogy is unpersuasive – not least because computer hardware does not deliver as software in the absence of a conscious user. This was the problem we encountered in the simpler example of a statue. It, too, can be multiply realized: in marble, yes; but also in clay, paper, or metal. The lump of material does not, however, count as a statue in the absence of a conscious subject able to see it and, what is more, see it for what it is, what it is intended that it should represent. It is not a statue for a chimpanzee leaning against it, or a slug crawling over it, and even less for another piece of marble colliding with it. The notion of something being realized in something therefore depends on separating an entity from itself – into realizing and realized features – and reuniting it with itself to classify it as an instance of something. This separation and union clearly needs precisely those higher-level states that are supposed to be explained, or at least deflated, by the notion of realization. I am, of course, talking about what-is being made explicit.

So much for weak emergence.

In the case of "strong emergence", there is downward causation from the whole, with its emergent properties, on its lower-level constituents. Moreover, the emergent properties "are not *deducible* even in principle from truths in the low-level domain".[7] In short, emergent entities exhibit behaviours that are not, and could not be, exhibited by the material from which they emerge. To this writer, it is difficult to distinguish the appeal to strong emergence from giving up on emergence as explanation.

It has been suggested that strong emergents, while they have radically non-physical features, may involve unknown lower-level processes that might in due course ("one fine day") be discovered. This belief is connected with the case for panpsychism, according to which there is no mystery about the emergence of consciousness in certain organisms because it was there all the time in the matter of which organisms were made, though for various reasons it is not observed. (We shall discuss this in the next chapter.) The emergent properties may exercise both downward causation – as when my thoughts that envisage possibilities shape my actions and hence the movements of my body – and horizontal causation – as when one thought leads to another.

For most philosophers, the passage from lifeless matter to organisms is an example of weak emergence. It is assumed that life is, or will be, explained by the behaviour of cells in turn explained by biochemistry, itself explained by chemistry. There is no need to invoke a "vital force" additional to those forces evident in the non-living physical world – gravitational, electromagnetic, and the strong and weak nuclear forces. The passage from organic being to conscious being, however, is in reality strongly emergent: it does not look susceptible to being explained by what is known of the processes seen inside the bodies of living organisms, even inside neural tissue.

Some philosophers make much of the distinction between diachronic and synchronic emergence. In the former, the base properties and the emerged properties appear in temporal succession; in the latter, they exist side by side. While one might expect that that which has emerged to be later in time than that which it has emerged from, the distinction is not clear cut. Few will dispute that lifeless material objects appeared in the universe before living organisms and insentient organisms before conscious subjects. There is an obvious diachronic succession. Our unholy trinity, however, reminds us that the Thing, with the properties of material things, the Organism with the characteristics of living things, and the Person with the features of a conscious subject, can, and indeed must, exist side by side. The entity in the mirror is a thing that exerts and experiences gravitational force and reflects light like any other material object; and it is the site of many biochemical processes that it shares with lowly worms. The Person, that is, is not a place where pebble-like entities have evaporated into spirits. We do not shake off our material or organic past – hence, alas, our mortality.

It is tempting to argue that it is the succession of *types* – "material object", "living being", "conscious subject" – that emerge diachronically while it is in the *token* or individual instances, that basic and emerged properties exist side by side. The distinction is not useful. After all, the token Raymond Tallis has a history of diachronic emergence *in utero* – as molecules are gathered up into an organism that itself develops towards personhood, or at least its possibility, and the emerged entity retains, as we have noted, the properties it shares with things and with organisms.

Based on the story so far, the Person in the mirror, being something of a sceptic, suspects that "emergence" has more explanatory promise than explanatory content. It certainly does not point to a mechanism. To what, then, does it owe this apparent explanatory promise?

As already noted, "emergence" usually implies something gradual – as when, in the history of the universe, persons emerge from things via organisms – a process that is granted billions of years. New kinds of things that come into being slowly may seem like business (almost) as usual. It takes the surprise out of change: each phase, each stage seems to be part of a continuous process permitting ontological creep. The following passage from William James is very much in tune with this intuition: "The demand for continuity has over large tracts of science proved itself to possess true prophetic power. We ought therefore ourselves sincerely to try every possible mode of conceiving the dawn of consciousness so that it may *not* appear equivalent to the irruption into the universe of a new nature, non-existent until then".[8] It is not, however, self-evident that slow emergence is more self-explanatory or easier to explain than sudden emergence.

One could be forgiven for suspecting that "emergence" has little explanatory power; rather that it is a description of change packaged with the reassurance that there is "nothing (unexplained) to see here".

2.3 EMERGENCE FROM WHAT? "EVERYTHING IS MADE OF ATOMS" 1: THE EVAPORATION OF ATOMS

> According to convention, there is sweet and bitter, a hot and a cold, and according to convention there is colour. In truth, there are atoms and the void. Democritus[9]

It may be argued that the problem to which emergence is an unsatisfactory response is self-inflicted. It is the result of starting with matter along with the assumption that, courtesy of the physical sciences, we know what matter is and what it might give rise to, the forms it may take, the behaviour that is to be expected of it. We shall look critically at this assumption in some detail in Chapter 4, but it is appropriate to examine it now in the light of the assumption that all is matter and, given the nature of matter, persons are ontologically particularly enigmatic. But there is an explanatory gap even before we get to the Organism and the Person.

While the thinginess of the Thing may be less puzzling than the beastliness of the Beast and the personhood of the Person, the scientific advances of the last century suggest that the matter of which the Beast and the Person are composed is densely packed with question marks. As we noted in the last chapter, the very idea of matter, including the macroscopic matter that we see when we look at ourselves in the mirror, has become deeply puzzling. In short, the jumping off point of any emergence story – what, in short, emergence is emerging *from* – is not very supportive. The starting point is as enigmatic as the destination and the journey between them. Notwithstanding some rather extraordinary transformations, the *atom* – or rather "atom-talk" – remains at the heart of the scientific account of the basic stuff of the universe. Consider this famous passage from Richard Feynman, Nobel Prize-winning pioneer of quantum electrodynamics. It appears early in the first of his *Lectures on Physics* (1963):

> If in some cataclysm all of scientific knowledge were to be destroyed, and only one sentence passed on to the next generations of creatures, what statement would contain the most information in the fewest of words? I believe it is the atomic *hypothesis* (or the atomic *fact*, or

> whatever you wish to call it) that *all things are made of atoms – little particles that move around in perpetual motion, attracting each other when they are a little distance apart, but repelling upon being squeezed into one another.*[10]

These "little particles" have undergone some strange transformations in the past century. Indeed, they had done so by the time Feynman was giving his classic lectures, not the least as a result of Feynman's own contributions to physics. The "particles" of particle physics have become ever more ethereal; they seem to be saving their skin by successive redefinitions. According to one, fairly standard, interpretation, they are quantized waves in a field that spreads out through the entirety of space and time.

Matter composed of elusive subatomic particles – fermions with half-integer spin seem to be the closest to self-subsistent entities – doesn't seem to be much of a platform from which to set off on any explanatory journey to understanding the relationship between the members of the unholy trinity in the Mirror. The various interpretations of quantum mechanics that Paul Dirac claimed to "explain all of physics and most of chemistry"[11] rival one another in what by normal standards would seem madness, a madness which, however, must be taken seriously because of the astonishing reach and precision of their predictive power and their role in making possible the technology that surrounds us wall to wall and vastly enhances our agency.

Atoms seen through the eyes of contemporary physics are not self-standing nanoscopic billiard balls bumping into one another. As James Ladyman and Donald Ross argued in *Everything Must Go*, "Physics has taught us that matter in the sense of extended stuff is an emergent phenomenon that has no counterpart in fundamental ontology".[12] Such as they are, most elementary particles even have to send out for their mass: they need (for example) "the Higgs field to gain masses and seed structure".[13] In "quantum fields, virtual particles bubble in and out of existence".[14] What is more, "the stuff of the elementary particles of the Standard Theory accounts for only 5% of the mass of the universe".[15] The remainder is currently unobservable dark matter that keeps dark energy for company. And it has been recently suggested that the material universe – which is supposed to have had the unpromising, if precisely defined, beginning, "as a hot speck of energy ... a hundredth billionth of a billionth of a size of a proton"[16] – may have had no beginning at all.[17] It would appear that the base state from which higher entities emerged did not itself emerge.

That is speculation. What is not speculation is that one widely accepted outcome of the quantum revolution is the conviction that there is nothing definite at the atomic or sub-atomic level in the absence of observation or, more precisely, measurement. As Manjit Kumar expresses it: the belief in an objective world whose smallest parts exist objectively in the same sense as stones or trees exist independently of whether we observe them was a throw-back to "simplistic materialist views that prevailed in the natural sciences of the nineteenth century".[18] The physicist Sabine Hossenfelder has put it very clearly:

> Matter is not made of particles. It is made of elementary constituents that are often called particles but are only described by wave functions. A wave function is a mathematical object which is neither a particle nor a wave, but it can have the properties of both. The curious thing about a wave function is that it does not itself correspond to something which we can observe or indeed something that travels in the sense in which a wave in water travels. Instead, it is only a tool by means of which we calculate what we do observe.[19]

And if that does not make pebbles and our own bodies sufficiently ethereal, Hossenfelder reminds us that it is only after a measurement has been made that the probability of what has been measured changes to 100 per cent. That is when "what-might-be" becomes "what-is".

The bottom seems to have fallen out of the most basic level of being; as Tuomas Takho expresses it, "Reality has no ultimate building blocks".[20] Takho's argument is that fundamental entities have to exist independently and there are elements in the Standard Model of the Atom – notably quarks – that seem ontologically interdependent.

The orthodoxy in theoretical physics is increasingly that ordinary sized bits of matter are *not* ultimately made up of nanoscopic bits of matter. Your material body – or the basic constituents out of which it is supposedly constituted – seems to evaporate before the gaze of physics more completely than Eddington's table. Solidity, location, size, all its modes of definiteness, of thisness, seem to be absent in those entities it is supposed to be made of.

Whether and for how long this will remain true is uncertain. The central point is that the notion that the atomic hypothesis has undergone such transformations that nothing recognizable remains in the idea of atoms, not even their indivisibility – which is what, etymologically, the term "atoms" ("*atmos*") means. (It is ironical that "The Atomic Age" began when it was recognized that atoms were *not* "*atmos*".)

At present once-indivisible atoms are divisible into sub-atomic particles. The list so far is 12 fermions (6 leptons and 6 quarks), 12 gauge bosons (Z, W^-, W^+, 8 gluons, and the Higgs boson). The menagerie continues to grow, there are virtual as well as actual particles, and their relationship to one another is fluid. At present, as we have noted, the top of the hierarchy is the spin-negative fermion. And those particles, far from being discrete, are so closely entangled that it is possible to interpret them as quantized field excitation states. Indeed, the dissolution of particles into fields may lead to their becoming purely relational, mathematical structures. As Ladyman and Ross expressed it in the passage quoted earlier, "matter in the sense of extended stuff is an emergent phenomenon that has no counterpart in fundamental ontology" – according to which "everything must go".[21] "Farewell" to indivisible and discrete nano-billiard balls. "Hello" to the discreteness of, for example, quantized space does not seem to make up for this loss.

We need, therefore, to modify Feynman's message to the Martians that all things are made of atoms by adding that "And nobody knows what the devil atoms are; nor, judging on present performance, are they ever likely to". The names of sub-atomic particles – "quarks", etc. – seem to mock any endeavour to break out of mathematical characteristics of the ultimate constituents of what-is. None of which is to discount the heuristic power in physical science of an homogenizing substance ontology, guiding inquiry notwithstanding that substance seems to evaporate as we dig into it. Or that (to quote Neils Bohr) "Everything we call real is made up of entities we cannot call real".[22]

The assumption implicit in atomism that ultimacy is identical with minute, discrete, indivisible entities seems, to say the least, vulnerable. At any rate, something outside of the reach of fundamental physics is required to explain how what-is can underpin the solid fellow – the Thing-Beast-Bloke – in the mirror. According to an increasingly dominant interpretation of quantum physics, that something is human consciousness. While the bottom seems to have fallen out of the most basic level of being, the top seems to have come to its rescue. The elementary particles that constitute macroscopic bodies assume definite states only when they are observed. Without a "ghost in the atom"[23] there is no atom. That ghost is the human observer.[24]

In any case, the challenge of making sense of the journey from the Thing to the Person has radically changed. Persons, it seems, may be implicated in the emergence of things from the ultimate constituents of what-is. Without an observer imposing a definite state on microscopic units of stuff and subsequently gathering trillions of them together, there is no body for the person to look at – or indeed (and this is the killer) to look *from*. It is, or so it appears, our individual or collective basilisk stare that freezes the world to solid bodies, including the body that supports the visual system that looks at itself in the mirror, the very possibility of the basilisk gaze. The Thing is part of a reality that does not exist independently of observation.

What it is about an *organism* that enables it to perform observations that solidify solid objects – how the body visible to itself came out of the material world as it is seen through the eyes of physics – is now even less clear. The lack of explanatory force delivered by talk of "emergence" is yet more obvious. The burden of explaining how a human body visible to itself emerged from a material world of lifeless material objects becomes heavier when so much of what has to be explained, and remains unexplained, is now located in the passage from ethereal "atomic" constituents to macroscopic objects. The most obtrusive mystery is how any solid object – the table, my body – emerges out of such slippery nanoscopic ingredients.

There is, in short, a good deal of "emergence" to be accomplished *before* we get to what is usually regarded as the starting place for the journey from material things to citizens: we first have to get to things. While it seems that persons require more explanation than organisms and organisms more explanation than lifeless material objects, the latter, too, seem to demand explanation and the mysterious journey from things to persons addressed by the idea of emergence seems to begin at its ending. Thinghood, it appears, requires personhood as much as personhood stands on, or grows out of, thinghood.

It may be a stretch (understatement of the century) to suggest that, in the absence of conscious subjects, there are no supra-atomic levels at which ethereal fields give birth to solid objects. This, however, is one way of responding to the privileging of the atomic (or sub-atomic) level in the material world: that conscious subjects are necessary to generate, and curate, any levels of being above that of the atom. Peter van Inwagen has argued that, in the last analysis, there are only atoms and persons (or at least conscious entities).[25]

If, on the other hand, we assert that the fundamental constituents of matter are *not* microscopic or not solely microscopic – so that the emergence of supra-atomic objects does not require explanation – we have to reject some of the basic precepts of physics that are central to quantum mechanics and the sciences dependent on it. If, on the other hand, atoms as "fundamental reality" are given a free pass – they do not have to explain themselves, other than the bare fact that they exist – then macroscopic objects are comparatively unexplained. As, too, indeed are the macroscopic spaces between them.

2.4 EMERGENCE FROM WHAT? "EVERYTHING IS MADE OF ATOMS" 2: AN EMPTY TRUISM?

The assumption that natural phenomena are entirely constituted out of microscopic fundamental physical components – that the material world is not a continuum of structureless gunk – deserves conceptual examination. Is Feynman's "Everything is made of atoms" perhaps merely the truism that "Everything is made of the smallest things that everything is made of"? Or even that "Everything is (basically) made of basic stuff – namely the stuff everything is made of"? Feynman's summary of what science tells us, after all, does not include a commitment to the actual characteristics of atoms as they are described in the Standard Model. The menagerie of particles with their distinctive properties and forces and the numerous constants (for example, the ratio of the mass of the proton to that of the electron) look to be contingent additions to mere existence. Ontologically privileging the smallest as revealed through the nanoscopic view is clearly problematic.

At any rate, it is difficult to say what "Everything is made of atoms" *does* mean, given that the atomic hypothesis no longer carries the traditional implications (maintained from Democritus to Dalton and thence to Boltzmann) that the fundamental constituents of matter are (a) discrete – being separated from one another by the void – and (b) indivisible. Quantum entanglement has put paid to discreteness; and atomic fission to indivisibility. Indeed, in view of the rapid transformations of the very idea of atoms, "Everything is made of atoms" has become, as I suggested in the previous section, "Everything is made of atoms – whatever they are – whatever they turn out to be". It looks dangerously close to "Everything is made of whatever everything is made of".

Ladyman's and Ross's *Every Thing Must Go* was devoted to mocking the idea of the basic constituents of matter as little particles engaged in "micro-bangings".[26] At the fundamental level there are only fields or structures, even mathematical structures. In short, the central intuition of atomism – that there is structure "all the way down" and that we never reach formless gunk – is saved only by extending the idea of structure to encompass mathematical structures, with stuff evaporating into a system of magnitudes.

As Ross Cameron has expressed it, the case against what-is being, at bottom, formless gunk "is that composition could never have got off the ground".[27] Without discrete pieces on the board, able to be arranged in differently ordered and disordered ways, there would be no basis for the variety of entities and states we see in the world. Even so, it is not clear that various arrangements of uniform fundamental elements could generate the kinds of variety we see in our world.

While the central dogma that everything is made of atoms remains a fundamental idea of basic physical science, it has little theoretical regulative power given the rich, speculative, complexity of theories of the atom and the 10^{500} "landscape" of string theories that are not susceptible to scientific evaluation.[28] Likewise, the claim that the universe is a single wave function could be translated as the idea that it *is* gunk – and not merely all the way down but also all the way up.

While it looks as if atomism – in the absence of a minimal definition of an atom as something indivisible and discrete – shrinks to the unsurprising claim that "everything is made of whatever everything is made of", there are two remaining features that have survived the revolutions of the past century. They are the assumptions that the true nature of the material world is as it appears at the *smallest* level and that at that level there is *uniformity*. Together they imply that everything is fundamentally – at the level at which things are seen for what they really are – the same.

Let us deal first with uniformity. It is counter-intuitive: it seems to block any explanation of the world of heterogeneous macroscopic objects such as stones and trees and the body of the man in the mirror that we seem to inhabit. If what-is boils down to sameness, how does it boil up to difference? More particularly, the lack of qualities – notably so-called secondary qualities – at the fundamental level makes the widespread presence of those qualities in the world we inhabit inexplicable. What, then, is the attraction of, the case for, the idea of uniformity at the most fundamental level?

In part it is based on a truism: "Everything is made of whatever it is that everything has in common". To gather everything into speech of manageable length, it is necessary to homogenize this totality. For entities to qualify as basic constituents they must be identical to one another so that they can be the constituents of everything. The fundamental physics that investigates basic constituents is contrasted with the special sciences that narrow their attention to classes of entities that have distinctive features – meteorological phenomena, living creatures, human beings, societies – that are limited to certain parts or aspects of what-is.

A Theory of Everything that will necessarily exclude anything that is not shared by Everything, runs the danger of course of becoming a Theory of Nothing in Particular and this may not be clearly different from A Theory of Nothing.[29] This may explain why such theories are willing to embrace impossibilities such as virtual particles created out of instability in a quantum vacuum, entities that are often the progeny of imaginary numbers, and negative probabilities.

It is therefore neither unexpected nor even a discovery that instances of types of fundamental particles (which are now subatomic) are identical with one another. If any differences are discovered between members of a type, this is a signal to dig deeper in search of an explanation which often takes the form of proposing *more* fundamental particles differently organized or combined. These more fundamental particles then take on the mantle of the basic constituents of matter, of what-is.

So much for uniformity. The Theory of Everything will necessarily be located at a level at which everything is like everything else. What level will that be? This brings us to the second surviving feature of atoms: *smallest-ness*. It is no accident that what are accounted as the real constituents of the entirety of what-is are so small – or small compared with us. (The atoms in the body of a 70-kilogram person are approximately $1/7 \times 10^{-27}$ the size of that body.) The smaller the entities, the less room there is for qualitative variation, indeed, for any distinctive qualities. If reality is that which everything has in common, it must be featureless; and sub-atomic particles, being the smallest possible entities, must have no scope for variation in features. Indeed, what features they have will be purely quantitative – such as charge or spin. To put this another way, what-is at the ultimate level will be drained of qualities: it will be as abstract as the charmless "charm" or colourless "colour" of quarks. Indeed, as we know from quantum mechanics, there is a level at which items lack basic parameters such as definite location and momentum.

The featurelessness of the fundamental level reaches its culmination in the reduction of what-is to a substanceless mathematical portrait, with the convergence between two claims: (1) that elementary particles can be fully described by mathematical formulae; and (2) that those particles *are* mathematical structures. Stuff evaporates to numbers, most clearly when virtual particles – particles that seemingly form out of nothing for a micro-instant and then return to nothing – are introduced "to balance the books".[30]

The reader might protest that, surely, it is more than a mere truism that pebbles, chaffinches, and citizens are, at a certain level, made of the same kind of stuff and that that stuff has properties that are described by fundamental laws that have universal application. It might, however, be said with equal conviction that it could have turned out that the stuff of which they are made is *not* the same, given that, at the level we experience them, they most obviously are *not* the same. Pebbles do not have self-growing feathers like chaffinches or thoughts like citizens. Something therefore is being said when it is asserted that these items are *fundamentally* the same – especially when "the same" is translated into aggregations of half-spin fermions – or (to namecheck the latest fashionable particle) axions.[31]

What, then, justifies the assumption that the *true nature* of pebbles, chaffinches, and citizens is to be found at that level where they are discovered to be fundamentally similar – indeed, self-evidently justifies them to the point where they are truistic? Namely, the two connected aspects of the truism: (a) the necessary generality of properties that apply to everything; and (b) the smallness of whatever it is that atoms are. Particles, to count as truly elementary, must be so small that they are almost featureless and there is no room for differences between them. There is room only to house properties that are so simple they must be identical.

To draw the conclusion from the universality of properties evident at the sub-atomic level that, notwithstanding appearances, this is where reality is to be found and consequently to embrace the (challengeable) assumption that things are ultimately the same, is then to be faced with the problem that exercises those who believe in the explanatory power of the notion of emergence: of how, despite being the same, they came to be seemingly so fundamentally different. If everything really is the same underneath, how does it come to be different and to behave differently on top, such that mountains block the view, birds sing, and citizens think?

The puzzle sits inside a perfect circle:

1. That which is truly fundamental is that which is most real – atoms are more real than chaffinches;
2. That which is fundamental necessarily lacks properties which distinguish it from other things at the same level;
3. The shortest route to being indistinguishable from other things is to have as few properties as is compatible with existence;
4. That which is smallest has least scope for variation and is therefore that which is closest to being featureless;
5. The constituents of atoms meet this criterion and hence are most universal and homogeneous.

We can now see that talk of emergence as closing explanatory gaps is doomed if we think of everything as "matter" and matter as being essentially atomic – or, more recently, sub-atomic – as when everything is reduced to axions. If we begin with essentially featureless sub-atomic particles we then have to explain the origin of a richly featured world out of near featureless constituents; of macroscopic variety and heterogeneity out of microscopic monotony and homogeneity.

If this way of thinking – such that everything seen aright will prove to be the same and that things will be seen aright only at the nanoscopic level – ever had anything going for it, what has happened in physics over the past century has put it into question. As atoms become ever more mysterious, elusive, complex, and contradictory, they also look less comprehensible: "Everything is made of atoms" seems to tell us less and less. As the model of the atom becomes increasingly mathematical, it becomes more unhinged from the world in which we pass our lives. And if everything boils down to that which has only charge, spin, and mass, how does it boil up to the rich, feature-filled, endlessly various world we live in?

Wait a moment, you might say. While individual atoms are homogeneous, they can give birth to heterogeneity by being gathered up in different quantities and in different ways – just as uniform bricks can be combined in different ways to create fundamentally different structures. Or just as, at a more fundamental level, sub-atomic particles can be combined to create atoms with different properties and those atoms can be combined in different ways to generate the almost limitless variety of chemical compounds with their distinctive properties. Unfortunately, the very idea of such combination has become problematic for several reasons.

The first is that the notions of space necessary to allow the difference between separation and aggregation and of time to allow sequences of events are under threat from contemporary quantum mechanics. Indeed, the entangled universe of quantum mechanics has suggested to some thinkers that the sum of things is a single indivisible unity. All objects are encoded in a universal wave function that describes a single, entangled state.[32] Far from everything being made of atoms, the universe is a single, quantum whole represented as a hologram.[33]

The second is the question of an ascent of scales from the nanoscopic to macroscopic. There are good reasons for thinking that scales are not inherent in the physical universe.[34] They are imposed by conscious observers, a fact which undermines the assumption that the ultimate truth of what-is will be found in its smallest components, given that observers are macroscopic.[35] This latter point is connected with the increasing suspicion that "fundamentality" is not a natural property and fundamental elements are not natural kinds. As Emily Adlam points out, "many of science's most important paradigm shifts have been tied to alterations in our understanding of the fundamental"[36] and that the levels of reality that are created only insofar as they are exposed by ever more intrusive experimental method cannot be thought of as fundamental.

Our suspicion that the starting platform for the "emergence" story is rickety seems to be increasingly well-founded. The problem of accounting for the infinitely varied world of macroscopic objects and entities such as organisms emerging out of a monotonous nanoscopic realm and of understanding how persons emerged from an insentient macroscopic one is self-inflicted. It is a result of embracing the initially ill-defined and subsequently obscure assumption that "Everything is made of atoms" or identical sub-atomic particles subject to unbreakable, universal laws, an assumption that (a) erases differences, and (b) locates reality at the smallest scale. That small scale is ultimately defined purely mathematically and, perhaps as a consequence of this, is running into the buffers – as we shall discuss in Chapter 4. This should make us question why we feel it necessary to find an explanation for the solid object that is RT's body as seen through his own eyes (and as lived in his own life) can have emerged spontaneously from the largely empty swarm of particles of an increasingly dubious and un-thing-like nature.

At any rate, the kind of emergence that would be required to build the potentially visible body that I see in the mirror out of its ultimate constituents seems at least as puzzling as the emergence of living, breathing, eating, organisms out of a lifeless universe or the emergence of conscious subjects, able to see themselves in a mirror, out of organic life. The gap between what physics considers to be the ultimate constituents of the material world and the potentially visible macroscopic body seems no narrower than between the Thing and the Bloke.

The idea that an entity is *really* the fundamental constituents that can be discerned in it is enjoyably criticized by Larry Wright. He challenges the notion that things boil down to their most basic constituents and it is that which defines what they are.[37] Wright imagines a person who looks at the list of ingredients on a carton of milk and, not finding "milk" on the list, concludes that the carton contains no milk and that there must be another carton which does contain the milk. The idea is that the reality of an entity is *really* its constituents if we consider that everything is made of atoms or of the most real sub-atomic particles – such as a half-negative spin fermions or axions – when we have only the one constituent. This prompts some to think that the world of heterogeneous entities is an illusion.

We (like chaffinches and pebbles – and cartons of milk) are not just what we are made of; otherwise, there would be no fundamental difference between macroscopic entities and atoms. The world would be a desert of atoms – though in the absence of a viewpoint not belonging to it, it would not even be gathered up into a desert. The atomic vision, which – as we shall further discuss in Chapter 4 – becomes a mathematical vision, drains the colour from the cheeks of being. Or it does if you think things are simply their ingredients. This then raises the question of how ingredients generate something – apparently equally real – different from themselves. That something begins with macroscopic material objects.

Which brings us back to what it is that allows macroscopic things to emerge from the soup of uncertainty that is the sub-atomic world seen through the eyes of quantum mechanics. If we are invited to believe that what-is at an atomic level is indeterminate until it is observed, then the idea of an arrow of emergence beginning with things and ending up as persons, is seen to be even more vulnerable than if we started with solid matter as instantiated in rocks, and pools, and winds.

At any rate, it starts to seem uncertain whether the Person lies at the end or at the beginning of the events that are the story of the universe that is typically ordered as

Material Thing → Living Organism → Conscious Person

There is the dizzying possibility, already hinted at, that it is the entities at the end of the sequence – conscious subjects – that clump hollow insubstantial sub-atomic entities into solid, substantial objects, while they themselves need to be clumped, perhaps even self-clumped, into such entities so that they might do the necessary clumping; that there are no macroscopic bodies without highly conscious agents and no highly conscious agents without macroscopic bodies.

And it is impossible to see how, and by what, this circle could have been drawn. By bodies using their bodies to manipulate other bodies (instruments) to discover that there are no bodies? If bodies are an illusion, how did they come about and what use would this illusion have in a world that is bodiless?

2.5 EMERGENCE WITHIN THE MATERIAL WORLD

This problem we have discussed of the starting point seems to have been overlooked by many of those for whom "emergence" offers a framework for an explanation of how persons arose in a world of insentient matter.

Many philosophers have cited examples of emergence *within* the material world to reassure us that the presence of persons in a universe that seems to have begun as a collection of material objects, energies, and forces can be accounted for in terms of those objects, energies, and forces. Ordinary, even commonplace, examples are intended to reassure us that there is nothing special, or specially mysterious, about the transition from the Thing to the Person. That which emerges, so we are assured, does so without breaking any existing laws or depending on new ones. Any new kinds of entities are composed of pre-existing stuff behaving as that stuff behaves. Emergence, even weak emergence, therefore, seems a goer.

A favourite example, among philosophers wanting to make emergence more respectable, of new kinds of things coming out of old stuff, of properties manifested by a system but not by its components, is the emergence of drops of water from molecules of H_2O. Here it is expressed by John Searle:

> An emergent property of a system is one that is causally explained by the behaviour of the elements of the system; but it is not a property of any individual elements of the system, and it cannot be explained simply as a summation of the properties of those elements. The liquidity of water is a good example: the behavior of the H_2O molecules explains liquidity, but the individual molecules are not liquid.[38]

The whole drop is more than, indeed of a different order from, the sum of its molecular parts. Let us reflect on this a bit.

Drops of water have properties such as shininess, capacity to reflect their surroundings, chilliness, warmth, transparency, surface tension – and, yes, wetness – that are not observed in individual molecules of H_2O, though those molecules still behave individually within the drop in the manner prescribed by the laws of fundamental physics. Indeed, while the surface tension of the drop is an expression of the principle that commands a population of mutually attracted molecules to crowd into the smallest possible surface, the properties of individual molecules are not changed. Additionally, in a falling raindrop, molecules are regimented to move in the same macroscopic direction – towards the earth – whatever random wriggling they may exhibit as the collective is falling. Their microscopic wriggling is subordinated to a macroscopic trajectory.

There could be no clearer, or indeed literal, example of downward causation. In addition to the thermodynamic influences that individual molecules are subject to and the overall gravitational influence on the drop as a whole, there are thermal micro-currents, intermediate in size, that will also dictate the movements of individual molecules within the drop. What is more, there is horizontal causation between wholes in the realm of macroscopic entities such as drops of water: drops may cause circling ripples in a puddle created out of drops. Those beautiful ripples are not evident at the molecular level.

There are at least three reasons why we should not consider this example as relevant to the question of whether emergence could account for the passage of the Person from the Thing – least of all weak emergence that aims to cross, or hush up, the explanatory gap.

The first is that it seems questionable whether emergence in the material world – for example of drops of water that have properties unknown in the realm of their basic constituents – can justify our feeling less perplexed by the emergence of the "I" of the person from the "it" of the body. After all, the difference between molecules of H_2O and drops of water is of a different – and (to put it mildly) less challenging – kind than that between material objects and conscious subjects.

The second is something we have addressed already: the well-founded suspicion that consciousness may play a (a non-causal) role in the transition from the microscopic to the macroscopic level. According to a dominant version of quantum physics, elementary particles may not acquire definite states in the absence of observers. Against this, it might be argued that, unlike atoms, *molecules* may not require assistance to acquire a location, a size, and other features such as velocity. If it were true then we are faced with the enigma of how atoms that are not definite particles unite in specific ways to produce molecules.

The third, and most telling, objection is that emergence of visible drops from individual molecules would not happen of its own accord if the default level of the material universe left to itself were that of the atom. If everything were indeed made of old-fashioned atoms, and atoms were the basic stuff of reality, it is not easy to see how that level is left behind and there is ascent to macroscopic stuff; how fundamental elements add themselves up to (say) drops of water; how such parts throw their lot in with wholes that are more than the sum of their parts, and acquire the additional – secondary – properties we have highlighted such as shininess, transparency, fluidity, viscosity, and warmth, that are not observed in atoms whose only properties are charge, spin, and velocity and which may even lack location and momentum.

We could be forgiven, therefore, for suspecting that the transition from the molecular to macroscopic level requires assistance from the very consciousness that Searle and others invoke to try to demystify with examples of emergence in the material world. That suspicion seems justified when we recall that at least some of the emergent properties just referred to are secondary qualities – entities we shall discuss in Chapter 10.

As has been recognized since Democritus, and reinforced by Galileo, Descartes, Locke and many others, these exist only in the mind of a perceiver and have independent presence only through their effects on such a mind. They do not, at any rate, seem to be accounted for by mere aggregation of molecules whose properties are confined to primary qualities that threaten to evaporate to pure quantities. While we may reduce the temperature of a body of water to a thermal quantity identified with the result of an interaction with a material object such as a thermometer, we cannot reduce warmth in this way. Warmth is an *experience* and not an intrinsic thermodynamic property of a material entity.

It seems justified, therefore, to conclude that consciousness seems to play a part in securing emergence within matter. The side-by-side existence of molecules and drops of water, of the micro and the macro, presupposes multiple scales of attention, none of which is inherent in the material world. Scales are not intrinsic to what-is; rather they emerge at the levels at which what-is is made explicit. "Macroscopic" and "microscopic" are not natural kinds. The boundary between them is not drawn by nature herself.[39] As Elay Schech has argued, any claim to the realism of physical theories, has to acknowledge that the data on which their truth claims stand depend on artifices of scale, and other idealizations that have no place in physical nature.[40]

Having said that, it may be a stretch to suggest that in the absence of necessarily scaled attention of conscious subjects there are no unitary entities of different sizes – for examples, mountains, pebbles, and centipedes – or that the difference between an amoeba and a person is merely a matter of the direction and grain of attention. It may be a stretch but, as we have noted, at least one respected philosopher – Peter van Inwagen – is willing to take it. Indeed, he goes further and argues that, since there is no universal or incontrovertible answer to the question as to when an assemblage of physical parts adds up to (one) object there are no such objects understood as self-individuating, self-defining entities. There are only atoms – and people (or, at least living beings) whose direction, grain, and scale of attention will determine what is gathered up into an individual object.[41]

There are, however, (at least) two problems with this rather startling claim. First, as we have seen, the status of "atoms" is no longer very clear: they are problematic building blocks of the localized objects differentiated by our attention. Secondly, it is difficult to know what to make of persons if there are no macroscopic objects in which they are embodied which, anyway, seems necessary for them to do the work of bringing atoms together into the illusions of objects. The grain, direction, and scale of attention seems to require an embodied – or more precisely an *am*bodied – subject, as we shall discuss in a later chapter.

What is definitely attention-dependent is the deliberate switching between scales as when, for example, a physicist burrows down to investigate the nature of the smallest components of his body – an inquiry guided by the over-riding principle that "everything" – that is every macro thing – "is made of atoms" so that we can see a pebble not only as a single entity but also as (a) a small part of a stony landscape, and (b) as a collection of 10^n atoms.

Little seems to remain of Searle's reassurance that the emergence of the "I" of the conscious person from the "It" of matter – via the intermediary of an organism whose activity at a cellular level boils down to (bio)chemical processes seen throughout material nature – is not after all mysterious because it is no more mysterious than the emergence of solid tables and shiny drops of water from atomic constituents that are neither solid not shiny. Emergence within the material world does not diminish the mystery of the relationship between persons, living things, and insentient matter because it is itself mysterious. Levels are no more part of the physical world than scales or the distinction between signals and noise:[42] They have to be introduced by conscious subjects if everything (else) is atomic or sub-atomic particles - or waves of excitation in quantum field. The molecular and the drop-of-water level at which different properties are observed are both on the far side of the very explicitness that has to be explained.

2.6 A BRIEF LOOK AT THE EMERGENCE OF ORGANISMS

At the beginning of the present discussion of "emergence" as explaining the arrival of persons in a universe consisting overwhelmingly of matter considered to be insentient, I set aside organisms because the lower boundaries between them and insentient material objects and the upper boundaries between them and conscious subjects are unclear. This move might, however, be regarded as absurd, given that the passage from a virus to higher mammals covers most of the territory separating material objects and persons. In this section I shall make what I hope will seem to be amends for what might appear an unjustified omission that unfairly weights the case against the claim that emergence has any explanatory force.

Allowing non-human, insentient organisms to reoccupy the gap between insentient material objects and human beings makes, or seems to make, available concepts that smooth the transition from the former to the latter. The most important are autopoiesis, unity, and function.

Autopoiesis is, as we noted earlier, a characteristic shared by bacteria, humans, and organisms in between, of self-assembly and self-maintenance. The active endurance of organisms (as opposed to the passive endurance of inert material objects) seems a step towards the self-curating of humans. What is missing, however, is the explicit, forward-looking, self-curating that is the central theme of human life.

The emergence of organisms seems, at least, to deliver a kind of supra-atomic unity different from that seen in the physical world. Yes, a drop of water has a unity that transcends that of its molecular constituents, but it is not self-individuating. It can be assimilated into a pool of water. As for a rock which may resist such assimilation, its relationship to its surroundings is not that of an *environment* with which it makes self-maintaining exchanges. Rocks do not look after themselves even in the way that single cell organisms do. The environment of an insentient organism may seem, however, to be half-way between the mere physical location of a material object and a world inhabited by a conscious subject, some at least of whose interactions with its material surroundings (and inside its innards) serve a biological function. As for function – this seems to span the gap between mechanism and purpose, perhaps narrowing the gap between material interaction and agency. Again, however, organic function falls short of purpose as we understand it in humans in whom it is explicit and delivered to a significant degree by conscious agency.

Evolution and its landmarks – such as the passage from uni-cellularity to multi-cellularity, gastrulation, the acquisition of a vertebral column, innervation, etc. – seem, however, to indicate intermediate steps across the gap between insentient matter and conscious subjects. Likewise, increasing complexity in which individual all-purpose cells are supplanted by organs and systems subserving the needs of a new whole. And the vast stretches of time over which these evolved changes take place seem to smooth out any jumps, dividing the gap into bite-sized chunks.

These, at least, are the intuitions that appear to support the general idea of emergence and to justify the hope of making sense of the origin of human beings in insentient matter. I think, however, I have already said enough to expose the idea of ontological sneaking for what it is. For what remains conspicuously unexplained is the emergence of sentience, of self-consciousness, of the transformation of physical surroundings into a *world* ("my world") that is *faced*; in short, the transition from what-is to the infinite varieties of that-it-is, of explicitness.

2.7 CONCLUDING THOUGHTS

> [M]usic and governments cannot be reduced to carbon, oxygen, metal, or some deeper alternative structure.
>
> Graham Harman[43]

The limitations of emergence as an "explanation" of the presence of phenomena different from those which can be accommodated by sub-atomic physics – or whatever is accounted, by fundamental science of fundamental stuff, as the ultimate constituents of what-is – are evident even when we think of something as basic as the ordinary macroscopic objects we perceive and engage with in the natural world. The transition from quantum fundamental reality to the classical world in which we live and have – and indeed are – our being lacks explanation. And that is before we attempt to address the transition from the Thing to the Person.

It is hardly surprising that when we appeal to emergence to account for the origin of consciousness in the insentient material world, we fail to make progress. The putative emergence of mind out of neural activity is nothing like the emergence of (say) a drop of water out of billions of molecules that individually lack the properties we associate with, and seem to find in, the drop of water. After all, it is difficult to separate the transition from individual molecules to collectives with properties not seen in molecules from the idea of a conscious subject with a fluctuating scale and direction of attention. This is certainly true irrespective of whether one considers either molecules or macroscopic drops as primary constituents. If the latter, then there is a looping journey from drops to molecules and back to drops.

This point about scale and direction of attention has been made with characteristic clarity by William James:

> [N]o possible number of entities (call them as you like, whether forces, material particles, or mental elements) can sum *themselves* together. Each remains in the sum, what it always was; and the sum exists only *for a bystander* who happens to overlook the units and to apprehend the sum as such; or else it exists as some other *effect* on an entity external to the sum itself. Let it not be objected that H_2 and O combine of themselves into water, and thenceforward exhibit new properties. They do not. The "water" is just the old atoms in a new position, H-O-H; the "new properties" are just combined *effects*, when in this position, upon external media, such as our sense organs and the various reagents on which water may exert its properties and be known.[44]

We should not assume that because drops of water boil down to many molecules those many molecules, gathered together by the forces of material nature will have the properties of drops of water, given that quite a few of those properties (shininess, warmth, etc.) are (mind-dependent) secondary qualities.

It is impossible to resist citing James' demolition of the notion of the block of marble realizing the statue – the example that we discussed at the beginning of the chapter:

> Aggregations are organized wholes only when they behave in the presence of other things. A statue is an aggregation of particles of marble; but as such has no unity. For the spectator it is one; in itself it is an aggregate; just as to the ant crawling over it, it may again appear a mere aggregate. No summing up of parts can make a unity of a mass of discrete constituents, *unless this unity exists for some other subject, not for the mass itself.* [emphasis added][45]

This should sound the death knell for the notion that emergence of higher-level properties takes place in the material world in accordance with its intrinsic properties. And if emergence in the material world seems to be consciousness-dependent, the emergence of consciousness must remain entirely unexplained. Self-evidently, beings in a mindless world cannot depend for their unified existence on consciousness. In short, the examples resorted to by writers such as Searle are far from defusing the problem of the relationship between the members of the unholy trinity.

Even more to the point, emergence within the physical world – drops of water out of molecules of H_2O – is nothing like the emergence of persons out of organisms – of the social "I" out of the biological "It". Persons have characteristics, such as awareness, self-consciousness, intentions, goal-orientated behaviour, values, etc., that are not seen in the realm of material things, not even where those material things are engaged in biochemical processes. More fundamentally – and central to the thesis of this book – the transition from molecules to drops of water is nothing like any putative emergence of explicitness out of a realm of insentient entities, the transition of what-is to that-it-is. A drop of water is no closer than the individual molecules of which it is made to being self-aware, world-aware, or to passing judgement on itself and others.

The intuitions, or undeclared assumptions, that make the idea of emergence of conscious subjects from matter attractive to some are (1) that matter was in place before there was life and there was life before there were conscious subjects; and (2) that matter is easier to understand than conscious subjects; that its properties can be appealed to as the basis of explanation rather than something that has to be explained. Once these assumptions are spelled out, they are exposed as open to challenge, and when they are challenged, it becomes clear that the appeal to emergence is merely a restatement of the problem of the relationship between things, beasts, and persons. It highlights, rather than closing, the explanatory gap between the physical world and people who experience, manipulate, inquire into, and think about that world. Nothing happening in what-is seems to explain how in persons "the universe" acquires inverted commas; how nearly 14 billion years after it was self-created, matter becomes the referent of a thought as "matter", bits of it come to matter to other bits of it, and the sum total of it acquires a name – the universe.

It is possible that the appeal to "emergence" is a surface dressing for a self-inflicted wound. The wound is the result of several assumptions: (a) that everything boils down to some basic stuff – let us call it matter; (b) that basic stuff boils down to atoms; and (c) that atoms and aggregations of atoms have properties that are nothing like the properties of conscious subjects. Weak emergentism proposes that macroscopic entities are the progeny of aggregations of atoms that, in virtue of being combined, acquire collective properties utterly unlike those they have as individuals. This seems like magic thinking. There is therefore an appeal to strong emergence which essentially throws in the emergentist towel and admits that, as Chalmers expressed it, the emergent properties "are not *deducible* even in principle from truths in the low-level domain".[46]

The classification of emergence theories into those that are "weak" and those that are "strong" should be a red flag highlighting a profound ambiguity in the very idea of emergence. This is admitted in an authoritative, and largely sympathetic, treatment of the history of the theories: "Clearly, emergence covers a wide spectrum of ontological commitments. According to some the emergents are no more than patterns, with no causal powers of their own; for others they are substances in their own right, almost as distinct from their origins as Cartesian mind is from body".[47] It is under the cover of this cloud of ambiguity that the attempts to close the explanatory gap seem to be delivering something. But "emergentism" is nothing more than "Turn-Up-Ism" or "It-Just-Turned-Up-Ism". Not only does it fail to account for the transition from what-is to that-it-is but also the origin of value, significance, and meaning – of matter mattering.

This may prompt us to reflect on what could count as a satisfactory explanation of the generation of conscious subjects in a universe that, according to standard physicalism, began (and overwhelmingly remains) as a collection of insentient, lifeless, matter; how exotic megafauna such as humans could arise out of the micro-entities that make up the universe. The answer to this question is unclear.

There are psychological, disciplinary, historical, and cultural contributions to the judgement that an explanation is satisfactory. They will emphasize logic, observation, intuition, etc., to different degrees. It is these that, to cite a problem sidestepped in this inquiry, determine whether or not instability in a quantum vacuum counts as a more satisfactory account of the origin of the universe than a Divine Creator. (For me, both "explanations" highlight rather than narrow the explanatory gap.)

A final thought. Emerged entities emerge into a universe composed overwhelmingly of entities that are (for the want of a better term) "unemerged". Conscious subjects, what is more, live their lives courtesy of insentient entities such as kidneys and hearts; and the scene of those lives is largely of a landscape of insentient material objects. Consciousness makes of those objects an *environment* which it faces and with which it interacts in different ways in order directly or indirectly to maintain itself. In short, emergence is always localized which raises interesting questions as to how a uniformly unfolding universe could give rise to such fundamental non-uniformities and to see how the ontological luck (or bad luck) of being a person rather than a stone should be so unevenly distributed. To say this is simply to highlight how the very notion of emergence does little to help us understand the mystery of the transition from things to persons.

It is against this background that I shall examine the most widely accepted approach to explaining the emergence of explicitness. I mean, of course, that of neuroscience, embraced by many contemporary philosophers who give the brain or nervous system a central role in the transition from the organism to the person, illuminating what-is with that-it-is; enabling the state of matter to open up the material realm as a world for a subject, for a being who mysteriously matters to itself. The brain, it is claimed, is not only the mediator of the expression of generally emergent properties such as consciousness but also the site at which an individual is born. This additional role ascribed to the brain is captured by William Hasker: "So it is not enough to say that there are emergent properties ...; what is needed is an *emergent individual*, a new individual entity that comes into existence as a result of a certain functional configuration of the material contents of the brain and nervous system".[48] The brain is the brewery of explicitness. Let us see, shall we.

CHAPTER 3
The received idea of the brain as the brewery of explicitness

3.1 THE MYTH OF "EASY" PROBLEMS OF CONSCIOUSNESS

Before I examine the claim that the transition from what-is to that-it-is, or to that-it-is-the-case, is delivered by the brain – more broadly, the nervous system or, more narrowly, parts of the brain – I want to challenge the distinction, particularly associated with David Chalmers, between the "easy" and the "hard" problems of consciousness. It illustrates the ease with which many manifestations of explicitness are overlooked.

The (supposedly) easy problems are "those that seem directly susceptible to the standard methods of cognitive science, whereby a phenomenon is explained in terms of computational or neural mechanisms", while the paradigmatic hard problem is "the problem of *experience*".[1] Another way of capturing the distinction is to contrast "functions and behaviour associated with consciousness" (easy problems) "with the experiential (phenomenal, subjective) dimensions of consciousness"[2] (the hard problem). An organism possesses the latter dimensions of consciousness when we can say, as Thomas Nagel expressed it, that it is *like* something to *be* that organism. This is true most incontrovertibly of organisms like you and me. It may be true of some vertebrate animals, but probably not of any plants.

The basic elements of mind, and the source of Chalmers' so-called "hard" problems, are qualia such as the experience of red, the smell of cheese, or the pain of toothache. For many philosophers, these present the most direct challenge to a physicalist account of consciousness – though a sense of irony might also be tricky. As we shall discuss in this chapter, nothing in neural activity explains how they are experienced by conscious subjects. Equally unexplained is the fact that our experiences are experienced as being had *by someone*: at the very least that they are connected, through a common bearer, with other experiences, and with thoughts and memories, and that in many cases they assume their place in a life-story or fragments thereof. They belong – at least tacitly – to a first-person being. For this additional reason they lie beyond the reach of third-person (or, strictly, no-person) objective neuroscience. Anyone can witness, and consequently equally partake in the fact that, some physical event occurred; but the experience of its occurring is private to an individual person or animal. Consider the difference between watching someone stub their toe on a table and experiencing the resultant pain.

So much for the source of Chalmers' "hard" problem of consciousness. The distinction between hard and easy problems is, however, deeply problematic; for the place where Chalmers draws the boundary between the two concedes too much to those who believe neural and computational science can explain consciousness. Chalmers is too ready to consider parts of the mind as being no more than physical way stations in the causal chain between sensory inputs and behavioural outputs.

In his most detailed development of the hard/easy distinction,[3] Chalmers' list of "easy" problems includes (1) our ability to discriminate, categorize, and react to environmental stimuli; (2) our ability to describe our mental states, to focus our attention, or deliberately to control our behavior; (3) how cognitive systems acquire and integrate information; and (4) the difference between wakefulness and sleep. This is puzzling. One would have thought that (receiving) information, (paying) attention, deliberating about one's behaviour, and wakefulness are things about which we can ask the question, "What is it like?". (So too, for that matter, is dream-filled sleep.) Indeed, if these mental features did not *feel like* anything, they would not be what they are supposed to be; and a difficult, indeed paradoxical, set of questions would remain about why they at least *seem* to feel like something.

The failure to see this curdles much contemporary discussion of the nature of consciousness and its place – and that of intelligence and attention and other higher-order manifestations of consciousness – in the larger scheme of things. Most importantly, it overlooks the way in which so-called "easy" aspects of consciousness are built on "hard" aspects. Crucially, for our present inquiry, they are all manifestations of explicitness. Chalmers' work, despite his reputation for pushing back on reductive trends, falls prey to the same mistake of overlooking the explicitness of all aspects of consciousness – not just so-called phenomenal consciousness.[4]

By saying that so much of consciousness is amenable, or potentially amenable, to a neural-computational explanation, Chalmers gives too much ground to those philosophers who believe there is no fundamental difference between a conscious organism and a mechanism, because they regard the conscious mind as no more than a machine for linking environmental inputs to behavioural outputs in the most effective way. This concession makes Chalmers' fundamental position that "consciousness is not physical" – something to which we shall return – more vulnerable than it need be. And it opens him to the accusation of inconsistency from committed physicalists, for whom phenomenal experiences, qualia, are illusions – merely ontological spooks left over from "folk psychology".

Chalmers' distinction between "hard" and "easy" problems of consciousness has not been universally accepted. The theory of "phenomenal intentionality", according to which all modes of intentionality, and indeed aspects of conscious behaviour, are derivative from the intentionality of phenomenal experience, is consistent with the position adopted in the present chapter.[5]

Nevertheless, it is easy to see what unites the elements of the mind rounded up in the pen of "easy problems" and consequently the attraction of this notion: they often, or usually, have behavioural correlates that are sometimes quite direct. Consequently, they are associated with events in the physical world, observable by anyone, so that at least some aspects of them can seem to be describable as purely physical, as part of the universal network of physical causation. And this, for some, means that they can be entirely reduced to and or translated into their behavioural correlates, with awkward mental elements being eliminated. The behaviour associated with, say, waving your hand could be simulated by a zombie – a hypothetical being often invoked by philosophers of mind – that *acts* exactly like a conscious person but is not conscious. Because the actual *experience* of waving is in theory not required for you to wave your hand, or (more precisely) for waving to occur, phenomenal experience must be regarded as a sort of accidental add-on. Even if the experience exists, it does not actually *cause* the behaviour of moving your hand but is a mere bystander to the event. It is "epiphenomenal".

Of course, even from the evolutionary perspective adopted by most of those who reduce the mind to a physical way station between sensory inputs and behavioural outputs, there are near-fatal difficulties posed by the idea that experience is a mere bystander. How could a trait incapable of affecting an organism's behaviour, and so its ability to survive and reproduce, be conserved, and developed through evolutionary mechanisms? Why would evolution generate the experience of pain if your body could simply withdraw your hand automatically without it? Automaticity does not require motivation by a conscious subject. Besides, the sequences of events that generate and maintain most organisms are not required to be "done" by an entity motivated to do them. Leaving that aside, the phenomenal experience would nonetheless still exist and would need to be accounted for and explained even if it were merely epiphenomenal.

In conclusion, then, there are *no* "easy" problems of consciousness, entirely amenable to investigation by objective, quantitative (neuro)science, because supposedly "easy" aspects of consciousness such as intelligence, the ability to focus attention, the capacity to access and report on one's mental contents, all require phenomenal consciousness – indeed, they build on it. While the phenomenal content may be less salient in the case of, say, the thought of an apple than in that of seeing an apple, some phenomenal content – such as hearing words in one's head (often characterized as "sub-vocalization") – is essential for the thoughts to be had as mental tokens of a specific nature. While intelligence may be expressed in action, those actions are steeped in – prompted by, guided by – perceptions, memories, and thoughts, all of which have phenomenal content.[6] It seems appropriate therefore to think of intelligence and other supposedly "easy" aspects of consciousness as the upper storeys of a hierarchy of awareness whose necessary ground floor is qualia-level phenomenal consciousness.[7] This is equally true of the transition from deep coma to wakefulness: there is no "what it is like to be" the former and there most certainly is what it is like to be the latter.

Every aspect of consciousness is directed towards that which it is not, to an "other" – that may be the conscious subject herself – that is its target. That is why there are no easy problems especially amenable to scientific investigation. As Stan Klein *et al.* have expressed it, "To understand consciousness one cannot transform it into an object of consciousness – that is into a non-sentient entity suitable for conscious apprehension".[8]

The myth that there are easy problems of consciousness which will be solved using the methods of objective scientific investigation, depends therefore on a failure to recognize that there cannot be *any* consciousness without phenomenal consciousness. Recognizing this should head off any tendency to assume that, if consciousness serves a function, it can be explained by, even reduced to, that function. *Every* aspect of consciousness is hard and carries with it the problem of explaining how it is that, for certain entities such as human beings, the world and themselves are explicit; that what-is is what-is-the-case.[9]

3.2 BRAIN AND CONSCIOUSNESS: THE STANDARD STORY

The most widely accepted story of how it is that some matter came to be aware of itself and of its physical surroundings, and how it came to matter to itself, is that this was courtesy of the development, in higher organisms, of a nervous system of a certain degree of "complexity" – whatever that means. The transition from what-is to that-it-is, from happenings to experience, from material objects to presence, is we are told, secured by nervous tissue.

There are many reasons why this "astonishing hypothesis" (to borrow Francis Crick's phrase)[10] has been so widely accepted and provided the implicit framework of so many theories of consciousness and the research motivated by them. Most importantly, the identity of consciousness and brain activity seems to be supported by a vast body of empirical, in particular neuroscientific, evidence.

We know that damage to the human brain impairs or removes aspects of consciousness and severe damage obliterates it. Localized damage may take away particular functions – aspects of perception, memory, and the ability to act competently on the basis of what is perceived or remembered. It would appear from this that there are parts of the brain that are essential to generate feelings of seeing, hearing, smelling, desiring, hoping, remembering, and thinking. This conclusion seems to be reinforced by the observation that brain stimulation can generate elements of consciousness ranging from simple tingles to complex, structured experiences, depending on which sites in the brain are stimulated. Most persuasively, recordings from cerebral tissue show correlations between neural activity and experiences of what-is, or is happening, prompting subjects to report that it is happening. The concept of correlation as used in this context is, however, not simple.

There is wide variation in the kinds of cerebral entities whose activity is correlated with consciousness. Robert Van Gulick lists: global integrated fields, binding through synchronous oscillation; transient neural assemblies mediated by the neurotransmitter NMDA; patterns of activation in the cerebral cortex modulated by the thalamus; and reentrant cortical loops – to name but a few.[11] An additional complication is the need to separate the kind of neuronal activity which determines the *level* of consciousness (wakefulness vs dream-filled sleep and dreamless sleep vs coma) from that which is thought to correspond to particular *contents* of consciousness: different sensory modalities (vision, hearing, etc.); various types of memory (working and long-term memory, procedural memory implicit in skills, declarative memory for facts and events, semantic and autobiographical memory), thoughts and other propositional attitudes; and the sense of self and of others as selves.

Simple ideas of correlation are even less easy to defend when we try to understand how *level* of neural activity shapes, or even (as when a loud sound or a pinch awakens us out of dreamless sleep) permits, consciousness; and the sense in which the level of consciousness is distinct from, rather than being merely the sum of, conscious experiences. Or when we consider that there is not a clear grading ranging from deep coma to full wakefulness. Awake-versus-dreaming is not on the same spectrum as awake-versus-coma; after all, dream-filled sleep may be "deeper" than drowsiness but in one sense more alert – though not to actual surroundings.

I have mentioned the empirical support for the correlation between neural activity and levels and contents of consciousness that comes from observing the effects of brain damage and brain stimulation and of the spontaneous brain activity associated with conscious experiences. These have been the subject of countless studies and I cannot do justice to the relevant literature which runs to many millions of scientific papers.[12] Instead, I want to highlight a couple of studies that seem to present the greatest challenge to those such as myself who are sceptical of the idea that the properties of the brain, or of parts of it, "explain" consciousness and who doubt that observation of the brain is the way to further our understanding of the transition from what-is to that-it-is.

There have been observations of the close correlation between clearly developed hallucinations and spontaneous discharges recorded, by means of electroencephalograms (EEGs), in waking subjects in parts of the cerebral cortex. Detailed studies, using intracranial stereo-EEGs have shown that subjects who experience visual hallucinations have discharges in the visual (occipital) cortex.[13]

Perhaps even more striking are the studies carried out in the wake of neurosurgeon Wilder Penfield's pioneering research.[14] Penfield was a leading figure in surgery for intractable epilepsy. The surgery was aimed at removing the source of epileptogenic discharges. Penfield was concerned not to remove structures in the brain associated with speech, memory and other vital functions. His patients were therefore awake during the operation and were asked to report what they experienced when he stimulated different parts of the cerebral cortex. In a minority of cases, there were quite complex experiences, including memories recalled from the distant past. Subsequent research has confirmed and developed Penfield's findings.[15] This seems to suggest that stand-alone neuronal activity is not only necessary but sufficient to generate phenomenal experiences; or even that the neural activity is conscious experience.

There are, however, several reasons for not jumping to that conclusion. The most important are that (a) the studies took place in waking patients – that is in patients who already had background consciousness independently of brain stimulation; (b) the events reported had already been experienced in whole or in part by the natural route; and (c) the experiences were mere fragments – they fell well short of a phenomenal field, even less of a waking moment.

So much for the association between brain stimulation in the waking patient and fragmentary experiences. Other studies that seem particularly relevant to the claim that neural activity is a sufficient rather than merely a necessary condition for phenomenal consciousness have been performed on patients who, for a variety of reasons, appear to be in a coma and yet have residual consciousness. Adrian Owen and his colleagues investigated seemingly unresponsive brain-injured patients and tested them for covert consciousness using the advanced brain scanning technique functional magnetic resonance imaging (fMRI).[16] They asked such patients in the scanner to imagine themselves either walking through their homes – a spatial imagery task – or playing tennis – a motor imagery task. About 1 in 10 patients were able to comply as measured by brain activity and, what is more, different parts of the brain were activated in the case of the walking and the motor imagery task. These striking observations would seem to be entirely compatible with the idea that consciousness is identical with activity in certain parts of the brain and different kinds of consciousness are correlated with activity in different parts of the brain.

Compatible with, yes. But decisive proof of? Perhaps not, given that the activity observed in the brain in response to the request to imagine different scenarios, clearly draws on previous experience – a point we have already made in relation to Penfield's studies. More importantly, the difference between localized neural activity and experience or imagined experience also remains. In part this is because of the complexity of consciousness – perception, memory, thought, reflections, intentions, judgements, and planning are all part of the mental equipment of creatures such as humans for whom what-is becomes that-it-is.

We shall come to some of these – notably thought and memory – in future chapters. For the present I want to focus on a more fundamental problem; one that is present in the neural account of all forms of explicitness. I shall touch on the actual problem relatively briefly because I have addressed it on several occasions elsewhere.[17] I shall devote more time to the attempts made by many philosophers and neuroscientists to brush it under the carpet.

The problem is this. Neural tissue is made of stuff whose properties are fundamentally the same as the what-is that simply is and does not declare, of anything else or itself, *that*-it-is. The physical events going on in nervous tissues – electrochemical discharges – are seen elsewhere in the material world without any hint of that part of the material world being aware of what triggers this activity or of itself as experiencing it. More to the point, most of the human nervous system in which such activity is seen – the spinal cord, the cerebellum, and even stretches of the cerebral cortex – are *not* associated with consciousness and experience of the world and oneself. There is little fundamental difference between what goes on in the cerebellum and the cerebral cortex or between those parts of the cerebral cortex that are and those that are not associated with consciousness. Nothing, at least, to account for the fundamental difference between being physically located in the material world (as is a body or a brain) and revealing that world to one's self as a result of being in it. Or to explain how it is that certain neural states in certain locations generate what it is like to be not just, or even, a brain, or part of a brain, but a person in a world; that these states would open up the brain to a world in which it is located and the owner of the brain to herself in that world.

The appeal to "complexity" does not help. There are no clear-cut criteria by which the unconscious cerebellum is less complex than the (supposedly) conscious cortex. The cerebellum has four times as many neurons as the cortex and its mode of connectedness is at least as complicated as that of the cortex. Moreover, it is not at all clear what it is about complexity (however defined) that would explain emergence of consciousness and the initiation of a journey that eventually results in matter putting itself into inverted commas.[18] In short, there appears to be nothing in nervous systems, however numerous their elements and however they are wired within themselves and with the bodies that surround and nourish them, to account for the irruption of explicitness, to explain the transition within a part of the material world from what-is to that-it-is, so that it becomes the contents and objects of perceptions, memories, beliefs, and thoughts.

A more conventional way of highlighting this difficulty is to point out that all mental entities that contribute to making what-is explicit have a property, not seen in the material world, of *intentionality* or "aboutness". Perceptions, intentions, emotions, memories, thoughts, and so on are *about* something that is outside of, other than, the brain, and the neural activity that is supposed to account for them.

It is difficult to identify what it is in neural activity that enables it to be *about* anything, even itself. Nevertheless neural theories of consciousness rest on the assumption that the brain, or parts of it, enable the (mediated) effects of objects and events on its material to somehow reach back to those objects and events and reveal them as the source of those events; that it can somehow create internal *representations* out of some of the causes of some of its activity and project them either back into a phenomenal space experienced as outside of the body of the perceiver or into the body itself as corporeal experiences.

We have already commented on the failure to find a sufficient difference between the small minority of neural activity supposedly associated with consciousness and the overwhelming majority of activity not thought to be associated with consciousness. There is the additional failure to see how differences in neural activity would be sufficient to account for the distinctive character of the experiences they are supposed to underpin or indeed to be. The differences between experiences of red and blue, of a red colour and a loud sound, or between a perception and a memory or a thought, are not reflected in comparable differences in events in the brain. Nothing in the homogeneity of the electrochemical discharges corresponds to the profound heterogeneity of the world that is experienced, remembered, or thought about.

Neuroscientists would point to sources of difference beyond the intrinsic properties of neural discharges: different locations, patterns, and quantities of neural activity. None of these, however, seems capable of delivering the variousness in question. Let us examine them in turn.

Location is defined by connections and, in the case of perceptions, connection to the relevant sense endings. Regarding the latter, we cannot find anything to account for the difference between seeing a red rose and hearing a dog barking by examining what goes on in the entrance halls of the nervous system because there is nothing of what goes on in the eyes and the ears that amounts to a colour or a sound or the difference between sounds and colours. These secondary qualities are not properties of the events in those structures or of the external events that trigger them. Of course, the retina responds selectively to light of different wavelengths and acoustic pathways to specifically vibrations in the air but these structures know nothing of colours or of sounds. Photons changing the shape of rhodopsin molecules in the retina, the responsiveness of the nerve endings in the cochlea, the firing of tactile endings in response to distortions of mechanoreceptors in the skin do not capture the nature of the colour orange, of a high-pitched sound, or of a feeling of pressure on the fingers. There is, for example, nothing red about the changes in rhodopsin or discharges in the occipital cortex – nor would we expect them to be, since secondary qualities such as colours are not properties of the material world. The light bouncing off an orange is not orange until it is perceived by a conscious subject. The notes of a blackbird's song are mere vibrations in the air until they are heard by a conscious subject. Since events in the sense endings do not account for the phenomenal content of individual sense experiences, they cannot account for the *difference* in phenomenal content between different senses.

Shifting the credit for the variety of mental contents causally upstream to receptors in the eye, ear, skin, etc., may relieve neural activity of the duty of underpinning mental experiences or the difference between those experiences. But it does so at a price: that of taking us to places – sense endings – where there is nothing corresponding to those experiences. The gap between changes in the shape of electromagnetically-sensitive rhodopsin molecules and the experience of orange is no less than that between neural discharges and visual experiences.

So much for the location. As for *quantity* of neural activity, there is nothing that seems to translate into quality. If the experience of orange really is identical with activity in certain neural pathways, then more of that activity would give a more intense or extensive experience of orange. In short, there might be quantity-to-quality relations. That, however, would be true only after there had already been a transition from material events to qualitative experiences. The idea of a threshold marking the transition between insentience and sentience does not map on to the difference between more and fewer material events.

The appeal to *patterns* of discharges in the nervous system does not seem to do much explanatory work. It is difficult to see how their variety could translate into the variety of experience. If (to use the standard analogy) neural activity were a digital representation of an analogue world, this would not explain how it is an analogue world that is experienced. Moreover, there is a well-founded feeling that patterns exist only for a spectator. The passage from William James quoted in the previous chapter is worth repeating: "[N]o possible number of entities (call them what you like, whether forces, material particles, or mental elements) can sum *themselves* together. Each remains in the sum what it always was; and the sum itself exists only *for a bystander* who happens to overlook the units and to apprehend the sum as such".[19] The patterns don't know they are patterns; they don't add themselves up and pick themselves out. After all, the discharges that constitute or realize a pattern are converging on the relevant site from elsewhere in the nervous system. The appeal to patterns, therefore, seems to presuppose the very consciousness that is to be explained.

Location, quantity, patterns of discharges – none of these seems to deliver anything that explains the rich variousness of the experienced world, never mind that it is experienced. And there is a further, more fundamental, problem for neurophilosophers. What happens in the brain does not look like that of which subjects are conscious. But how could it? What would neural discharges have to look like or be like in order to look like experiences and capture the differences between them? We cannot expect the experience of orange to be realized in neural activity that looks, or in some sense looks like, orange. The taste of cheese cannot be realized in neural discharges that taste like cheese. The most compelling reason for this conclusion is something to which we have already alluded: secondary qualities are not intrinsic properties of material objects or events. And neural activity is composed of material events.

The common response to this objection to neural accounts of phenomenal consciousness is to assert that neural activity doesn't have to be "like" what it is about. It does not have to replicate it. Yes, it is a *representation* of that of which the subject is conscious but that representation is neurally "encoded" and, as in the case of codes with which we are familiar, there does not have to be a phenomenal similarity between what is encoded and the coding. (We shall return to the idea of the mind as neurally encoded representations in Section 3.7.)

Why, then, does the notion that the intracranial giblet has the property of making the extra-cranial world, and indeed the owner of the cranium, explicit seem plausible, or even self-evident, to so many who think, write, and research about this? In part this is because the idea of the connection between what happens in the relevant parts of the brain and what it is that the brain makes explicit has a seductive simplicity. And the pathway to accepting this has been facilitated by endless repetition since it was first suggested by Hippocrates. And there is, of course, evidence, earlier referred to, of the effects of brain damage and brain stimulation that strongly suggest that the brain is a necessary condition for consciousness and all the higher mental functions that we possess. I say "necessary", but this does not mean "sufficient". In the seemingly uncloseable gap between necessity and sufficiency, the mystery of the transition between what-is and that-it-is, and the supposed role of the brain as the brewery of explicitness, remains intact. The brain, looked at through the objective gaze of which science is a paradigm expression, is as apersonal, as unhaunted, as any other item in the world.

3.3 NATURALIZING INTENTIONALITY 1: CAUSAL THEORIES OF PERCEPTION

The feature that makes consciousness such an awkward customer for philosophers of a physicalist persuasion to deal with is intentionality. Because, as we discussed in the opening section of this chapter, intentionality is present, directly or indirectly, in all aspects of consciousness there are no "easy" problems of consciousness. It lies at the heart of the "that" of "that-it-is" and "that-such-and-such-is-the-case" and other modes of explicitness. There have been many attempts to naturalize intentionality.

The most familiar is the causal theory of mind whose most apparently straightforward element is the causal theory of perception. The apparent explanatory power of causal accounts is a striking demonstration of how easy it is to overlook the unique and inexplicable nature of explicitness. Here, as elsewhere, explicitness takes itself for granted. The strangeness of the assumption that what is made explicit is made of the same stuff, has the same fundamental properties as, that which makes it explicit, is overlooked. One of the attractions of the causal theory is that it seems to provide a criterion of truth applicable to perception. A perception is true when it is about an entity that actually causes it.

The causal theory is best expounded through a simple example: my perception of the cup in front of me. Why do I see the cup, why does it become explicit in my consciousness? Because, so the story goes, light bouncing off the cup enters my eyes, excites visual pathways, and ultimately causes activity in my visual cortex. What is happening in, or on the surface of, the cup amounts to a sustained, standing cause, that produces sustained standing effects in my visual cortex. Those sustained effects are my seeing the cup as a stable object before me.

Not even the most ardent causal theorist would think that this is the whole story. Even in the case of something as simple as seeing an object in front of me, consciousness is not passive. There is work to be done to transform what is sensed from a pattern of light into the intuition of an object that transcends what is revealed in sense experience; is of a certain kind; and is located in a certain quarter of "out there". There is, to use the standard jargon, a lot of "top-down" work to be done by the mind-brain – though this is construed as more neural activity. This, however, is not our concern here.

Making the causal theory more sophisticated will not mitigate the fundamental error of thinking that a causal interaction between two material objects – a cup on my desk and a brain in my head – mediated by light is sufficient to make one object, admittedly a rather special object – aware of the other. While causation may (if one accepts a pre-Humean realist account that allows causal interactions to exist independently of the mind) account for how the light interacting with the cup, generates neural activity, it does not explain how the neural activity forms a gaze that looks out and has the cup as its intentional object, explicitly present, and at an explicit distance from the observer.

Causation may explain (if it explains anything[20]) light getting into the eye and tickling up the brain but not the gaze looking out, alighting on the object and revealing *that it is*, even less that it is *present to me*, that it is in front of me. This is captured by Geoffrey Warnock in his critique of the causal theory: "[O]ne might want to represent seeing Socrates rather as something I do to him, than as something he does to me – as more like netting a fish, than being hit by a cricket ball".[21] Unfortunately, in correcting one error, Warnock makes another; for perceptions are themselves no more causes – netters of fish – than they are effects. What is more, in the case of proximate senses, such as touch, the material interaction is two-way, while the direction of awareness is one way: I am aware of the object I am touching while the object, unless it is the body of a conscious subject, is not aware of being touched.

There is a further problem with causal accounts of perception; namely that, as already alluded to, what is perceived does not exist in the material world. There are no colours, sounds, smells, or tastes. Even more fundamentally, light in the absence of eyes is not luminous: it is merely electromagnetic energy and does not reveal the objects it interacts with. Likewise, the motion of molecules does not create the feelings of warmth and cold associated with different temperatures. In short, causal interactions within the material world could not generate the kinds of differences, illustrated by the experiences we have of light and dark, of that which is revealed and that which is hidden.

There is yet another problem, arising most obviously in the case of distant senses such as sight, which is that causal chains, if one believes in them, have no natural beginning or end.[22] As the activity in my visual cortex reaches back along its causal ancestry, there is no material reason why it should alight at the cup. It could theoretically pause at events in my retina, and thus fall short of the cup, or go beyond the cup to the source of the light that illuminates it.

There are apparent answers to this objection, the most obvious being that my perception of a cup is woven into my sense of my own body, my willed and meaningful interaction with the cup and the context in which I am observing it, and so on. (This is discussed in Chapter 5). It is not, however, clear that this will entirely overcome the problem of dividing a continuum of events that in theory stretches back to the Big Bang, into (a) a source of perceptions – the perceived object, and (b) a place where the perceptions land – the perceiving subject. In short, it will not explain why, within the causal continuum, there is a beginning (the events in the surface of the cup that are supposed to be responsible for its being potentially visible) and an end (the events in the brain supposedly accounting for the perception). The suspicion that such beginnings and ends are themselves the children of perception – and so cannot account for the beginnings and ends that characterize perceptions linking conscious subjects with the objects of which they are conscious – seems well-founded.

The suggestion that objects of perception are picked out as "affordances" that populate the environment with which subjects-as-agents have to engage does not address the division of the material continuum into a world that is faced and a subject that faces it – of what-is into that which matters and the individual to whom it matters.

Even if this difficulty were overcome, there is a yet more profound problem. The claim that an event (or a cluster of events) in the brain is *about* an object because it is caused by events in that object would license the conclusion that causal connectedness is sufficient to ensure aboutness. We are, it seems, invited to believe that for something to be *about* an object or event it is enough that it is an *effect* of an event in that object. This risks suggesting that the entire universe, throughout its history, in the presence or absence of conscious organisms, would be engaged in making itself explicit. Aware of this difficulty, some neurophilosophers (an increasing minority) have seen the notion that all causal interactions generate information not as a problem for, or even a reduction to absurdity of, the causal theory of perception, but rather as a solution to be embraced. This view – paninformationalism – is one we shall discuss later in this chapter.

Hold on, a more conventional neurophilosopher might respond, it is not suggested that *all* causal interactions between objects would generate awareness in one object of another. Not all causal interactions with the body or even with the sense organs deliver experiences; even fewer an awareness of their causes; and fewer still of the objects that host the events that are the causes. This claim would be supported by reference to the special properties of neurones woven into complex nervous systems. As we have seen, however, it is not clear from neuroscience why a small minority of ionic currents in the brain could become the very stuff of consciousness. The appeal to complexity – even if we could clarify the concept and separate it from anthropocentric criteria such as our own difficulty of grasping the nature of an entity or the length of the description of its salient features – does nothing to explain what it is about events in the nervous system that make them not only to refer backwards to their intermediate causes – events in an object that is located "out there" – but in doing so also to reveal an object as existing independently of the experiences of them; that is to say to reveal the object as an *object* with all that that implies – including the fact that it is "other than me" and "more than my experiences of it". We can (perhaps) say what makes the nervous system of one organism more complex than another (more neurones, more connections, more modes of firing together and separately) but not in what salient respect it is more complex than unconscious organs such as a kidney. Nor can we say what it is that enables a nervous system to exist at, or adopt, the scale at which it can *be* its complexity.

In sum, the causal theory of perception (of mind, of consciousness) does not explain the very specific circumstances in which an event, object, or state of affairs becomes explicit as something that is present to someone, as part of their ongoing, lived situation.

3.4 NATURALIZING INTENTIONALITY 2: TRACKING THEORY

Tracking theory is a variant of the causal theory of perception.[23] It is another attempt to naturalize the passage from what-is to that-it-is, and hence to make the transition something that the material world could cook up without assistance. It reduces the "aboutness" of consciousness to a co-relation between states of the (material) brain and states of the (material) environment, such that what is going on in the brain *tracks* what is going on in the environment. "Tracking" theory merges, or (less kindly) conflates, a causal connection with a correspondence relationship or a relationship of appropriate covariance between consciousness on the one hand and states or items in the environment on the other. This correspondence, so it is claimed, delivers the key aspects of intentionality, of aboutness – namely, carrying "information" about the contents of the environment. By this means, intentionality is reduced to an entirely natural relation between brain states and states of the environment.

It will be evident from our earlier discussions that causally driven tracking will not do. If causal relations were sufficient to deliver intentionality the entire universe would be in a permanent state of being "about" its previous states, irrespective of whether there were any conscious subjects on the scene. Because this absurd consequence seems obvious, the question arises as to why anyone gives tracking theory the time of day; why the fact that processes in two different objects are coupled, being causally connected and/or co-variant, should be sufficient to make what happens in the one an experience of what happens in the other, so that the former is about the latter.

The answer becomes apparent when we remind ourselves of the slipperiness of the terminology used by tracking theorists and other naturalizers of intentionality. Causal relations are conflated with correlations or correspondence relations, which are then elevated to the status of "information".

Consider an example of tracking in the natural world: the image of a passing cloud in a puddle. The image tracks the movement and matches the shape and colour of the cloud: what happens "in" the puddle "tracks" what is happening in parts of its environment. The image in the puddle, however, is clearly not a consciousness of the cloud: puddles do not perceive clouds – or, indeed, anything else. Strictly, the "image of the cloud" will be an image only for a third party – namely a conscious observer, who will supply the necessary intentionality connecting the phenomenal appearance of the puddle with the phenomenal appearance of the cloud. The conscious subject, what is more, has an attunement of attention that will pick out from an unfolding world a subset of objects and/or events that will be selected as contents of consciousness. Co-relation, that is to say, has to be identified and, even then, is not enough to ensure that an event in one object is conscious of an event in another object, even less of the latter object.

What's more, it is not always clear what is correlated with what. Even if we were to confine a moment of consciousness to perceptions, it is difficult to think of its contents as the sum total of what is going on in the material world around the sensory apparatus of the embodied subject.

While tracking theory – like the standard causal theory – gives an unsatisfactory account even of elementary perceptions and their intentional projection to entities other than themselves, such as when we see a flash of light as being "out there" or coming from "out there", its failure is even more profound in the case of more complex elements of consciousness such as memory and thought. And it has nothing to say about how it is that we humans are *subjects*, offset from, or facing, or explicitly surrounded by, a world of which we have knowledge that presents itself as being objective.

Even less do causal-tracking accounts of intentionality address those special kinds of elective, quality-controlled experiences that form the basis of science and elaborate its external view on nature and nature's laws: namely, measurements. It is difficult to see how a scientific observer whose consciousness is a passive effect of items in the world would acquire capacity to envisage, plan, execute, register, and collate such observations. But we are here anticipating the preoccupations of the next chapter.

Tracking theory hovers on the borders of the paninformationalism that, as we shall discuss in the next section, is ubiquitous in contemporary attempts to make sense of consciousness. The claim that tracking is correlation, that correlation is information, and information is an intentional relation overlooks what is required for correlation to become, or be a source of, information. For an event B that correlates with event A to become information about A, we need B to refer back to A; and for this we need a conscious subject. Correlation cannot, therefore, be the foundation of the intentionality that is the distinctive character of the mind of the conscious subject. It does not provide a naturalistic account of intentionality.

It is clear that appealing to causal connectedness and/or tracking brings us no closer to understanding how the interaction between the brain and the world outside of the cranium generates a conscious subject facing a world made explicit as his or her environment. It is too much to expect the movement of ions in neural tissue, however closely correlated with what is going on "out there", to deliver what is needed. Many philosophers who are aware of this but are reluctant to give up on materialist naturalism look to ways of giving the brain less work to do. In order to do this, they embrace two "pantasies": paninformationalism we have already touched on; and panpsychism that is becoming increasingly popular among philosophers of consciousness. These are the subjects respectively of the next section and the one following it.

3.5 TAKING THE STRAIN OFF THE BRAIN 1: PANINFORMATIONALISM

The misuse of the word "information" is ubiquitous. Addiction to this word, such that it is used in places where it has no right to be, is understandable because it is a solvent dissolving many otherwise insoluble problems. If consciousness is "information" and information is present in every event in the universe, consciousness doesn't require any more explanation than does the unconscious remainder of the universe. It does not have to emerge: it was there all the time.

Thus, we arrive at paninformationalism, perhaps the most determined effort to overlook, or bypass, or demystify explicitness. It overlaps with panpsychism and many paninformationalists are panpsychists. However, they deserve separate treatment, and panpsychism will be discussed in the next section.

For a bold, bald statement of paninformationalism, it would be difficult to trump David Chalmers, who raised fewer eyebrows than he should have done when he asserted that "[W]herever there is causal interaction, there is information ... One can find information states in a rock – when it expands and contracts, for example – or even in different aspects of an electron".[24] It is not very clear in this example what the information is *about*. Let us assume a rock passes from State 1 to State 2. Is State 2 *about* State 1? If it was assumed that State 1 caused State 2, this would be analogous to the causal theory of perception, where an event in the brain is about its causal ancestor, the event that causes the event in the brain. It is, however, not clear how a state of an object can, of itself, cause a successor state, if the state is supposed to encompass the entirety of the rock. Nevertheless, this seems to be a requirement if the information is supposed to be about the rock as an object – as when perception delivers a full-blown object. In any case, it remains unclear how events in the material world can, in the absence of a conscious subject, amount to any kind of information.

More recently, Chalmers has offered a more complex view of information:

> [I]nformation (or at least some information) has two basic aspects, a physical aspect and a phenomenological aspect. This has the status of a basic principle that might underlie and explain the emergence of experience from the physical. Experience arises by virtue of its status

> as one aspect of information, when the other aspect is found embodied in physical processing.[25]

And he has further modified the claim that causal interaction is information in this recent passage: "What matters for the emergence of experience is not the specific physical make up of a system but the abstract pattern of causal interaction between its components".[26] Referencc to "the abstract pattern of causal interaction" raises the question of what, or more likely who, abstracts the abstract pattern.

Leaving this aside, the assumption that causal interactions are information and certain patterns of causal interaction elevate information to phenomenal experience, seems to permit philosophers and others to approach consciousness with the tools of natural science. It is for this reason that Chalmers' voice is but one of a swelling choir of philosophers, physicists, biologists, and many other scientists, singing from similar hymn sheets.

The view that causation is revelation, or that causal interaction between two samples of what-is or two states of a single sample will transform one of them into an intentional object for another as subject, is close to an even more radical version of paninformationalism advocated by many thinkers for whom the entire universe is a computer and everything that happens can be understood as information processing, by which the universe "computes" its future state from its present state. Information, we are told, is more fundamental than matter, energy, space, and time.

The view is particularly associated with the pre-eminent physicist John Wheeler who postulated that "every it – every particle, every field of force, even the spacetime continuum itself – derives its function, its meaning, its very existence entirely – even if in some contexts indirectly – from the apparatus-elicited answers to yes-or-no questions, binary choices, bits".[27] The assertion by Basil Hiley, also a physicist, that "one tends to regard the quantum potential as arising from a field that is more like a field of information than a physical field" reflects a growing orthodoxy that all physical processes are information processes.[28] Hiley quotes none other than the legendarily tough-minded Richard Feynman as saying that "he thought of a point in spacetime as being like a computer with an output and an input. The point would have a memory for all the fields and particles that are possible and would act like a computer".[29] The philosophers Ladyman and Ross concludes from their reading of science that:

> Reference to transfer of some (in principle) quantitatively measurable information is a highly general way of describing any process.

> More specifically, it is more general than describing something as a *causal* process or as an instantiation of a *lawlike* one. If there are causal processes, then each such process must involve the transfer of information between cause and effect; … and if there are lawlike processes, then each such process must involve the transfer of information between instantiations of the types of processes governed by the law.[30]

We can cross the explanatory gap between the physical world and phenomenal consciousness, it seems, without doing any particular explanatory work.

If paninformationalism appears to you somewhat topsy-turvey, you are not alone: "it is not the laws of physics that determine how information behaves in our Universe but the other way round. The implication is extraordinary: that somehow information is the ghostly bedrock of our Universe and from it all else is derived."[31] We are to believe, it seems, that what-is is always and everywhere translating itself entirely into that-it-is, without the assistance of conscious subjects. Or that that-it-is even *precedes* what-is. The sum total of what-is is not merely floating in, or dissolving in, a sea of information: it *is* a sea of information. To accept this is to embrace the idea that information – usually thought of as the *product* of informing and being informed – far from being a sophisticated manifestation of consciousness is everything everywhere. This doesn't leave much work for the brain to do. In short, one wonders whether, with respect to consciousness, it has any work to do, any distinct contribution to make.

The road to paninformationalism is paved not by scientific discoveries or advances in understanding. Rather it is the product of an increasingly lax use of the term "information". The *Oxford English Dictionary* gives the everyday meaning of information as "knowledge communicated concerning some particular fact, subject or event; that of which one is apprised or told; intelligence, news". Information understood in this way clearly presupposes consciousness – that of sentient beings in possession of, passing on, and potentially receiving information.

A new understanding of "information" – both wider and narrower – was introduced into scientific, technical, philosophical, and general discourse in the 1940s by Warren Weaver, Claude Shannon and others who were looking for ways of quantifying what was transmitted by telecommunication systems. Information in this specialized technical or engineering sense was the resolution of *uncertainty* regarding a range of possibilities resulting from the successful transmission of a signal. The measure of information transmitted was the quantity of uncertainty resolved. A device transmitting a succession of letters of the alphabet resolved uncertainty as to what the next letter would be. If each transmitted letter was equally probable, being one possibility out of 26, its transmission would carry between 4 and 5 bits of information.

This definition places the conscious transmitter and receiver of information – who would after all be required to *entertain* the possibilities that the message resolved – off stage. She is, however, still indispensable. A telephone system without such players creating and using it would be a rather improbable configuration of matter. Shannon and Weaver were aware of this and warned that "information" was being used "in a special sense and must not be confused with its ordinary meaning. In particular, *information* must not be confused with meaning. In fact, two messages, one of which is heavily loaded with meaning, and the other which is pure nonsense, can be exactly equivalent from the present viewpoint as regards information".[32] A "message" that consists of one letter out of a random string of letters would carry more information in the engineering sense (4 or 5 bits) than the Yes/No answer to the question "Is my beloved dead or alive?" (1 bit).

As Shannon and Weaver were at pains to point out, what is missing in an exchange between phones isolated from conscious subjects making use of the phone is *meaning*. It is this meaning that turns events into information in the sense that matters. If my phone was struck by lightning and this triggered an automated message to be sent to a friend who had lost his phone, there would have been no exchange of information because the "message" would not have been meant and its meaning would not have been received. This is sufficiently obvious, just as it is (or should be) obvious that a pocket calculator does not make calculations, a coffee-making machine make coffee, or Deep Blue play chess. In none of these cases is the entity an agent.[33]

Nevertheless, the subsequent fortunes of the concept of information, such that it is now ubiquitous in psychological, neuroscientific, biological, computing, and more broadly scientific, discourse, framing questions and for some thinkers seeming to solve them, have been built on this conflation of the engineering and the ordinary (that is to say, full) meaning of the term. The marginalization of consciousness – of conscious communicators – in the engineering sense of information has seemed to some to license the idea that information does not require conscious subjects at all.

It should hardly be necessary to point out that a telephone conversation is not a conversation between two telephones but between two people using the phones. That it is necessary is illustrated by this account of Shannon–Weaver communication theory: "A communication system is composed of an information source (a person or *thing* that selects a state or message from a set of possible messages ... and the destination (which can be a person or a *thing* interpreting or using the message)" (emphasis added).[34]

"Person or a thing?" Once the concept of information is cut free from the idea of a conscious someone being informed (or wanting to be informed) by a conscious someone doing the informing (or wanting to do it) we are on the road to what under normal circumstances would be dismissed as absurdity. What is more, in the absence of a framework defined by the interest of a subject, information is not even quantifiable – the original aim of 1940s information theory. Where there is no conscious subject, events not only fall short of being information in virtue of lacking meaning, but they also fall short of being information in the Shannon–Weaver sense of being quantifiable as yes-no, 1-0 bits.

As we shall discuss in Chapter 9, it is only within the context of *experienced uncertainty* that an event counts as information and only in the context of a *framed uncertainty* that what "actually" occurs fulfils one out of a definite range of possibilities and delivers a certain quantity of information. Otherwise, there are neither uncertainties nor frames of reference to quantify the information content of an event understood as an outcome; in short there is nothing corresponding to information even in the impoverished engineering sense.

An event that actually happens counts as the end-point of a passage from a range of possible states to a definite state only if possibilities are entertained in advance – and (it hardly needs saying) only conscious subjects can entertain possibilities – or, in virtue of entertaining possibilities, experience the uncertainties that information may resolve and hence transform events into information.

While information is typically something transmitted by one conscious subject and received by another, we can derive information from entities that have no informative intent. The most obvious examples are natural phenomena in which any state of the world, in virtue of being typically co-related with another state, may seem to inform an observer as to a future state of things. "Clouds mean rain", "tracks in the forest mean deer", "spots mean measles" – and so on. Of course, these entities do not of themselves, in the absence of a conscious subject, *mean* anything: they are what they are rather than signs of an as yet non-existent future or, indeed, of a once-existent past.

Take the example of the rings in a tree trunk. Yes, to the informed gaze of an individual presented with a stump, they indicate the age of the tree but in the absence of such a gaze, they do no such thing. The tree is not in a constant state of declaring how old it is. The material world, insofar as it makes sense, leads us to have largely correct expectations. Our knowing gaze informs us as to what is going on and what will, most probably, happen next. Of course, natural entities – forests, clouds – are not informants. It is the conscious subject, with her experience, who transforms what-is into that-it-is and the latter into a source of information about what is likely to happen.

All of which seems sufficiently obvious. Why on earth, then, can we be so easily seduced into thinking of information as something that can be generated in the absence of conscious subjects? When information is made only potentially available and is generated by subjects who are not present – in books, notes, signs – it seems as if there is a period in which information can exist in the absence of either a conscious transmitter or a conscious receiver. Such information is typically described as being *stored.* I write you a letter on Wednesday and you receive it on Friday. Forty-eight hours pass between my writing the letter and your picking it up, a period during which its informational content seems to exist in the absence of a conscious subject. In fact, it is only potential information during that time. If, after I had written the letter, I had forgotten to post it and it had subsequently been destroyed, it would not have had any period as the virtual information I intended to communicate to you. In other words, that period of virtual information is a retrospective description looking from the standpoint of the information being received. In reality, there is no information in the absence of a conscious, interpreting subject.

The concept of "stored" information has been extended in parallel with the expanding capacities of an endless variety of ubiquitous technologies that have been invented in the last century. Magnetic tape, gramophone records, compact discs, hard discs, and the iCloud have all been characterized as means of "storing information" which can be accessed by suitably qualified individuals using the relevant machinery. Calling what is on magnetic tape "information" even before it is accessed may seem to be justified because it was inserted by a conscious subject wishing to convey intelligence to others, though the originating subject may be many steps and many devices and processes distant from the place where the what-is is experienced as information. It is around here the "information" starts to be detached from conscious subjects. The failure to recognize that what is in a device is only *potential* information for a recipient licenses an expansion of the concept of information to encompass states of material objects even in the absence of informees and informants.

We should be discouraged from extending the reference of "information" over the entire material world by reminding ourselves that, in the absence of conscious subjects, there is no basis for the distinction between an event that is the information, those events that constitute the transmission of information, and those events that count as the reception of information; or between the information, the receiver or recipient of information, and (if required) the transmitter or source of information. Paninformationalism drains specific content from the notion of information; consequently, in the paninformationalistic world, the difference between signal and noise is lost. Without this difference there cannot be a subset of events that are signals and another subset that is noise – and, of course, between the knower and the known.

Paninformationalism may be a madness too far for some (including, as will be obvious, yours truly), but it makes lesser extensions of the widening of the notion of information comparatively respectable. These lesser extensions are sufficient to secure transport across the brain-mind, or material world-mind, barrier and to make the brain as a conscious machine at least in principle possible.

Consider the triggering of ionic currents in neurones by electromagnetic energy. There seems to be an insurmountable barrier to securing the transformation of such energy into consciousness of light – into sight or visual knowledge. The barrier is overcome if *both* the energy impinging on the brain-machine *and* conscious experiences are the same stuff: information. This is made even easier if all conscious experience is information and minds are "information processors".

If what comes into the brain is information, and what goes on in the brain is information, and consciousness is understood as information, then the explanatory gap between the material brain and the conscious subject, between a piece of what-is and the illumination of a world as a that-it-is, is closed. The brain's job as the brewery of explicitness is done for it; it has no special role in making what-is explicit. Unfortunately, by making things so easy for the brain, paninformationalists remove pretty well anything specific that organ brings to the party.

There are additional problems with consciousness seen as information in the respectably scientific technical sense of "the resolution of uncertainty" – as when the brain is characterized as a "Bayesian engine". The brain, according to this theory, is "a statistical organ of hierarchical inference that predicts current and future events on the basis of past experience".[35] To the contrary, consciousness is not simply the resolution of uncertainty: it has phenomenal content at many levels – perception, memory, thought. The experience of red is not reducible to the selection – or realization or actualization – of one out of a range of possibilities, of possible colours: it has intrinsic as well as differential, relational phenomenal content. The global nature of experience – as in sunbathing or indeed moment-to-moment world-bathing – is not the passage of awareness down a succession of discrete forking paths. Sunbathing, in short, is not "sunbayesianing". And, as already noted, possibilities must be *entertained* before consciousness-reduced-to-information can resolve them. In short, uncertainty Bayesian or otherwise *presupposes* consciousness, presupposes explicitness. It would be an egregious form of magic thinking to suggest that successive states of the universe are either global or local resolutions of uncertainties generated by their prior states.

Paninformationalism is one of the inescapable consequences of an understanding of "information" that conflates what happens in machines – radios, telephones, computers – with what happens in the (conscious) people who use them to transmit and receive information. Before there were conscious observers, clouds would not mean rain, nor spots measles; even less would there be events created to mean or to transmit meanings, such as the typing that I am engaged in now. The spots that mean measles do not *mean* to mean "measles". The emergence of conscious subjects was necessary before causally or otherwise correlated successive events could count as information about their predecessors. Any event or state of the world may be informative but only in the presence of a consumer informed by it: it is not itself information.

We should acknowledge that, except in conscious beings, "information" is only *honorary* or *potential*. And, as we have already seen, it is important to use the notion of "stored" information with care. This caveat applies most obviously to "stores of information" in nature. But this is equally applicable to "information" "stored" in devices such as books and computers: their state or their contents would cease to be "information" or even potential information in any meaningful sense if all humans capable of reading their contents vanished. To put this another way: there cannot be such a thing as prosthetic explicitness.

The original meaning of "information" is connected with a rather high level of consciousness, when what-is is registered and articulated by one conscious subject in response to her experience or transmitted from one conscious subject to another. It is a peg above sensations and perceptions, though its possibility is built on a base of phenomenal consciousness. This pecking order is overturned when everything that happens in the universe is granted the status of information. And it is also turned upside down in one of the most cited of current theories of consciousness: the so-called integrated information theory (IIT) particularly associated with Guilio Tononi.[36]

According to IIT, the brain, instead of being responsible for the transformation of what-is into something like information, gets information for free. Consciousness, however, has to be worked for. It is *built up* in the brain out of the right kind of integration of information. Consciousness emerges when the amount of information held collectively by neural networks exceeds a certain threshold signified by the Greek letter "phi".

Rarely has topsy been so turvey: instead of information being a manifestation of a higher-level consciousness capable of sharing experience, "consciousness emerges from the way information is processed within a 'system' (for example, networks of neurons or computer circuits) and that systems that are more interconnected, or integrated, have higher levels of consciousness".[37] Information is not specific to the brain: a simple photodiode or a thermostat will contain information and hence be conscious to some degree. Indeed, Tononi argues that IIT implies that "consciousness is a fundamental property possessed by physical systems having specific causal properties".[38] As such, it will not be confined to brains, or even to biological systems.

According to IIT, "contents are conscious (rather than unconscious) when, and only when, they are incorporated into a cause-and-effect 'complex' (where a complex is a subset of physical systems that underpins a maximum of irreducible integrated information)."[39]

What counts as the "whole" as opposed to the "parts" and who or what is keeping track of the sum total of the information in the whole and counting it as exceeding that in the parts is not clear. All of this, however, seems irrelevant: if information has no necessary connection with the consciousness of conscious subjects, then it is not clear why "integration" of information should generate or explain such consciousness. It is even less clear why the input of only some parts of the nervous system should contribute to the totality that is integration when consciousness is created; why the cerebellum and the spinal cord are not listened to. There is the additional problem of how inputs can be added up without losing their identity, retain the distinctiveness of what they individually bring to the party while not being lost in the unity of wakefulness.

It is hardly surprising that recent empirical research has failed to confirm that predications from either IIT or its rival global workspace theory have been confirmed.[40] There is no evidence of sustained synchronization between different areas of the brain corresponding to the transition to conscious experience predicted by IIT; and the neural activity observed in conscious experiences does not closely correspond to the "broadcasting" predicted by GWT.

Nor is it surprising that an impressive number of leading figures in neuroscience regard IIT as "pseudoscience", as evidenced by a letter, with over 100 signatories, recently published.[41] What is more, the axiomatic foundation of IIT has been pretty conclusively demolished.[42] There are, it seems, many reasons for saying "Fie!" to Phi.

There are other versions of the informationalist approach to consciousness in the brain. Among them is the emphasis on re-entry circuits being central to the transformation of neural events into consciousness. This seems to be influenced by the idea that if the door of a house is knocked hard enough, the furniture will wake up; or if the brain bangs hard enough on itself, it will rouse itself.[43] There seems no ground for believing that loops in neural circuits will awaken neural circuits to awareness of that which is other than themselves and hence to a world that surrounds them. The belief is an expression of the seemingly ineradicable intuition that that-it-is could be ignited by bits of what-is interacting with one another resulting in one participant in the interaction being made explicit to itself and locating other participants in a world that is its world.

The doctrine of "paninformationalism", which holds that wherever there is a physical state there is an informational state and wherever there is phenomenal consciousness there are informational states, is clearly too high a price to pay for keeping alive the hope of an explanatorily satisfactory account of the brain as a consciousness-making machine. It is also self-defeating. For the terminological wriggling that closes the gap between human beings and conscious machines also closes the gap between conscious machines and unconscious ones and between unconscious machines and the rest of the stuff of the universe. And we lose the sense of information as something that carries significance; something that "thatters" and hence matters. More broadly, if consciousness boils down to information and everything else boils up to information, the gap between subjects and objects, selves and their world, is lost and there would seem to be nothing metaphysically distinct about neural citizens and the stuff of which they are made. And that, utterly fundamental, distinction is what neural theories of consciousness are supposed to explain.

Paninformationalism is the most determined effort to overlook explicitness. Or, at the very least, it is the most striking expression of the tendency of explicitness to look past itself. It cannot accommodate the distinction between information, the subject who is informed by it, and the informant. Or, indeed, the difference between an object, a perception of it, and the sense organs that make it possible. Or the asymmetry of perception such that I am aware of a cup but the cup is not aware of me.[44]

3.6 TAKING THE STRAIN OFF THE BRAIN 2: PANPSYCHISM

One rather desperate response to the failure to find anything in the brain that can account for its apparent role in generating the consciousness of the conscious subject is to claim that it does not have to generate consciousness at all because consciousness is present in *all* material objects, great and small. Consciousness, so the argument goes, is a ubiquitous feature of the natural world, a fundamental stuff or force analogous to mass, energy, or electromagnetic force. There is consequently no need to explain how consciousness emerges in sentient organisms courtesy of ionic currents in the nervous systems. It didn't *emerge* at all because it was there all the time. Thus panpsychism: the view that mind or consciousness is present in the basic stuff of the universe.[45]

This jaw-dropping claim hardly seems to square with the facts of experience, according to which the stuff of the universe is overwhelmingly insentient. The defenders of panpsychism point out, however, that, when we observe or, indeed, experience the natural world, we do so from the outside. All that observation reveals to us about material objects, and the events and processes they are caught up in, is what they *do*, rather than what they *are*. The most sophisticated observations at the cutting edge of science give access only to their extrinsic, relational, dispositional, or mathematical properties. This is equally true of observations in biology, chemistry, and physics. In this sense, all stuff is dark stuff. Alternatively, as Chalmers has expressed it, "science reveals the structure of the physical world but not its intrinsic nature".[46]

There is, however, one place where we are not limited to an outside view, one piece of matter that we do know from within in virtue of *being* it – namely ourselves. And, if our selves are our brains, then we uniquely know our own brains from within: we have access to its intrinsic properties. In this one case, we can see that its properties include consciousness: the conscious mind is an aspect of the intrinsic nature of the brain.

From this, so the rather astonishing argument goes, we are justified in concluding that consciousness is an intrinsic property of matter beyond the brain – indeed throughout the universe. As Philip Goff has put it: "[T]he simplest hypothesis concerning the intrinsic nature of matter *outside of brains* is that it is continuous with the intrinsic nature of matter *inside of brains*, in the sense that both inside and outside of brain matter has an intrinsic nature made up of forms of consciousness".[47] Since consciousness is a property of all matter, though it is usually hidden from our gaze, and it is present in the fundamental elements of matter – electrons and such-like – we do not have to scratch our heads as to how ionic currents in the brain under the scratched surfaces of our heads could make unconscious matter conscious. Consciousness is already present in the stuff of which brains are made.

What is special about our brain is not that it (uniquely) generates consciousness but our privileged relationship to it, a relationship that gives us access to its intrinsic nature which, as is the case with all matter, includes consciousness. I know my brain from within, and thus have access to its true nature, in virtue of being it. The problem with the claim that conscious experiences are identical with certain brain states – namely that the least one would expect of two items that are in fact identical is that they should resemble each other – can be avoided. Brain states are conscious experiences, or the basis of conscious experiences, seen from without; and conscious experiences are the interior of brain states.

This solution to the mind-matter, mind-brain, problem gives rise to more problems than it claims to solve.

First, it does not explain how we uniquely know brains from within, in virtue of being our own brains. It does not account for the selfhood we have, or are supposed to have, in relation to our brains. It does not explain why our brains have such a relationship to themselves. Why first-person beings should arise courtesy of brains – such that there is an I who *am* that brain which then has privileged access to itself – is unexplained. The identity of a material object such as an electron, a stone, or a mountain, with itself is fundamentally different from the way that I *am* myself. (We shall discuss first-person being in Part II.) The way Raymond Tallis is himself is fundamentally different from the way a pebble is itself – not the least (but not only) because the latter's status of being identical with itself depends on this being made explicit by creatures such as Raymond Tallis.

The problem of explaining the emergence of consciousness in an unconscious world is therefore replaced by the equally, or even more tricky, problem of explaining the emergence of first-person being in an apersonal world, of "amming" in a world of being, of explaining how there emerge items that are conscious of being conscious in a seemingly unconscious world. Merely *being* object X is not, for the overwhelming majority of objects in the universe, sufficient to know object X from within; even less is it sufficient for X to "am" itself, to have access to itself and to discover that it is conscious – a property which it then recognizes as general and, indeed, generalizable to all other beings in the universe. Why would having, or rather being, a brain confer this epistemological privilege, this ability to nip round the back of the veil of interactive properties to the intrinsic nature of things?

That which is made explicit to itself in virtue of "amming" it – a conscious, embrained subject – rather than merely being it, doesn't seem to be explained by a brain, given its apparent status as a piece of matter whose key events are material processes similar to those seen elsewhere in the material world (including those that take place in, say, the spinal cord which is not associated with consciousness). If being X is not sufficient for X to "am" X, thereby giving it access to its own inner nature, and through this to have a window on the inner nature of what-is, that is hidden from the gaze of objective science, even less is it sufficient to explain the consciousness of consciousness that is supposed to be delivered in virtue of my being my brain.

Panpsychism does not thereby bring us any nearer to understanding why the brain "ams" itself; why I taste the entity that I am, while a cup of tea does not taste itself. Or why the brain – according to neurophilosophers – "ams" itself while a kidney or the rest of the body does not am itself.

Secondly, being a brain that knows itself from within is not the same as knowing an outside world in which the brain is physically located. Ordinary phenomenal consciousness is transitive consciousness *of* a realm outside of the brain – a realm defined (according to materialists) by that which directly or indirectly interacts with the brain. While the interactions between the brain and the world outside of it may generate events in the brain, there is nothing, in the consciousness supposedly available to all matter, corresponding to the intentional reach of phenomenal consciousness to events and objects that are causally upstream of the brain. If consciousness is everywhere, it is not at all clear why it should, courtesy of a brain, discover itself as being anywhere in particular, with that particular anywhere being defined by that of which it is conscious courtesy of the brain. Panpsychism, therefore, has nothing to say about how consciousness, supposedly intrinsic to all objects, uniquely in a brain reveals not just itself but a *world*.

Even less does it offer any explanation of the origin of the difference between intransitive consciousness – such as an itch – and transitive consciousness such as my awareness of a tree over there, along with the realm in which it is located that is explicitly other than myself as a conscious subject. The further transition to third-person mental entities such as thoughts and items of knowledge also takes us further from consciousness as an intrinsic property of the brain. Nor does it account for the objective knowledge of a shared world, the kind of knowledge we share with others, that we collectively build into a thatosphere that, among other things, ultimately enables us to talk about brains.

We can unpack this point slightly differently. If the matter of which the brain is composed comes already equipped with its own consciousness, it is difficult to see where or how it can make room for consciousness *of* – of the things that impinge on it from within and outside of its body. The supposed intrinsic consciousness of its material stuff seems at best intransitive, whereas human consciousness is overwhelmingly transitive, being about something other than itself. (We shall return to the contrast between intransitive and transitive consciousness in Chapter 5.)

Making consciousness a universal property, therefore, so that it is present even in electrons, must require jettisoning its fundamental characteristic of being conscious *of* – of itself and/or of a world. Phenomenal content, intentionality, propositional form, will be erased in a fundamental stuff that is ubiquitous at the basic or microscopic level. More specifically, panpsychism cannot deliver explicitness, if only because "panexplicitness" would be a self-contradictory state of universal explicitness that makes nothing in particular explicit. Or of everything making itself explicit to no particular subject.

The challenge of explaining the transitivity, the world-revealing nature, of human consciousness has sometimes been characterized as that of accounting for the emergence of a macro-conscious self-aware subject from micro-conscious elementary particles or other less fundamental constituents. And many, perhaps most, panpsychists accept that while material objects are made of elementary microscopic components with experiences of some kind, they also accept that the overwhelming majority of such objects (stones, clouds, trees) don't have (unified) minds: they are not selves in worlds like you and me.

And so, we return to the problem of explaining the seemingly close, privileged, unique, association between the brain and the consciousness of fully formed conscious subjects, the revelation of a world around an "I", an implicit "me here" facing an explicit "out there", an individual who cares for herself in a world that matters to her. The gap between atomic consciousness and full-blown conscious subjects like you and me is as wide as the gap between matter and mind, brain and consciousness, that panpsychism tries to close.

The claim that brains have full-blown minds, rather than being simply consciousness dust, because their parts are organized in a certain way, leads to the obvious problem of trying to understand what it is about brains that confers this faculty on them. What kind of organization would deliver what is needed? Reference to, for example, "fusion" in the brain of micro-level subjects into a human mind or a self only adds an extra layer of opacity.[48] It certainly does not explain why the same brain can be in different states of consciousness – waking or asleep – or be associated with different contents of consciousness depending on its location as a material object in a material world. Nor, of course, why a brain has a consciousness beyond the capacity of a stone, a kidney, or a spinal cord. Or why there is a "what it is like to be a person with a brain" but not a "what it is like to be a stone" or "what it is like to be a spinal cord". Nor how the brain enables its own atomic soul dust to add up to conscious subjects with memories, agency, thought, a sense of responsibility, and a life history that it embraces as its own.

We have arrived at the most discussed of the problems acknowledged by panpsychists: the so-called subject combination or subject-summing problem. This is the conundrum of how micro-level consciousnesses add up to a mind or a human subject. The analogy with the combination of, say, molecules of H_2O into a drop of water does not hold up because twinkles of consciousness do not seem amenable to adding up in the way that molecules do. There is a built-in closed-off-ness of elements of the consciousness of a human subject. My consciousness of the apple in front of me does not combine with another's consciousness of a snowstorm in the arctic or even that of a person standing next to me looking at the same apple as I am. Beneath this concern is the not unreasonable intuition – discussed in Chapter 5 – that experiences have to be "had" by, and uniquely attached to, an experiencing subject. And the example of molecules of H_2O adding up to a drop of water can only seem to deliver a hint of explanation if we overlook the fact that the different scales – molecular and macroscopic – are imported by a conscious subject, as discussed in the previous chapter.

Subject-summing remains a formidable barrier to accepting a panpsychist "solution" to the problems of human consciousness. According to Basil Hiley and Paavo Pylkkänen "There is nothing like it to be a single electron but when electrons and other elementary particles and fields are arranged in the right hierarchical structure (e.g., that of the human brain) phenomenal properties in a full sense emerge from the underlying protophenomenal ground".[49] This sounds more like a restatement of the gap to be closed than an account of how it might be closed. What "the right hierarchical structure could mean" – other than that which solves the problem – is unclear.

Combining large numbers of micro-consciousnesses into a subject does not seem likely to generate anything more hospitable to the transitive, world-revealing, consciousness of human subjects than the microscopic elements that go into the mix. Irrespective of the scale of the elements whose individual consciousnesses are combined – atoms, intracellular organelles, cells, etc. – it is difficult to see how the result of their combination would be an organized conscious subject facing a highly structured world evolving in a law-governed way rather than pandemonium – or some kind of consciousness mush.

Panpsychism, even if it solved, or justified bypassing, the enigma of finding mind in matter and the additional problem of allowing mind to have causal properties, would not address the challenge of explaining the origin of highly structured, indescribably complex, individual conscious subjects, more or less sealed off from other subjects, who, in virtue of their perspective on the world, in part given to them, in part constructed by them, inhabit particular worlds in which they live. It is not at all clear how the aperspectival, intransitive mind dust of elementary particles adds up to a coherent conscious subject. Or, as it is put by Miri Albahari (who defends an idealist panpsychism against materialist panpsychism), how combination would be possible if the component conscious elements were envisaged "as encased in observer-independent material bodies that are located in mind-independent space".[50] It is, as she says, "very hard to imagine how microexperiences belonging to different microperspectives could combine to produce microexperiences that associate with a single macroperspective".[51] At the very least they would have competing, or even incompatible, perspectives. The appeal to "phenomenal bonding" seems to restate rather than contribute to solving the problem.[52]

The ascription of consciousness to elementary particles, as well as raising the question of how they can pool their resources to become a unified conscious subject, also raises the question of what on earth they could be conscious of. Of themselves? In which case what would it be like to be an electron? Of the electron next door? Of the atom of which it is a part? Or of a micro-world? How far would that micro-world extend? Why should an electron be enworlded? Even if the consciousness of an electron were intransitive, like a nano-proto-itch, how would transitive consciousness arise? And there remains the question already referred to of the competing scales realized by the objects of which it was a part – sub-atomic, atomic, molecular, and so on upwards – if such scales exist independently of conscious subjects with specific interests. (Which, of course, they don't.)

One response to these many problems is "cosmopsychism", according to which the entire universe is a single consciousness, or at least displays psychological properties as a whole.[53] For some philosophers, cosmopsychism is consistent with certain interpretations of theoretical physics, according to which "the fundamental building blocks of reality are not particles but *universe-wide fields*, and that particles are simply local vibrations within those fields".[54]

Cosmopsychism generates the opposite problem to the subject-summing problem of building macroscopic human subjects out of the nano-twinkles of sentience had by elementary particles; namely, that of unpacking the consciousness of the cosmos into individual subjects each having their own perspectives, worlds, destinies, preoccupations, and life stories. This is the so-called "decombination" problem. What would the single, universe-wide consciousness be conscious of? What would it have as its intentional object(s)? The universe? If yes, then we would have to account for the limited scope of the consciousness of individual subjects: the world that seems to be anchored in the first instance on our bodies, supplemented by the shared realm created out of the community of minds of which we are epistemological citizens. And if the intentional object of cosmic consciousness were the (entire) universe over (all) time, it would have no definite object because events, processes, and states of affairs would cancel each other as the universe evolves: that which goes up usually comes down. Even if the Second Law allowed for a vector of change over time, the object of consciousness would be a blur.

If consciousness were genuinely cosmic it could not sample the universe in time slices, since relativity theory has taught us that there is no such state as "the (entire) universe at time t". Such foliation is observer dependent. What's more, the universal consciousness of cosmopsychism would be as intransitive as that of an electron, because there would be no space for the otherness, the outside-of-me, of intentional objects.

All of these problems arise even without the cosmopsychism-related claim – elaborated by Philip Goff – that the universe has an overall purpose.[55] It is difficult to see what that purpose could be. Purposes as they are usually understood are typically plural, local and anchored in the ever-changing needs, viewpoints, preoccupations, responsibilities, and hopes of individual conscious subjects. They are usually short-lived or at least of finite duration, and extinguished by fulfilment. They are often conflicting – an especially pronounced barrier to unification. Even when the idea of purpose is expanded to that of *telos* – most obviously in the case of whole organisms or organs that serve the needs of organisms – it cannot be further expanded to the entirety of being, not the least because purposes are plural and typically competitive. Predator and prey could not be seen as having a common purpose. The idea that they jointly contribute to the fulfilment of a purpose of which both parties are unaware – that they respectively flourish and die for the sake of the progress of the universe, the evolutionary pilgrimage of what-is – raises the question of what purpose that process serves, given that even its most advanced product, human beings, seems for many to lack a sense of unwavering, overall purpose or be able to imagine what kind of purpose they serve by living and dying as they do.

What, anyway, could be the purpose of the universe as a whole? To last longer, get bigger, or to please a God whose judgements – projected from those of humanity? – seems to be uncertain and even conflicting. Purposes typically exist only insofar as they are entertained, and they are typically entertained when they are not yet fulfilled. Cosmopsychism does not seem able to accommodate the gap, at the heart of purpose, between what-is and what-might-be; between how things are and how they could be; between the successive phases of the fulfilment of a purpose.

In short, the cosmopsychic answer to the combination problem results in over-combination – and consequently leads us to a place as remote from the kind of conscious subjects we are as the micro-consciousness of standard panpsychism or, indeed, the passage of ions through the semi-permeable membranes of nerve fibres.

The most serious problem with panpsychism is one to which I have already referred; namely that it is (to put it mildly) something of a stretch to extrapolate from our own inner experience of our material brains – or indeed one brain, our own – to the properties of the entire material universe; that we can get to the heart of matter by being a piece of it. Being something – first-person "amming" it – does not give you third-person knowledge of that thing; even less does it reveal the universe in which that thing is located. Knowledge, after all, has to be corrigible, and being something – as I am my brain according to panpsychism – is not corrigible.

More damagingly, being something – for example Raymond Tallis' brain – does not deliver the free gift of knowledge of its fundamental nature. The very idea that I know one item – my brain – through and through because I *am* it, raises the question of how much of it I know and what form that knowledge takes. Being a particular brain does not make one a sufficient authority on it to conclude from that brain's being conscious that consciousness (or proto-consciousness[56]) is a property of the entire material world.

The passage from supposedly first-person experience of one object to knowledge of a general nature of the stuff of which it is composed is long and difficult: third-person, objective, general knowledge of such a wide scope is not a free gift that comes from being a particular entity. A crucial step on that path would be the classification of our brain as a piece of matter and that consequently justifies the belief that it is fundamentally the same as the stuff of electrons, chaffinches, mountains, and the remainder of our bodies. "Amming" a brain, even if it were self-explanatory (which most certainly it is not), would not give us access to the intrinsic nature of the remainder of the universe. Just as being a piece of matter does not, of itself, give us insight into the nature of matter.

So much for the astonishing claim that we can get past the veil of appearance in virtue of being a part of what is behind the veil. At the very least, this leaves unexplained the origin of the localized "am" (and "that-I-am") in a universe of mere "is". Why should "amming" take place in a brain as opposed to in a pebble, a leaf, or an entire planet? Why does a brain "am-being" itself in addition to "is-being" itself while, so far as we know, other material objects – including our organs such as our hearts, kidneys, and spinal cords – simply *are*, without embracing that fact as their identity. Nothing that we can observe in the brain accounts for its being the place not only where what-is becomes that-it-is but where that-it-is becomes that-I-am or the what-is becomes the I that I am.[57]

If, in short, we are to presume that it is the distinctive (objective) nature of our brain that enables it to be conscious of itself (and thus delivering the data on which panpsychism stands), we have to identify what that nature might be and explain why it might deliver what is needed. This seems to be no less of a challenge than the one facing materialist neurophilosophers trying to find the basis for consciousness in the distinctive objective characteristics of neural activity. Avoiding the difficulty of explaining why consciousness should be associated with (certain) activity in (certain) parts of the brain by asserting that "consciousness is everywhere" seems only to compound the difficulties faced by a neuroscience of consciousness. The claim that the mind is a basic ingredient of the entire natural world does not narrow the explanatory gap between the seemingly insentient material world and the feeling of joy or irritation or between cheese and the taste of cheese.

Let us therefore return to the question of what it is that happens in the brain that might account for its apparently crucial role in the generation of explicitness.

3.7 THE SEDUCTIVE METAPHOR OF REPRESENTATION

The metaphor of the mind as "representational" and the connected idea that its representations are "realized" in patterns of neural activity exemplifies the endeavour to make consciousness comprehensible as part of the physical world by squeezing out the distance between what-is and that-it-is – in short, explicitness.

My contention is that talk of "representation" bypasses the fundamental mystery of explicitness because it does not acknowledge that *there is no representation without presentation*. Representation cannot account for presentation – for presence, or the givenness of the given. Representational theories of mind, therefore, bypass what is central to the conscious mind.

At its simplest, the representational story is about the intentionality of conscious contents: a perception, memory, or thought is, so we are assured, *about* an object or an event or a state of affairs in virtue of its *representing* the latter. It qualifies as a representation by having something fundamentally in common with it; by, in some concrete or abstract sense, *replicating* it. The world is present to us by being in some sense mirrored in our brains.

Representational accounts of the mind have become increasingly complex because, on account of their intuitive attractiveness, they have been embraced and developed by many philosophers, neuroscientists, computer scientists, and cognitive psychologists. In the latter half of the twentieth century, when it was at the height of its popularity, the representational theory was narrowed down to the more precise claim that the mind is a *symbolic* or *encoded* representation in neural pathways of what surrounds the subject and impinges on her brain and that what happens in those pathways is analogous in important respects to computer processing. Adam Rostowski (who opposes representationalism) has expressed it as follows: "Inspired by developments in computational theory and technology, 'cognitivism' took the lid off 'the black box' of the mind, modelling its inner workings through its formalization of cognitive processes as algorithmic operations executed over internal representations".[58]

According to the representational theory, what happens in the subject – or her brain – counts as a representation because it in some important respects encodes and hence indirectly replicates features of the perceived object. This seems intuitively acceptable because what qualifies as a representation in the extracerebral world – in photographs and oil portraits as well as natural images in reflecting surfaces – does so in virtue of its sharing properties with the represented object. The image of a person in the mirror "represents" that person because it has (for example) the same shape, or size relative to the objects around her, or colour, as the person. The face in the mirror looks like the face. Or, as in the case of a scaled-down, black-and-white, portrait, it has the same proportions of parts as in the original, even though it is profoundly different. The fundamental principle – that something of the original is *replicated* in the image – is nevertheless upheld.

Relatively little may need to be preserved of the original in the image if there is a disposition to see the former in the latter: I may see faces in clouds, which otherwise have very little in common with faces. What is conserved may be quite abstract. Hence the attractiveness of the idea that the passage from what-is to that-it-is for a conscious subject is secured by the entity that is made explicit being replicated in some form in the brain of the subject.

Advocates of the representational theory of the brain-mind are, of course, aware that neural activity is nothing like the world it is supposed to represent. There is no isomorphism between brain events and the contents of consciousness and the world of which it is conscious. And, of course, there are no secondary qualities in neural activity. Colourless neural discharges don't look like my grey-bearded, bottle-hardened face replicated in the mirror. Indeed, the mismatch is more profound. After all, not only is neural activity not pink but neither is the light which impinges on the brain. What we perceive as one colour rather than another is the response of the brain to light of a particular frequency.[59]

The idea of replication is therefore widened to include some kind of *encoding* – analogous to the encoding that takes place in a computer. If what I see in the mirror is encoded in neurally realized computations in the brain, triggered by light from the object landing on the retina, the experience of the object need not directly match any aspect of the object. Just as the electronic patterns that are associated with the generation of an image on a television screen are nothing like that image, so the encoded "images" in the brain are nothing like the object that is registered through those images. The activity that corresponds to the experience of yellow does not have to look like, even less to be, yellow. Some kind of replication, however, is retained. For example, the pattern of neural activity preserves an aspect of the *pattern* of light that underpins the visual appearance of the object. The homogeneity of neurally encoded images of widely different objects – rainbows, stones, faces – should not, it is argued, be a matter of concern. After all, language is homogeneous – it is composed of a narrow range of sounds or letters – and it still can point to, or bring into presence, a rich and varied world.

There is an obvious problem with this view of the cerebral realization of perceptual and other modes of consciousness. Encoded representations do not deliver consciousness of objects without *de*coding. Decoding is necessary in order that homogeneous neural activity here in the brain becomes a heterogeneous reality out there in the world. It is not at all clear what form decoding would take. Translating neural activity into other neural activity would scarcely deliver what is needed. The translation of what is going on in the electronic circuitry of a television into an image ultimately requires a conscious interpreter looking at the screen, gathering up the pixels into a picture: the couch potato staring at the screen. And the same, of course, applies to language. Verbal expression captures the richness of the world only because listeners and readers are able to translate what they hear or see into the *referents* of what is said or written because they have experienced those referents, or their components – individually and collectively – directly through their own perceptions or indirectly through the reported perceptions of others and have been taught the meaning of the words that refer to them.

A cycle of encoding followed by decoding does not therefore meet the challenge of understanding how neural activity that does not look like objects of perception or their appearance to us, delivers us a portrait of our surroundings. This is connected with a deeper problem of the representational theory of mind: that of ascribing to the brain the capacity to translate what-is into the *appearance* or *presence* of what-is, into that-it-is, in virtue of replicating some aspect of it. There is nothing in replication *per se* that would ensure that the replica would *represent* an original, however "faithful" the replication.

Let us take a very simple example: the literal three-dimensional replication of an object, a *reduplication*. Consider a cup on a table. Another cup, identical in all respects, is placed next to it. Of these two cups, it is not possible to say that one of them represents the other; that the that-it-is of the one is *of* the what-is of the other. Indeed, unless they are used in this way – as signs of each other – by a conscious subject, this relationship does not arise. Cup A may be like Cup B, but neither is an image of the other in the absence of a conscious subject. The same applies to mirror images. They do not of themselves represent, refer to, make explicit, that which is mirrored in them: without the assistance of conscious subjects, replication does not deliver experiences. After all, entities that were mirrored in other entities before there were conscious subjects – as when clouds passed over pools on the earth billions of years ago – did not acquire the status of images. It takes a conscious subject harvesting and interpreting the light to see an image of cloud on the surface of a pool. No amount of reference to arcane processes such as "encoding" and "decoding" can or could find that subject in the brain.

The representational theory of mind or consciousness has become less popular of late, though probably for the wrong reason. Or rather for a reason that does not dig deep enough, because it overlooks the explicitness that is the essence of consciousness.

It has been most clearly challenged by the enactivist movement in philosophy which, over the last 30 or so years, has generated a vast and often rich literature. The representational theory of mind, it is argued, makes the conscious subject materialized in the brain too much of a spectator – quite unlike the busy engaged individuals we are. Our "absorbed coping" is not at a distance from the world with which we are interacting in pursuit of our goals; and the representational theory of the brain-mind seems to isolate the brain from the rest of the body and its environment.[60]

All of this is true; but it does not get to the bottom of the error of representationalism. The right reason for rejecting representational accounts of mind is that they overlook the need for prior *presentation* for there to be *re*presentations and this requires, rather than constitutes, a mind.

Let me clarify this. A representation must be a *re*presentation of something that is (already) present. Consciousness is first and foremost about presence; awareness of what-is *that* it is. Representation of objects is a secondary, mediated presence that depends on a prior capacity to make things present. Representation would not deliver awareness of objects were the image itself not experienced as being present. My body and its image in the mirror are both present as I look in the mirror. The image does not have intentional reference to my body except when I am conscious of the image. If I were unconscious, the image of my comatose body in the mirror would not be a presentation unless some other conscious subject came upon the scene.

The belief that replication is sufficient to make that which is replicated *present* as an object of consciousness is as absurd as the suggestion that an echo can turn vibrating air into a sound.

The claim that phenomenal consciousness, with its intentionality, is generated by (coded) replication which mysteriously becomes representation is, among other things, a variant of the "tracking" theory of consciousness which we discussed in Section 3.4. If intentionality were simply replication, then it would not be possible to get things wrong. It is only when the reflection of the cloud in the puddle is experienced and consequently interpreted as an *image* of a cloud by a conscious individual that the image becomes a piece of truth. Replication could not generate this normative (true vs false) dimension.[61]

Just how adhesive is the upside-down notion that presence or presentation could be generated by *re*presentation is illustrated by the seriousness with which the higher order theory (HOT) of consciousness is taken. The key element of this theory is the claim that "the phenomenal character of a [conscious] state is determined by the properties that the relevant meta-representational state ascribes to it ... A mental state is conscious in virtue of being the target of a certain kind of meta-representational state".[62] In other words, consciousness depends on the representation of representations; or presence depends on third-order presence of the presence of presence.

I don't want to spend too much of the reader's time on this wrong tree up which so many philosophers have barked – given also that the tree is now barked up less frequently. I visit it only to highlight what is perhaps the greatest mystery surrounding the unholy trinity of Chapter 1: the very fact that I *see* the mirror image. The transformation of what-is into that-it-is is not comparable to the replication of my face or any other entity in a mirror – with or without any amount of encoding.

For (perhaps) 13.7 billion years, light has travelled through the universe from sources to surfaces that either absorb or reflect it. For 99.99 per cent of those years, the light was not received in the way that it is received by entities such as you and me. For you and me, the light is luminous and illuminating: it reveals both itself and some of the objects that it encounters in its journey to the eye. The visibility of items in the world, the revelatory power of light, is a recent phenomenon, confined so far as we know to a minute part of the universe.

3.8 COLLAPSING THE MIND: FUNCTIONALISM

There seems to be little progress, or even hope of progress, in closing the explanatory gap between what happens in the realm of material entities and the fact that some entities are conscious of some of those happenings – and of themselves to whom, in whom, or around whom, those happenings happen.

As we have seen, attempts to naturalize the most awkward aspect of consciousness – intentionality – have proved unsuccessful.

One way of dealing with the truth, frustrating to materialists, that nothing in the brain accounts for the transition from what-is to that-it-is is to minimize the significance of the transition. Until the advent of the pantasies described in Sections 3.5 (panpsychism) and 3.6 (paninformationalism), the commonest expression of this was the endeavour to squeeze consciousness into a vanishing gap between input from sensory receptors and output of behaviour, thus dispensing with any inner space belonging to the mind opening up in the brain.

This is the essence of functionalism: mind is what mind does; what the mind does is what the physical brain (physically) does; mental states are identified with certain causal roles connecting brain activity with visible or invisible behavioural or physiological responses. What makes "a mental state of a particular type does not depend on its internal constitution, but rather on the way it functions".[63] "The way it functions" refers ultimately to the visible behaviour or invisible bodily adjustments it prompts.

The research programme of functionalism had as its over-riding purpose that of uncovering continuous causal chains connecting material entities or energies impinging on sense endings with neural activity, and thence with motor or other effector responses constituting behaviour. At no stage is there a transition of what-is to that-it-is, even in seemingly pure that-it-is such as pain or thought. Pain, for example, can, it is claimed, be translatable without remainder into a set of responses – physiological reactions, or pain behaviour such as crying out, withdrawing a limb, saying "Ouch!", reporting that "I am in pain", or seeking the help of a doctor, or an increased probability of such responses.

While this crude behaviourist version of functionalism is less popular among philosophers than it was, it is worthy of mention because it highlights the tendency to exclude or marginalize explicitness even from the portrait of the mind. In pursuit of this goal, functionalists endeavour to close any gap between the material events that impinge on subjects and what those subjects get up to in response to, or anticipation of, such events.

This project is very much alive and well in some versions of the aforementioned idea of cognition as being embodied, such that the elements that constitute a cognitive system are not confined to the brain but include the rest of the body and even its environment. Accordingly, the conscious organism does not have to represent the world in order to interact with it. Instead, the organism and the relevant parts of the environment and the environment together comprise a single, coupled system.[64] This is a direct assault on something central to our human being as that of an *un*coupled animal, so that we can be agents acting upon nature from a virtual outside – something we shall discuss in Chapter 7.

The case against functionalism has been framed in different ways but the most common approach has been to point out that there is rarely a direct, even less a simple, correlation between stimuli, sensory experiences, and behavioural responses. The seemingly straightforward, elementary example of pain – which may prompt withdrawal from the source, a cry of "Ouch!", a consultation of the timetable of the public transport system that will take me to the health centre, cancellation of an appointment, or a decision to "just put up with it" – illustrates this. Sensory experience – or more complex mental contents such as memories, thoughts, and propositional attitudes – are very loosely connected with types of behaviour, especially when the latter are characterized in physical terms.

More to the point, the precise phenomenal content of experience may be irrelevant to behavioural output. The lack of a tight connection between phenomenal experience and behaviour is highlighted in the "inverted qualia" thought experiment, according to which we may imagine two people experiencing the same colour differently.[65] Jane experiences red as you and I experience green and she experiences green as you and I experience red, while John experiences red as you and I experience red, and green as we experience green. However, John and Jane both respond the same to traffic lights and call the colours by the same name.

Thus, there is a fundamental dissociation between phenomenal experiences and the behaviour they prompt. This is possible because all that is needed is the right kind of connection between material events such as traffic lights changing and drivers stopping. The phenomenal experience seems to have little or nothing to do with any particular causal events being actually experienced or with a particular stretch of a causal chain being experienced in a certain way or being associated with a specific kind of experience.[66] After all, secondary qualities such as colours do not have a place in the material world. The examples of advanced computers and of insentient organisms responding to inputs demonstrate how much function can be delivered without anything being made explicit in, or to, the entity in question or without there being anything it is like to be that entity and to have events happening within it as experiences. In short, there is a dissociation between the conscious contents of mind and any functional connections between what may be classified as the sensory input and what may be classified as the behavioural output of a human subject. Functionalism cannot accommodate phenomenal consciousness.

There is a computational version of functionalism that for some philosophers still has life. According to this version, the organization relevant to consciousness is computation. The mind is, for example, "a virtual machine", defined thus by Jennan Ismael: "A virtual machine is a generic word for a functional duplicate of a real or hypothetical machine made not of mechanical parts, but of virtual components".[67] The link between sensory input and behavioural output takes the form of a virtual machine. Virtual machines are machines created mainly by programmes running on other machines.[68] Mind is "software" implemented in the "hardware" of the brain. The "soft" in "software" makes it mindlike.

It is not clear how, without an already existing mind (indeed intentionality), there could be any basis for the differentiation between or separation of hardware and software – between the brain and itself; between the solid, soggy brain and the virtual machine that is realized in it. Nevertheless, Ismael asserts that "[T]he right way to think of the relationship between mental activity and brain activity is *implementation*: not correspondence, none of the much simpler, visualizable relationships that philosophers have sometimes thought have to hold if the mind is part of the physical world" (emphasis added).[69] This does not really help to support the aim of absorbing the mind into the physical world. While this restores something that has been lost in functionalism, namely the collapse of the gap between input and output, what fills the restored gap does not seem at all mind-like. Most importantly, there is nothing corresponding to phenomenal consciousness. What is more, the idea that the hardware "hosts" or "instantiates" the virtual machine remains to be clarified and, once clarified, explained.

While software may do things not evident to external inspection this does not bring us any closer to narrowing the gap between a piece of what-is (the brain) and that-it-is or what-it-is-that-is (mind). It is, after all, the phenomenal consciousness of conscious subjects that creates the algorithms that shape the functioning of the computer. What goes on in the computer can be seen as the operation of, or realization of, algorithms ultimately only in the light of phenomenal consciousness.[70]

The functionalist endeavour to squeeze out, or tame, or naturalize mental entities by gathering them up into a continuous causal path between external events impinging on the nervous system and the behaviour that results, fails. Which is precisely what we would expect. The assumption that the passage between states of what-is could somehow generate explicitness is clearly unfounded. At any rate, the continuous causal connectedness between sensory input and behavioural output where, as Klein has put it, movement is "caused – not intended – by neurally instantiated, cranially located structures linking input with output"[71] hardly captures anything of what is going on when human agents engage with a world that is *their* world on behalf of an entity that is themselves. (This is something to which we shall return in Chapters 5–7.)

3.9 CONCLUDING REFLECTIONS

It seems clear that, without the metaphorical or simply careless use of terms such as "information" or "representation", the brain, understood as a material object made of the same stuff and subject to the same physical laws as other material objects, seems to lack the capacity to secure the transition from what-is to that-it-is – never mind to "that-I-am", the conscious subject who receives, correlates, makes sense of, and consciously acts on, that which she makes explicit as the "reality" within which she is located, takes account of, and of which she is a part. If intransitive experiences such as itches resist explanation in neural terms, transitive experiences – such as your seeing the text you are currently reading – are even further from the reach of neural explanation. It is unclear how the brain gets itself out of the way in order to permit itself to be that in virtue of which portions of what-is are allowed to declare that they are and in some difficult-to-specify sense – discussed in Chapter 10 – see what it is that they are.

For these reasons, it seems that what it is like to be Raymond Tallis and to have the experiences that locate him in his body and that body in his world, is not what it is like to be a brain – even his brain – with its objective properties, notably the ionic currents passing through it. "Presence" is not a property of material objects, whether they are cups that are perceived or brains that neurophilosophers believe do the perceiving: it is not an intrinsic property of either Raymond Tallis, insofar as he is a material object, or of the objects that are present to him. There is nothing in body or brain seen through the eyes of physical science that can explain how it is self-presencing or world-presencing.

The seriousness with which neural causal, neural representational, and neural tracking theories of intentionality, have been taken is evidence, if evidence were needed, of the elusiveness of explicitness – how its essential nature is missed.

To reject causal, representational, or tracking theories of intentionality is not, of course, to deny some essential connection between the object of consciousness and the subject who is conscious of it, mediated at some level through the body of the subject. If there were no such connection, it would be impossible to see how individual subjects with their distinctive trajectories through what-is, and their lives lived in individual worlds, could be accounted for. The connection is not, however, a connection of the kind that is seen in, or ascribed to, the realm of what-is. The transition from what-is to that-it-is or from what-is to that-I-am is utterly different from the passage from one phase of, or one location in, what-is to another phase or location of what-is.

Yes, in order to perceive a cup, I have to be in its vicinity but the distance of intentionality between my perception of the cup and the cup itself is not identical with, or even comparable to, that between my head, my retina, or my visual cortex, and the cup; or to the distance that is cancelled when I reach for the cup to pick it up. The space crossed by the light from my cup to my eyes is not comparable to that crossed by my gaze to the cup. This is obvious from the fact that the "over there" of the cup is a spatial separation that is *at the same time* both crossed and held open by my perception of the item in question.

While this is most obviously true of visual perception – where I am related to the objects that are explicitly "over there", outside of me – it is equally the case with all modes of object awareness. When I pick up a cup, the spatial gap between my body and the cup is closed but the subject-object difference/distance between me and the cup remains. This is reflected in an asymmetry: while I sense the cup, the cup does not sense me: the intentional relationship between me and the cup is not reciprocated. The difference or distance between an object and its material representation – for example my body and its mirror image – is not the same as the difference or distance between an object and its being present; between its being and its presence.

This point about the special kind of distance associated with consciousness is especially clear when we consider modes of consciousness other than perception, such as thought or memory, where the intentional objects are not present and may indeed have never existed. If it is not possible to see how the brain could be a place where (for example) electromagnetic radiation is transformed into a medium of perceptual revelation – of itself and of objects it falls upon – it is hardly going to be possible to see what it is about the material properties of the brain that enable it to transform what-is into that which is suffered, known, remembered, thought about, sussed out, puzzled over, got wrong, and valued as good or bad. Or an awareness that certain things do not exist or are not possible or unlikely.

If there is nothing in the matter of the brain that would account for its being the place where the material world acquires the property of standing outside of itself, even less is it able to explain how the organ can identity itself as "the brain" and think about "the brain" as the putative seat of consciousness including the thoughts it is having about the brain. The neuroscientific endeavour to construct the conscious subject from the apersonal, aperspectival perspective of objective quantitative thought is effectively to attempt to create a personless, mindless, theory of persons and their minds.

A connected point is made by John Shand. The difficulty of explaining how the physical could explain consciousness, he argues, is epistemic – "that of defining the physical as those features that can be known objectively, coupled with the contention that only those objectively known properties are real". It is this "that makes the explanation of subjective consciousness in terms of physical properties not just hard but impossible".[72] His solution to the hard problem is simple: "to hold that the world is indeed all physical but have the physical no longer defined as what may be known only objectively and hold that some physical properties may be known only subjectively".[73]

This, alas, is point-missing. To include in the physical that which is known only subjectively is to overlook the fundamental difference between what-is and that which makes it explicit – as the contents of consciousness that are directly or indirectly about it. What is more, objective observation of the brain is downstream, is built on, the explicitness that begins with basic sentience. Hence the fundamental obstacle to explaining conscious experience by physical science: its observations – howsoever mediated by science – are never entirely divorced from phenomenal consciousness, though science may set it to one side. In order for there to be science, there has to be something other than its objects: consciousness is that something. Hence there cannot be a science of consciousness.

The failure of objective knowledge to reach beneath itself to its own basis – to know how it is possible – is, of course, entirely unsurprising. The very idea of a scientific or other explanation of consciousness – of consciousness digging beneath itself – using tools created through the collective cognitive activity of conscious beings seems impossible. The hard problem remains unsoftened.[74]

As was argued at the beginning of this chapter, the hard problem is not escaped as we ascend from phenomenal, what-it-is-like, consciousness to the higher levels of cognitive function. Unfortunately, Chalmers' most recent iteration of the hard/easy distinction is no less vulnerable than earlier ones: "Intelligence isn't a matter of how a system *feels.* What matters are the objective processes in the system and the behavior they produce ... As a result we have a much better grip on intelligence than on consciousness ... While intelligence is a matter of objective behavior, consciousness is a matter of subjective experience".[75] On the contrary, intelligence cannot be separated from phenomenal consciousness, even less from consciousness period. If anything, it is "Consciousness+". It could not function except in a sea of phenomenal awareness, building on and going beyond it, as it deepens and extends explicitness.

The eyeless gaze of verbal knowledge – when we think about, indeed reflect on, and investigate, the material world (as when we use the word "light" to signify light) – seems yet more remote from what happens in matter than the gaze of the outer eye, but it ultimately builds on what is experienced courtesy of that eye and other senses.

The link between intelligence and behaviour highlighted by Chalmers in the passage just quoted – "intelligence is a matter of objective behavior" – does not justify separating it from (phenomenal) consciousness, from subjective experience. Nothing could be more complexly conscious than the observed behaviour of humans navigating their way through their own lives in a manner that they consider appropriate in the light of what they think or feel they are or would like to be or become. The execution of bespoke actions in the context of bespoke intentions directed to a bespoke future must be illuminated by consciousness at many levels and at every level there are elements of phenomenal consciousness. (We shall discuss this further in Chapter 7). While the navigation of a missile towards a moving target can be accomplished without consciousness, the navigation of an individual through all the spaces, physical, conceptual, and social, that separate her from her goal requires consciousness and self-consciousness – if only because that goal is not an existent something that is approached but a possibility that may or may not be realized. This is particularly clear when the goal is something abstract, complex, not defined in physical terms – as when (for example) I behave in such a way as to try to improve my relations with a colleague, fulfil the plan to have a holiday in the sun, qualify as a doctor, or raise the standard of the services provided for stroke patients.

If we insist on extending the idea of intelligence to encompass whatever shapes adaptive behaviour, this may backfire: it can be widened to include the complex processes seen inside individual cells. This is what follows if we separate intelligence from consciousness. We should, however, think of intelligence as the upper storey of a hierarchy of awareness whose ground floor is phenomenal consciousness: it is a faculty exercised by a creature who knows what she is doing and why she is doing it. We cannot, therefore, escape the problems that face us when we think about the emergence of consciousness in all its manifestations and try to understand the role of neural activity in making things explicit. All such problems are "hard" problems.

The temptation to evade such problems by invoking panpsychism and its sibling paninformationalism is understandable. It is easy to sympathize with William James' giving up on emergence: "If evolution is to work smoothly, consciousness in some shape must have been present at the very origin of things. Accordingly, we find that the more clear-sighted evolutionary philosophers are beginning to posit it there".[76] If the most likely candidate for the brewer of explicitness – the brain – does not deliver this, then nothing can. It must have been there all the time.

While this may seem to address one problem, it generates a multitude of others – including accounting for the special role of the brain in our lives as conscious subjects. And if consciousness truly is spread throughout the universe, then we need to separate our idea of it from what hitherto we regard as its defining characteristics – intentionality, phenomenal content, propositional awareness, connectedness with a self. So there then remains the problem of understanding how it acquires these properties. The gap between atomic consciousness (or indeed cosmic consciousness) and a conscious subject seems no narrower than that between matter and consciousness as usually understood.

Nothing of what has been said in this chapter is intended to defend the idea that the mind is a parallel realm of ethereal stuff tucked away in some localized, metaphysical inside, even less to support the case for substance dualism. It is hardly necessary to rehearse the many arguments that have been mounted against Cartesian ontology. It seems appropriate, however, to make a couple of points.

The first is that there is a default sense of "real stuff" which matter fulfils. This being so, anything endeavouring to qualify as real that does not have the characteristics of matter – with solidity, space-occupancy, and push-and-shove interactions with other entities like itself (such as our bodies) – will seem unreal or less real, notwithstanding that, as we have already seen, matter itself seems to have become rather less substantial in the eyes of those for whom micro-physics gives us the last word on its intrinsic nature. (We shall return to the evaporation of matter before the gaze of quantitative science in the next chapter).

The second is more closely connected with our fundamental thought: that mind, distributed among conscious subjects, as that in virtue of which what-is becomes that-it-is, has elements and instances, but it does not exist parallel to matter. Explicitness is not a stuff. If "mind-stuff" were competing with matter-stuff for the accolade of substantiality, this would get in the way of its being just revelation. The duality of the location of perceptual consciousness – *in* the subject and *of* the object – reminds us that it is not located in the way that material objects and occurrences are located.

This is most obviously true of thoughts, whose intentional objects are typically not present, are often of a general nature, and may not exist. *Where* a thought takes place – if that notion makes sense except with respect to a location borrowed from the body of the thinker – does not determine what it refers to. It highlights how mental entities are not located with respect to that which they are about. A perception is no more at a definite distance from the perceived object than a thought about Paris is at a definite distance from Paris.

Consciousness, then, is not a stuff; neither is it a quasi-physical relation. It is that in virtue of which stuffs are made explicit such that there are relations. That is why the brain's location inside our body does not enhance its claim as to being the location of the mind. It remains a piece of what-is and its connections with other pieces of what-is do not add to its case for being that in the world in virtue of which what-is becomes that-it-is. As with mind and consciousness, explicitness is not a stuff, a thing, a process, a succession of events; rather it is that in virtue of which what-is is stuff, things, processes, events for a subject.

In the chapter that follows, we shall extend our critique of endeavours to provide a scientific account of the mind as rooted in the brain – to a broader critique of the claim of science, in particular physics, to be the last word on what-is, and how it overlooks its own nature.

CHAPTER 4
What-is as how much: cutting measurement down to size

4.1 METAPHYSICS UNDER SIEGE FROM SCIENCE

Notwithstanding the failure of neuroscience to account for consciousness, or for the transition from what-is to that-it-is, many thinkers, including quite a few philosophers, believe that the spectacular advance of science has discredited any independent contribution philosophy may make to advancing our understanding of the fundamental nature of what-is. Metaphysical inquiry, we are now told, should be guided by physics. The special sciences, notably biology, may have something to add; indeed, as we shall discuss in Chapter 10, biology can seem to justify radically revisionary views of the nature of the natural world. That said, fundamental physics is the place to which philosophers inclined to scientism usually turn.

If philosophy has any role in guiding our thoughts about the fundamental nature of things, it is merely secondary, that of commenting on science, helping us to understand what scientists are up to, and trying to make sense of the relationship between the world as it is experienced in everyday life and the profoundly counter-intuitive, even unintelligible, accounts of reality as it is described in cutting-edge science. Trying, for example, to make sense of Eddington's Two Tables.

The view that anything philosophy has to say about fundamental reality should be based on science is expressed with exemplary clarity by James Ladyman and Donald Ross:

> Any new metaphysical claim that is to be taken seriously should be motivated by, and only by, the service it would perform, if true, in showing how two or more specific scientific hypotheses jointly explain more than the sum of what is explained by the two hypotheses taken separately, where a 'scientific hypothesis' is understood as an hypothesis that is taken seriously by institutionally *bona fide* current science.[1]

They dismiss philosophy that does not have this as its aim as "Neo-Scholasticism". The philosopher's role – if any – is that of a Lockean humble under-labourer clearing away the litter from the path to truth – stuff often dropped by other philosophers. Physicist Carlo Rovelli has allowed that philosophy may have something to say but he believes that "we need to adapt our philosophy to our science and not our science to our philosophy".[2] Whether the scientist would welcome even this modest contribution is dubious. As Richard Feynman said (or is said to have said), science needs philosophy as birds need ornithology.

Philosophical "science-cringe" is not universal, even among scientists. Some prominent physicists – Lee Smolin springs to mind – have argued that physics needs philosophy to help it out of some of its present difficulties. And some philosophers have mocked colleagues who embrace the latest science as the last word on reality as "Bunsen burner realists".[3] Nevertheless, many accept the authority of science in what have hitherto been traditional areas of philosophy: ontology (what kinds of beings are there); epistemology (the basis of and limits to our knowledge); metaphysics – the fundamental stuff of the world, the nature of space, time, and causation; and what kind of entities we human beings are. In short, what-is in the most general sense.

In this chapter, I want to put science in its proper place. At the heart of the critique that follows will be belief that science's claim to ultimate authority on what-is is that it overlooks itself – the explicitness of which it is a manifestation – and the nature of its procedures (notably measurement) which have a tendency to reduce what-is to quantities. My argument does not, of course, deny, even less take for granted, the awe-inspiring precision, and the staggering predictive, explanatory, and practical power of science.

Regarding precision, a couple of examples will do. On the microscopic scale, calculating the magnetic moment arising out of the intrinsic properties of spin and charge of the electron give results that agree with measurement to 14 decimal places. And on the macroscopic scale, there is a consistent, seemingly robust account of the first moments of the universe. According to the inflationary theory: "The universe began as a hot spot of energy, a tiny seed a hundred billionth of a billionth of the size of a proton" which then inflated exponentially, "by twenty-six powers of ten in each of three spatial dimensions in 10^{-32} seconds ... a hundredth of a thousandth of a millionth of a billionth of a trillionth of a second".[4] Even if this proves to be wrong, the precision of the error is breathtaking.

The practical power of science is expressed in the extraordinary application of the most esoteric research in creating the technology that has transformed our everyday life and vastly enhanced our individual and collective agency. To repeat an example that cannot be repeated often enough, because it illustrates how we take for granted the magic powers that science has granted to us: I can speak to a friend in Australia 10,000 miles away without raising my voice, using a device I can carry in my pocket which sends out signals that negotiate the curvature of the earth by bouncing off satellites circling the planet. These powers are beyond the dreams of magicians of the pre-scientific era. Our contemporary habitat is a wall-to-wall artefactscape of electronic technology that seems to validate at least some aspects of some of the most esoteric of theories in physics. Daily life depends heavily on thousands of interacting technologies that increase our life expectancy, health expectancy, comfort expectancy, and fun expectancy – as well as adding to the threats that we face individually and collectively. It has been estimated that by 2000 30 per cent of US GDP was based on inventions made possible by quantum mechanics.[5] The computer on which this is being typed, the music playing in the background, the artificial light illuminating my study, and the regulation of the heating that frees me to be a thinking mind on this winter day, rather than a suffering body, make it difficult to challenge the pragmatic truth of the ways of seeing that lie behind them.

So "putting science in its place" does not mean (absurdly) questioning its incredible theoretical and practical power or denying that, notwithstanding the small issue of its intelligibility (of which more presently), quantum mechanics is possibly the most successful theory of all time. As John Wheeler has expressed it, "quantum theory in an everyday context is unshakeable, unchallengeable, undefeatable – it's battled tested".[6] Rather, "putting science in its place" is a rejection of the claim that its success has rendered superfluous traditional – usually mocked as "armchair" – philosophical inquiry into the fundamental nature of things, as it is a cognitive relic of a pre-scientific past. The failure of science described in the previous chapter to make sense of our nature as explicit animals, is inseparable from its nature as objective, quantitative inquiry. And there is a more intimate failure: to make sense of itself.

4.2 PROBLEMS WITH PHYSICS AS METAPHYSICS

Whether or not physics is in crisis,[7] physics-as-metaphysics is – or should be – in crisis. It is widely acknowledged that the two most powerful (and seemingly metaphysically salient) theories (general relativity and quantum mechanics insofar as it is applied in developing the Standard Model of the atom) are at odds with one another:

> In general relativity, events are continuous and deterministic, meaning that every cause matches up to a specific local effect. In quantum mechanics, events produced by the interaction of subatomic particles happen in jumps ..., with probabilistic rather than definite outcomes. Quantum rules allow connections forbidden by classical physics ... Relativity gives nonsensical answers when you try to scale it down to quantum size, eventually descending to infinite values in its description of gravity.[8]

The most determined attempts to reconcile relativity and quantum mechanics have generated world pictures that, outside of science, would be dismissed with laughter. What is more, many theories – most notably string theories – are not susceptible to empirical testing even in principle and for some physicists this means that they do not count as science.[9] This may suggest that general relativity and quantum mechanics represent the termini of their respective cul-de-sacs or, worse, that fundamental physics might be barking up two wrong trees, albeit mighty Californian redwoods, which have delivered more than we are entitled to expect.

The confusions of fundamental physics are also manifest in the wildly different, equally unsatisfactory, attempts to make stand-alone sense of quantum mechanics, which we touched on in our discussion in Chapter 2 of the dogma that "everything is made of atoms". Niels Bohr attempted to deal with the paradoxes associated with the dynamic collapse of the wave function in response to measurement by embracing idealism – something to which we shall return. The "hidden variables interpretation" of quantum mechanics tries to dispose of the seeming entanglement of paired elementary particles, to avoid the charge of faster-than-light transmission of signals, and "spooky interaction" between the particles. According to this interpretation, a particle is not a particle or a wave but is always both, with the field, or "pilot wave", guiding the particle. Essentially, it is argued that particles have definite values all the time and this is obscured by the influence of yet to be discovered variables. Unfortunately, the hidden variables have remained stubbornly hidden.[10]

The implication of quantum theory – that the values of variables such as position or momentum of certain sub-atomic particles are not settled prior to measurement, that many probable values are superimposed – has been dealt with in a variety of ways. There is the Ghirardi–Rimini–Weber theory, according to which collapses of superimposed multiple values to determinate values happen spontaneously, so there is no need to ascribe magical properties to the act of measurement or the observer. These spontaneous collapses are so numerous that superposition of states will not be visible at the macroscopic level: the microsuperposition will not turn into a macrosuperposition. And the collapse associated with measurement is the result of incorporating the microscopic system being measured into a macroscopic complex system, through hitching the former to a measuring system. Finally, it has been claimed that the wave function collapses that occur in response to measurement generates an infinity of universes.

All these interpretations have passionate adherents as they come into and go out of fashion. And there remains the suspicion that the interpretation of observed limits to the precision of measurement as an imprecision of being is illegitimate. A sceptic (and perhaps the present writer falls into this category) might see this as being analogous to blurred vision being projected into the world such that the latter is understood to be intrinsically blurred; or seeing "is" as "is-ish".

None of this is entirely encouraging for anyone who wishes to ground metaphysics in physics. The promissory note that there will soon, or one day, or eventually, be a Theory of Everything that will, among other things, reconcile quantum mechanics and general relativity is, after nearly a century, getting close to its expiry date. The judgement of Richard Feynman – one of the leading quantum physicists of the twentieth century – that "anyone who thinks they understand quantum physics doesn't understand quantum physics"[11] speaks for many physicists who prefer to focus on the practical exploitation of the fundamental equations – advising those who want to understand quantum physics just "to shut up and calculate".[12] As for philosophers (among whom I number myself) who can't calculate, they should just shut up.

The suggestion that quantum mechanics, while lacking ordinary intelligibility, has "mathematical intelligibility" offers little consolation. As William Seager has expressed it, "Although the attainment of mathematical intelligibility is impressive and desirable, it comes at the cost of abandoning mundane intelligibility".[13] And that cost is very high – notwithstanding that "The step towards mathematical intelligibility begins with the observation that mundane intelligibility can be represented mathematically".[14] John von Neumann, the mathematician, theorist of quantum mechanics and father of the computer, famously said to one of his students "Young man, in mathematics, you don't understand things. You just get used to them".[15] Which fits with Seager's observation that "As we develop faith in mathematics, the need to coordinate the mathematical representation with some intuitively intelligible system falls away".[16] We get used to talking of impossibilities such as infinity, the square root of minus one, negative probabilities (yes, *negative* probabilities), and other such characters. But that still does not reconcile us to Niels Bohr's observation that "Everything we call real is made up of entities we cannot call real. If quantum mechanics hasn't shocked you yet, you haven't understood it yet".

We may conclude that little or none of the life-changing, world-changing impact of fundamental physics, its unbelievably accurate predictions and their coherence, and their applications, depend on its being intelligible, nor that the truth of those aspects of physics which are central to its powers may be regarded as being metaphysical or as having definite metaphysical implications. For all that quantum mechanics is "battle-tested" the relevant battles – prediction and application – do not seem to depend on an interpretation of the fundamental nature of items, such as the wave function, central to the science. John Horgan's characterization seems apt: "Quantum mechanics is a black box; from it we get extraordinarily precise predictions, but we have no idea what happens in the black box".[17]

The closer physics gets to metaphysics, the less successful it is. Physics at its cutting edge is a metaphysical car crash – illustrated by the fact that (as we shall discover) quantum mechanics both reifies non-entities such as probabilities[18] and also de-reifies the fundamental elements of genuine things such as tables and chairs to entities that, as we have seen from Bohr's assertion a few lines back, are in some elusive sense unreal.

The privileging of stories that are mathematically satisfactory or even "beautiful" has bred certain habits – for example postulating new particles when the maths gets stuck – that result in a world picture that is at best unimaginable but at worst self-contradictory. Sabine Hossenfelder has argued that the pursuit of mathematical beauty has resulted in fundamental physics losing its way.[19] At the very least, the succession of profoundly different world pictures – Newtonian physics, general relativity, various iterations of quantum mechanics – suggests that science is a work in progress, with no end in sight.

As if this were not trouble enough, there is an unresolved argument as to the status of the entities that lie at the heart of fundamental physics, not the least those "virtual particles" that live (or die) in the shadow of the uncertainty principle, and dark stuffs – dark energy and dark matter – that are mobilized to balance the books. A century after its proclamation, the Schrödinger wave function, which is central to quantum physics, remains enigmatic.

Three profoundly different interpretations remain in contention. Eddy Chen summarizes the rival interpretations as follows:

> *Instrumentalism*: The wave function is merely an instrument for making empirically adequate predictions.
> *Epistemicism*: The wave function merely represents the observer's uncertainty of the physical situation. The collapse that is supposed to take place when a measurement is made, such that the wave function is replaced by a definite value, is not a physical event; rather, it is an advance in our knowledge of the system.
> *Realism*: The wave function represents something objective and mind-independent.[20]

This last – realistic – interpretation is fraught with problems as to whether or not the entities it deals in are real. The claim that (for example) the position and momentum of an electron are not only resistant to being measured at the same time but also that the values of these parameters are not determined simultaneously has been thought by some (most famously Bohr) to be a decisive demonstration that they do not exist prior to measurement. After all, it is difficult to imagine a particle existing but not having a position – being there but being nowhere.

It is interesting that Chen ends his defence of a realist interpretation of the wave function, with this *in*conclusion:

> Perhaps the lesson of quantum mechanics is that the wave function does not fit into any familiar categories of things; it is a new kind of entity. *Perhaps it is neither ontological nor nomological.* In that case, the wave function has its own category of existence that is distinct from anything we have considered. In other words, the wave function is ontologically *sui generis.* (emphasis added)[21]

The sentence in italics seems to undermine any hope of understanding what the wave function is and, what is more, also to contradict the claim in the final sentence that the wave function is a kind of being.

Yemima Ben-Menahem has pointed out that not all quantum properties seem to depend on measurement:

> [S]ome properties, such as electric charge, or being a boson or a fermion, do not superpose, and are represented in QM as independent of observation. Thus, although QM alters our conception of some

> classical properties, it does not deprive us of objective properties altogether. The set of properties QM takes as real is indeed different from the set taken as real in classical mechanics, but it is not empty.[22]

In short, even if QM is anti-realist, it is only patchily so. Indeed, even if they are not precisely measurable at the same time, the co-existence of conjugate properties suggests that there might be a common substrate for them, which would correspond to a certain ontological robustness.

Jonte Hance and Sabine Hossenfelder adopt an instrumentalist position: "[Q]uantum mechanics is a mathematical machine. Into this machine we insert some known properties of a system we have prepared in the laboratory. Then we do the maths, get out a prediction for measurement outcomes, and compare the prediction to observation".[23] Essentially quantum theory is just a mathematical tool to predict the results of measurement and the effects of interventions. So perhaps we should, after all, just shut up and calculate. Those of us, that is, who (unlike the present author) are capable of calculation.

The point is this: the jury is very much out on all the fundamental issues in quantum theory. For this reason, it does not seem as if the Standard Model of the atom based on quantum mechanics – the nearest physics comes to a comprehensive statement of what-ultimately-is – has much to contribute to metaphysics. The extraordinary predictive and practical power of physics is not rooted in, or dependent on, anything like metaphysical truth. That power has, after all, been exercised in the absence of any consensus as to the metaphysical implications of physics. The unresolved arguments about the idealist vs realist interpretations of quantum theory or entanglement vs hidden variable explanations of "spooky action at a distance" have not got in the way of the ever more astonishing applications of quantum mechanics. It would seem that there is little relationship between any putative metaphysical world picture based on quantum mechanics and the extraordinary advances in quantum-based technologies that are transforming everything in our lives from shopping to geopolitics.

There is another problem facing physics-as-metaphysics: the vast, seemingly unbridgeable, and certainly unbridged, gap between the world as we experience it in the serious business of everyday life and the world as it is described in fundamental physics. We illustrated this in Chapter 1 with Eddington's Two Tables. The claim that this gap demonstrates how our everyday reality is an illusion – developed to self-contradiction in Donald Hoffman's "case against reality" discussed in Chapter 10 – does not take account of the fact that the world of the physicists who do the physics is inescapably that of common-sense reality. If the universe were as described in the leading interpretations of fundamental physics, the actions that have generated fundamental physics would not have been possible. The laboratory, the kit in the laboratory, along with physicists – with their definite actions, ambitions, training, collaborations, long-distance travel, sustained thinking – who populate the laboratories, and the engineers and architects who built them, are closer to the world of classical physics than to the world as portrayed by quantum mechanics. Without self-individuating, solid objects, located in space and time, it seems difficult to envisage how physics could proceed. As Eddington expressed it, "It is true that the whole scientific inquiry starts from the familiar world and in the end must return to the familiar world; but the part of the journey over which the physicist has charge is in foreign territory".[24]

That physics cannot explain the possibility of the discipline of physics – conducted by macroscopic physicists living their lives and practising their profession in a folk physical world – is another reason for challenging any claim of physics to being *en route* to a Theory of Everything or the only route to a comprehensive account of reality. (This is something to which we shall return.)

Admittedly, this argument could be applied with equal force to the revisionary metaphysics pursued by some philosophers. Philosophers who, for example, deny the reality of time will still (to borrow an example ascribed to G. E. Moore[25]) need to explain what they mean when they say that they had breakfast before their lunch. The same criticism does not apply to descriptive metaphysics which simply aims to describe the implicit metaphysics of everyday life and the way in which it does, or does not, cohere.[26]

We should not at any rate build metaphysical castles on the manifestly temporary foundations of physics, in which successive theoretical frameworks challenge their predecessors – in the case of quantum theory undermining them to the point of leaving nothing standing, so that the notion of a progress towards reality as an asymptote is more than a little optimistic. There is of course a rejoinder: the history of metaphysics as it has developed independently of science is not even a work in progress. The conflict between materialists, idealists, dualists, neutral monists, seems never-ending and no nearer to resolution.[27]

Nevertheless, the claim of physics to have supreme metaphysical authority remains (to put it mildly) open to challenge. Granting this authority is an unjustified projection from the physical sciences' entirely justified kudos as the shaper and sustainer of our modern world. The fact that we can (for example) use quantum-based technology massively to extend our capacity to communicate with our fellows does not demonstrate that the theoretical foundation of that technology has exposed our everyday understanding of the what-is around us as fundamentally wrong or that philosophy has nothing to contribute to our understanding of the nature of what-is.

I ought to end this introduction with a declaration of interest. If fundamental physics provided a definitive and unassailable account of reality, my lack of mathematical attainment would debar me from either judging the case for this account or indeed understanding many aspects of it. It might be argued that my relative innumeracy disqualifies me from entering into the discussion of physics-as-metaphysics. And suspected that my hostility to the very idea that physics is the last word on metaphysics is the resentment of a eunuch at an orgy. But drawing this conclusion presupposes the very thing that is in question: namely, that quantification through measurement – which has led to fundamental physics – is the quickest, truest, indeed the only, path to the truth about what-is.

I shall contest this, at the very least to head off the objection that it is not possible to assess whether fundamental physics is our best and brightest hope for true progress in metaphysics, without being equipped to make sense of, and judge, the mathematics, which requires a sure grasp of Lagrangians, gauge symmetry, vector analysis, Hilbert space, and so on. To argue that a relatively innumerate individual such as myself has no right to assess the role of the quantitative sciences in making metaphysical sense of what-is would be to prejudge the outcome of the debate as to whether measurement is the one true path to reality and that that reality is mathematical. What follows challenges this assumption. I shall argue that physics cannot transfer its well-earned authority in creating a mathematical portrait of the natural world and the practical power that has followed from this to the question of the fundamental nature of things.

Ultimately physics-as-metaphysics rests on an assumption that one can answer the question of what, fundamentally, there is by drawing on essentially quantitative theories; that what-is boils down to "how much" based on unitized counting or measurement – of mass, length and time and, more recently, of forces and particles, and their connections portrayed in equations. My challenge to the belief that physics is the superior, even the only, path for metaphysics will focus on the implicit belief that measurement gets the measure of all things – a position that we might term "the apotheosis of measurement".

The inquiry will be motivated by the belief that the so-called "measurement problem" that causes such consternation in quantum mechanics has its origin in the extraordinary, unnatural nature of measurement (even of what we are inclined to describe as "ordinary" measurement) and the associated marginalization of consciousness in scientific observation.

4.3 BACK TO THE BEGINNING: MEASUREMENT

Measurement is the ground on which natural science stands. Lord Kelvin's famous *profession de foi* has become widely accepted in the century or more since he proclaimed it: "[W]hen you can measure what you are speaking about, you know something about it; but when you cannot measure it, when you cannot express it in numbers, your knowledge is of a meagre and unsatisfactory kind; it may be the beginning of knowledge, but you have scarcely, in your thoughts, advanced to the stage of science".[28] This claim was expressed more succinctly recently: "If we can't measure something, we can't know its true nature".[29]

The case for measurement being the royal road to unveiling reality – for believing that, as Wilfrid Sellars put it, "In the dimension of describing and explaining the world, science is the measure of all things, of what is that it is, of what is not that it is not"[30] – has been strengthened, as we have noted, by the spectacular transformations the applications of quantitative science have made to our lives. Whether or not "Measurement began our might",[31] it is certainly true that "In scientific measurement lies the magnification of our might".

While, as quantum physicist David Deutsch has pointed out "There is no ... logically necessary connection between explanatory power and truth",[32] it seems difficult to believe that quantitative, and increasingly mathematical, science, which has vastly increased our explanatory power and power period, hasn't got something fundamentally, or at least, importantly right about what-is. But it does not follow from this that the "what" of "what-is" ultimately boils down to "how-much"; that numbers or numerical relationships, capture the nature of things and of happenings.[33] While science may be the measure of all things, measure may not necessarily tell us what all things are in themselves.

What is the case for questioning the intellectual hegemony of the quantitative – what we might call "the rule of the ruler" – and for challenging the claim that, since quantification enables us to engage more successfully with what-is, what-is is most accurately portrayed as collections of quantities, or (more broadly) mathematical structures? If we are going to put physics in its proper (admittedly elevated) place without running foul of Hilary Putnam's "no miracle" argument according to which realism is the only view of science that doesn't make its success an inexplicable miracle, we need to reflect on the nature of measurement that underpins the quantitative portrait of the world.[34]

Measurement begins and ends with macroscopic objects called human beings who are also conscious subjects and in communication with a vast community of other conscious subjects. As such, it fits awkwardly into any physicalist picture of what-is. Indeed, as we shall discuss, "the measurement problem" of quantum mechanics (which we shall discuss later in this chapter) may be a long-delayed comeuppance for the failure to see measurement for the unnatural event that it is; for physics overlooking the world in which physicists operate and, indeed overlooking physicists themselves; looking past all that is necessary for the physical world to be portrayed (whether correctly or incorrectly). This is a particularly spectacular example of overlooking the explicitness implicit in our having a world of which we are aware and to which we are related as (among other things) an object of knowledge and inquiry. Overlooking explicitness and macroscopic bodies – both are necessary for its existence – are major omissions for a discipline that would claim to be on the road to a Theory of Everything.

4.4 MEASUREMENT AND THE MARGINALIZATION OF PHENOMENAL EXPERIENCE

Philosophers of science tend to focus on the ascent from observation and measurement to laws, equations, and the postulation of forces, fields, and fundamental particles. The key steps on the lower slopes of science, the transition from the informal awareness we have of the world around us to observation (elective quality-controlled experiences, sometimes by appointment), and thence to measurement, have attracted less attention. The first thing to be said about measurement is that it is active: it is the culmination of a journey from passive seeing to looking, to active purposive scrutiny, and thence to expert observation made by those trained in the necessary skills.

It is easy to overlook this special nature of measurement. Ernst Mach, a great scientist, and a hugely influential philosopher of science, argued that there was no fundamental difference between ordinary experience available to any creature with a nervous system and scientific observation: "Scientific thought arises out of popular thought, and so completes the continuous series of biological developments that begins with the first simple manifestations of life".[35] The observations central to scientific inquiry are remote from the mere "panning round" or gawping of other non-human creatures, even those with complex nervous systems. Scientific observations are closely prescribed as to their topics (pre-defined by a particular question and, more broadly, by an existing tradition of investigation) and methodology. The discrete actions that are scientific measurements may be very complex and have a vast hinterland. Not only may they require requisitioning, setting up, and inventing, requisitioning, and using instruments of varying degrees of sophistication, but they may also involve travel to facilities and cooperation with others.

While there is no right and wrong way to have an experience, there most certainly is a right and wrong way to carry out a measurement, so the journey to the measurement may require expertise in relevant techniques and broader training in the surrounding disciplines. There are rules governing, and standards of, measurement; quality controls, ruling out mistakes and the corrupting effect of noise, that ensure reproducibility within and across individuals, and within and across disciplines and cultures. Such standards developed in science feed back into everyday life via the bureaus and other institutions that uphold those standards. As science becomes more sophisticated, measurement and its objects become increasingly theory-laden. The very idea of the accuracy of measurements ultimately rests on a nexus of other measurements, and discovered laws, and physical constants.

As we saw in the case of Mach, philosophers of science often underestimate the wide gap between our experiencing the world as organisms in living our daily lives and scientifically investigating it, the distance between informal awareness and a requisitioned, quality-controlled, experience mediated by expertise and apparatus, that has a hinterland of a community of minds.[36] Contrary to what Mach claimed measurement is not just a sensory experience. Nor is it merely just a succession of physical events – a point that has escaped some scientists and philosophers of science. There is, for example, a fundamental difference between the events that are involved in making a measurement and the event that is the result of measurement, such as the assumption of a position by a pointer. The latter is a datum that is to be harvested. And there is a fundamental difference between the events constituting a measurement and other events that take place in the vicinity of the measurement that have nothing to do with it.

We do not have to travel to CERN to encounter the strange, even unnatural, nature of measurement. It is evident in something as simple as measuring the length of a table. The subject as measurer withdraws from the flow of her experience that delivers, for example, a furnished room as her surroundings, a parish of her *lebenswelt.* She reaches for a tape measure (after rummaging in a drawer), isolates aspects of the table (length and width) that could not exist by themselves, applies the tape and observes those numbers on it that correspond to the location of the corners of the table. She then transfers the numbers to a sheet of paper where they are stored as the frozen outcome of her action: they might survive even the destruction of the table. The length and width of the item are then multiplied together to deliver the surface area of the top of the table, separated from the table. The surface area is now available to be transported to a shop where a tablecloth of suitable size can be bought.

The differences between ordinary perceptions and quantitative observations, the standing back from the flow of random experience when something is measured, increases as measurement for every day, practical purposes evolves in the direction of scientific observation. In the latter case, measurement is quality-controlled, and mediated through instruments of increasing complexity. What is more the entity upon which the measurement is made is treated as representative of a whole class of entities. This is subsequently justified by further meassurements on other instances of entities judged to be qualified to belong to the same class. This is measurement in pursuit of generalities, patterns, trends, correlations, and even causal connections.

At its most basic level, a measurement represents an interruption of, an intervention in, an awakening out of, a passively received flow of experience in order to focus on, and look beyond or through, the object underpinning the experience. The experience is not dissolved in the unfolding story of sensorimotor engagement with the world. Measurement, that is, is a formalized *action* that begins with the posing of a question, ends with an answer, an outcome, a result, which represents the transformation of an aspect of the given into a datum, of what-is into quantitative data *about* what-is.

Just how much of the experience of the subject or subjects during the performance of the measurement is set aside is often overlooked. This is true even when the measurement is something as simple as determining the size of a table. When I extract the unitized length of the table from the table, I ignore most of the contents of my sensory field: not only the experience of the place in which the measurement is made, but also those experiences that arise from the table within and beyond the borders being measured, and most of the features of the measuring device.

When therefore I say, "I measure a table", this is not strictly true. I measure a parameter of a table. My measurement may focus on, for example, its length. In order to do so, I must ignore its width, height, weight, density, colour, surface patterns, and so on. Such selectivity applies equally to the device I use to make a measurement. I ignore the colour of the tape measure, the cloth of which it is made, its fraying edges, even the colour of the numbers printed on it. Indeed, I abstract from the characters on the tape the pure number – 25 or whatever – whose size, colour, typography, are all irrelevant. The resulting datum, or data-point, is effectively drained of phenomenal appearance. When a table length is measured, the interaction that generates the measurement reduces the object and the measuring device to abstractions of themselves.

After all, a ruler, a laser beam, the length of my thumb – which can all be used to measure length – have little in common. The outcome of a measurement is a *result* only when it is reduced to a number. Most of the appearance it has – as written symbol, a position on a voltmeter, flashes of light on an oscilloscope screen, clicks on a Geiger counter – is irrelevant; or rather it is to be looked *through* to its referent.

Thinking of measurements merely as "experiences" therefore overlooks what is distinctive about them, how strange they are. There is the world of difference between, say, on the one hand looking at a table, noticing that it is in the room, and on the other measuring its size or weight; between my seeing that it is "over there" and my measuring the distance separating me from it.

As the measuring gaze widens in pursuit of generality, it becomes more purely mathematical and manifest reality is progressively burnt off. The link between the measurements and physical items that are measured is broken. While the surface area of the particular table measured in order to determine the appropriate size of a tablecloth is connected with an individual object, the measurements that underpin scientific experiments have no essential relation with particular objects. They are about classes or types of objects or events – entities as exemplars, that are taken to be essentially identical to, or equivalent to, each other. A number of examples of the relevant entities are, for example, weighed and their weights are averaged. The examples are assumed to be typical of the class to which they belong, and their average weight is extrapolated to an indefinite number of other entities of the same class.

That is why it is misleading to think of a measurement, as Mach did, as just an *experience*, even a special one: it is, at best, a *de*-experienced – de-localized, de-individualized – experience. This is captured by J. Acacio de Barros and Carlos Montemayor who point out that the observer in quantum mechanics is not "a phenomenally-conscious observer, because measurements are not entirely determined by merely appearance properties of experiences, but rather by conscious interventions in an environment by a rational agent with specific goals that have unique theoretical meaning".[37] Purposive de-experienced experiences begin with more homely measurements in science and even homelier measurements in daily life, such as weighing out the ingredients of a cake or measuring the size of a table. In short, much of what Acacio de Barros and Montemayor say of quantum measurement applies to commonplace measurements in everyday life.

I have already mentioned in passing another distinctive aspect of measurement: it is that, unlike an experience, it can, in the form of a datum, be transported elsewhere. The length of the table can be written down in order to be communicated to someone else or to a future self. It is liberated from a location in space and time, including that of the object that has been measured and of the device that has been used to measure it.

Man, the Measuring Animal is uncoupled from the material world, indeed the biosphere, in a way not seen in other living creatures. The science that delivers the technologies that serve our needs, protect us from threats, fulfil some of our hopes, is possible because we can distance ourselves from those needs, threats, and hopes. So much for Mach's understanding of science as completing "the continuous series of biological developments that begins with the first simple manifestations of life".

The "dephenomenalizing" or "deperspectivalizing" nature of measurement makes it seem to be on the side of the object – as the conscious subject is pushed to one side by the intimate relationship between objects – between the tape and the table. It amounts to treating it with a kind of respect for its intrinsic nature – notwithstanding what we shall discuss in the final chapter about the ultimate inaccessibility of the properties of the "thing in itself". Measurement is different from noticing how the thing looks to me or from merely pushing and shoving it around or otherwise using it. It is, as it were, allowing it to speak for itself and amounts to a partial shedding of the egocentric viewpoint that one has as a conscious subject. You and I may disagree over the phenomenal appearance of the table: you may think it rather small, and I may think that it is quite big. What we cannot disagree over is its measured size. If we do disagree it is because one of us has got the measurement wrong – on that we shall agree.

Admittedly, although experience transformed to a measurement is dephenomenalized, the subject is not, however, entirely "got out of the way": there is still a residue of *interaction*, and it is between a conscious subject and a material object. That residue will come to haunt quantitative science at the fundamental level, as we shall see.

Consensus over measurement – how it should be done, agreeing on the results – has played a major role in social coherence: in regulating the exchange of goods and services, assessing productivity, determining and apportioning just reward, and drawing up and delivering on contracts.[38] In this respect, science as we know it is a relative latecomer in the exploitation of the possibilities of measurement. One account of the origin of the geometry that ultimately delivered the general theory of relativity is that it provided a way of assessing areas of land when they had to be reapportioned after boundaries had been erased by the annual Nilotic floods. It may be apocryphal, but it is a reminder of how measurement was woven into the fabric of daily life long before it was crucial to the advance of natural science. It was tied to solving local practical problems and addressing quarrels over apportionment. Even so, the principles underlying it had universal application.

There are two aspects, therefore, to the strangeness of measurement. One is its distance from the ordinary modes of noticing, the immersion in a bath of perceptions, that constitute our ordinary consciousness. The other is its distance from the processes that take place in the physical world that is the object of measurement.

One marker of that distance not yet mentioned is the fundamental asymmetry between the measuring device and the item that is measured. Yes, measurement is an interaction between objects, but the participants have different roles. When the tape measure is applied to the table, the table is measured not the tape measure. The result of the measurement is that the table is so many feet and inches, not that the tape measure is so many tables, or fractions of a table.

Those different roles are, of course, assigned by a conscious subject acting in accordance with conventions to which she consciously subscribes. A tape measure in a drawer does not do any measuring: it is not measuring the entities it is lying on nor, of course, itself. And, while the roles of the measuring instrument and the measured object can be reversed – so that, for example, I can use my foot as a unit to measure distances or use a ruler to measure the size of my foot – that of measurer (the conscious subject) cannot be interchanged with that of the other two parties. This may seem obvious, but some writers would like to extend the role of measurer to that of material objects. Consider, for example, this passage from Hossenfelder's *Existential Physics*: "A measurement in Quantum Mechanics does not require a conscious observer. In fact, it does not need a measurement apparatus. Even tiny interactions with air molecules or light can destroy quantum effects and so update the wave function".[39] This may be a consequence of a failure to differentiate between the dephenomenalization of the experienced world that is evident in the way the *outcome* of measurement is expressed and the conscious processes that are mobilized in making a measurement. To see measurement as a purely physical event is to overlook much that goes into making a measurement – the foreground of a conscious subject (the observer) and the hinterland of a community of minds (who underwrite the science, technology, and significance of measurement). It overlooks the fact that it is a *performance*. Contra Mach, measurement is remote from biological experience, but it does not leave experience behind. To imagine that it does is to confuse what lies within and behind measurement with the dephenomenalized reality that is the portrait of the world woven out of measurement. Physical processes in the absence of conscious subjects are not the answers to questions – even less answers to general questions.

I have teased out the complexity of what happens when a measurement is made, even one as simple as translating a table into an entity so many feet and inches long, not because I believe the reader needs instructions in how to pull off this elementary feat but to highlight its remoteness from ordinary observation, from the sense experiences we were born to have. Of course, this distance between measurement and ordinary perception has widened hugely with the advance of science. Scientific measurement involves instruments that are themselves the children of measurements, designed in accordance with general principles established on the basis of countless earlier measurements. Those measurements are often indirect, as when I measure the intensity of a current by looking at the position of a pointer on a meter. By the time we reach fundamental physics, measurements are made on entities that are unobservable – and of uncertain ontological standing – via measurements performed on observable entities that are quantitively connected with them by theories of a high level of abstraction.

In contemporary science, the extent to which a measurement may be distant from the flow of ordinary background experience may be vast, as instanced by the 30 years between the dream of the Large Hadron Collider and the experiments carried out to detect the presence of the Higgs boson. The CERN studies involved the cooperating and converging consciousness of many thousands of people – theoretical and experimental scientists, engineers, maintenance workers, administrators, accountants, clerks, employees of grant-giving bodies, to mention but a few of the communities involved. Very few of the experiences had by scientists working at CERN amount to actual measurements which they had been employed to make possible.

The outcome of measurement – irrespective of whether it is the product of a tape measure applied to a table or the mobilization of a Large Hadron Collider – is located in a space increasingly remote from that occupied by the objects and events that are directly encountered, including those that have been measured, and indeed the space occupied by, or more precisely inhabited by, the body of the measurer. The data obtained at CERN are relocated on pages and computer screens, and read by consumers, far from the legendary laboratory.

The act of measurement lies at the heart of the Copernican revolutions of science that will ultimately displace the subject from the centre of her world, locating her as a small item in a universe. It is the key step in human awakening from the *umwelt* granted to every organism to an articulated, shared world-picture of a kind that only humans entertain.

This said, there is an inescapable residue of experience in measurement. Much is, however, hidden by its apparent irrelevance. The conscious subject may be buried but, as we shall discover as measurement reaches its asymptote, she won't lie down. The so-called measurement problem of quantum mechanics is a reminder of the problematic nature of measurement: that it is, and it is not, about the consciousness of the measurer.

4.5 THE HEART OF MEASUREMENT: UNITS

While we think of measurements as being "objective", of being on the side of the object, and as an expression of our capacity to get ourselves out of the way, there is a sense in which the objects, too, are sidelined by measurement. When a table is reduced to an object of a certain length, it has lost most of its intrinsic properties as well as those revealed in the phenomenal experiences of perspective-bound observers. After all, the length of a table – or indeed the speed of light – cannot exist by itself.

This marginalization of the object is compounded by the fact that measurements are cast in *units* that will have been developed elsewhere and related to standard entities, such as a bar of metal in Paris, remote from the place where we employ them. When a table, a person, and a rock are all described as 2 metres long, they are homogenized: a similarity is asserted across profound differences. Measurement that reduces entities to so many units is an exotic mode of quantification: a "how much" that goes beyond the "how many" of ordinary counting. Counting the number of pebbles as 6 does not answer the question of "how much pebble?" For that we need standardized units so that the response to the question of "how much pebble?" is (for example) 5 kilograms or 5,000 grams. We accept that a kilogram of feathers is the same quantity as a kilogram of lead.

Focusing on units highlights the exotic, unnatural nature of measurement. This is put in italics when we remind ourselves that some of the earliest measurements involved treating parts of our own bodies as measuring devices and as sources of primordial units – so-called anthropic units. Feet, paces, inches (thumbs), cubits (forearms), fathoms (the length of outstretched arms) reveal a rather peculiarly alienated, impersonal, relationship with one's own body. The measurer treats her own body (a) as *any b*ody – a standard body, (b) as a material object on a par with other material objects, and (c) as a *tool* stripped down for the purpose to its own size.

A "mile" may begin as the distance covered by a Roman soldier in 1,000 paces, but it is not, of course, a particular Roman soldier, and evolves into a distance covered by any average adult in 1,000 standardized paces. It is lifted from a particular journey in order to be applied to any journey. Ultimately it becomes a disembodied, purely geometrical, distance abstracted from any actual or possible journey, or indeed any actual or possible separation between material entities, as when the distance between the Earth and the Sun or between one atom and another is measured in multiples of units of length.

This is an act or stance of humility that is almost scandalous. Using a body part to determine the size of another object is a striking expression of the ontological democracy we shall discuss in Chapter 5. The body part is reduced to its length, to a property it shares with any material object. To pace out a distance is to disenchant walking of its primordial purpose – that of getting from A to B in order to engage with whatever is at B.

While the use of a tape measure does not involve anything as scandalous, there is still the fact, already noted, that the measuring tool and the measured object are both reduced to one of their properties – length – that they have in common in order that one should measure the other; that the length of one should be used to measure the length of the other and the unitization of the spatial extension of the tape should be transferred to the spatial extension of the table. It is a meeting of abstractions.

There is another important difference between the body – or body part – as a measuring tool and, say, a tape measure, in addition to standardization that comes from the fact that the units written on tape measures, unlike forearms, really are of equal size, as opposed to being, accidentally, of roughly equal size. It is that the units are clearly distinct from – though written on – the measuring device. This is highlighted when, as is typically the case, "centimetres" and "inches" are printed on opposite sides of the tape. While numbers printed on the tape are features of it, the units that are counted are not properties of the tape. This is because units are not intrinsic properties of anything; rather they are *ratios* – between one object, or kind of object, seen as an exemplar and another. This is well expressed by Max Born:

> The foundation of every space and time measurement is laid down by fixing the unit. The phrase "a length of so and so many meters" denotes the ratio of the length to be measured to the length of a meter. The phrase "a time of so many seconds" denotes the ratio of the time to be measured to the duration of a second. Thus, we are always dealing with ratios, relative data concerning units which are themselves to a high degree arbitrary.[40]

It is very much to the point that there is this further mediator regulating the size of the units which are used to translate magnitudes into numbers, such that "big" becomes 6,000 units and "small" 0.0006 of a unit. The item to which the term refers is invisible. It is not something clear and present – such as the thumb exploited as an inch or the physical gap between two marks on a tape measure.

In the case of units of mass, the item in question was, from the eighteenth century until 2018, *Le Grand K* – a particular lump of metal kept under lock and key in an underground vault in Paris, underpinning the durability, precision and replicability of measures of weight.[41] With the need for increasing precision, the difficulty of curating the paradigm kilogram – which lost about 150 micrograms in 130 years – and the seeming circularity of verifying "the mass of an object that itself defines mass"[42] became problematic. The bar of metal in Paris has therefore been replaced by a *definition* rather distantly related to any perceptible entity – namely Planck's constant or 6.626×10^{-34} joule seconds This invisible fourth party – joining the measurer, the measuring device, and the measured object – is the ultimate basis of the calibration of measurement instruments, and of the quality control of measurements.

The absent or hidden fourth party is a further reminder of how a quantified magnitude is an abstraction – a "taking away". The taking away, as we have already noted, lies at the heart of even the homeliest measurement. We say that a table has a length – as if that were separable from it – but that length is clearly not stand-alone. Except in our discourse, the individual properties of an object cannot be isolated from its other properties in the actual world. No object could have just a length, or just a weight, or just a velocity, or even a combination of them. Even less could it have a surface area generated by multiplying two abstractions - for example a length by a breadth, with the former being discovered or reported separately from the latter. Something else is necessary for length, weight, or velocity to be instantiated in the object.

Any tendency to think that individual properties exist independently comes from our giving measurements a kind of substantiality by recording and writing them down as a unitized result. We extract the length from the table and then project it back onto the table as one of its properties: we say, "the table is six feet long". There is the implicit idea that there is a bit of stuff, the material of which the table is made, in which all the quantitative parameters we can extract from it have a local habitation, are instantiated in, possessed by, it.

At any rate, numbers fall short of actuality; magnitudes cannot exist by themselves, although they may do on the page, as when I report that the table is 6 feet long and 3 feet wide. While the magnitudes may point to how things might behave, it is the things, not the magnitudes that behave. And this is equally true of shape. There are circular, triangular, oblong things (even though they may be reduced to drawings) but there are no circles, triangles, oblongs existing independently of things.[43] Even a circle in a geometry textbook requires stuff – printer's ink – to exist and, as such, it is not a (perfect) circle. For the same reason, we cannot draw a genuinely one-dimensional line.

It will be evident that there is nothing natural about units, though they are necessary to extract how-much, expressed as a number, from what-is. Nothing is of itself so many standard x's: a tree is not intrinsically so many centimetres or a different number of inches or angstrom units. Nor are there even scales in the universe – as we discussed in Chapter 2. Nothing is, of itself, "very big" or "very small",[44] though it may be bigger or smaller than another object with which we choose to compare it. A tree may be big compared with a sapling and small compared with a landscape gathered up in a gaze from a vantage point.

It may seem unnecessary to be reminded of this. That we sometimes need such a reminder arises from our propensity to use measurement metaphors to describe complex, coordinated natural processes. We find, or think we find, clocks everywhere in nature – from our own bodies with their circadian rhythms, to the entire universe whose law-governed activities are seen to add up to a vast clockwork machinery.

Einstein was acutely aware of the unnatural nature of measurement:

> One is struck by the fact that the theory [relativity] ... introduces two kinds of physical things, i.e. (1) measuring rods and clocks, (2) and all other things, e.g., the electromagnetic field, the material point, etc. This, in a certain sense, is inconsistent ...; strictly speaking, measuring rods and clocks would have to be represented as solutions of the basic equations (objects consisting of moving atomic configurations), not, as it were, theoretically self-sufficient entities.[45]

The problem is serious: "Nature does not have to settle whether a given mechanism counts as a 'clock' in order to determine how it should behave. A term like 'clock' unlike 'light ray' or massive particle cannot appear in the statement of any fundamental law".[46] Clocks do not measure the time nor rods length: it is *people* who use clocks and rods (however indirectly) to measure time and length – and who extract quantity from time and length from material objects. Events that take time, or occur at a certain rate, do not of themselves *time* other events. The transition from "time" to "timing" is an extraordinary example of how we make things explicit. Out of this arises the distinction between temporal location, temporal order, and temporal duration.

We can summarize the burden of this section in one sentence. Physical science, which is based on measurement, cannot accommodate measurement.

4.6 THE PROBLEMATIC RELATIONSHIP BETWEEN MEASUREMENT AND CONSCIOUSNESS

We have emphasized the unnatural nature of measurement. Nevertheless, conscious subjects, as producers and consumers of measurements, tend to be sidelined. It is measurement's guilty secret. There are reasons for the marginalization of consciousness in the way we think of measurement.

We have already noted how most of the experience of making a measurement is excluded from the result. In becoming a datum, the observation that the table is 6 feet long – never mind of an elementary particle that it has such and such a position or momentum – is stripped of all of the experiences that went into making, and subsequently registering, it. The extraction of the length of the table by the length of the tape measure drains most of the actual or potential experience of both these items, even before unitization – the reduction of what is observed to abstract ratios. What is more, visible qualities of the resulting datum, of the record, are irrelevant: they are looked through, as opposed to looked at. It doesn't matter what is the colour or size of the number on the table of results.

The exsanguination or dephenomenalization of the observed is taken even further when, for example, the datum takes the form of a pointer reading, a bright spot on a screen, or a point in a graph gathering up a succession of data in a pseudo-spatial line or curve. At the macroscopic level, the position of a pointer is not itself an image of the state of that which is being measured. Rather, it is an encoded representation of that state – pointing to something that is not at all like it – for example pressure, temperature, or a time interval. The height of a line on a thermometer is nothing like warmth or cold or even temperature as a standing condition that determines that ice melts or water freezes. The transition from 1°C to -1°C as reported in a text is nothing like freezing. And the transformation of a pointer position into a reading, a result, a datum – into information – could not be unpacked from its experienced physical properties; rather it depends on the complex background of conventions that makes the position a sign of a quantity of a parameter.

And so, to the third reason for the marginalization of consciousness in the idea of measurement. The transformation of the interaction between a measuring device and a measured object or event into a measurement, datum, or reading, is mediated by knowledge, understanding, expertise, and by belief in many things such as the accuracy of the measuring device and the hinterland of theory which supports it. All of this draws on the sedimented consciousness of a community of minds instantiated in the consciousness of the individual performing the measurement. This sedimented consciousness is remote from phenomenal, here-and-now, awareness. It is therefore easy to overlook – as is the increasingly theory-laden nature of measurement and, indeed, the nature of what is measured. The history of the metre and the kilogram has a profound subsoil of custom that draws on many aspects of the community of minds that is everyday social being.

This third point will be particularly relevant when we try to understand the interaction between consciousness and the physical world in the so-called collapse of the wave function: it is not clear whether we are thinking of the individual mind of the conscious observer, the community of minds to which she belongs, or mind as a kind of stuff.

The fundamental point is that, while measurement delivering a datum is a set of material events, it is not just that. The position on a clock face, the location of a pointer, that are consequences of the way things are set up and what is done in the set-up, count as *outcomes*, as arrivals, as measurements, only when they are registered by a conscious being who makes sense of them. A pointer does not point to a quantity except in the presence of an informed mind that makes the connection between the pointer and the number on the screen it points to. Temporally ordered events, as in the succession of positions of the hands on a clock, do not "time" other temporally ordered events without the transformative presence of a suitably informed or trained observer. For such an observer, what happens takes place in a time frame only because he or she frames it in time, using one set of changes to time another set, thus delivering "duration of" and "rate of" changes or second-order observations of the rate of change of the rate of change. Physical events occur in time but they do not of themselves amount to timing – whereby one lot of events times another. There is no measurement without that awkward entity, that gate-crasher in any physical theory, the the individual or collective conscious human subject who turns a parish of what-is into the scene of an experiment with calibrated disturbances as inputs and subsequent disturbances as outputs, as *results*.

The formalization of measurement and ever more sophisticated measuring equipment can make it seem that the subject can be absent from measurements and that the values they generate are properties of the physical world – a key step on the path to the mathematical idealism mentioned just now which we shall visit in Section 4.9. This illusion is reinforced when sophisticated measuring devices are able to get on with much of the business of measurement without human involvement. Indeed, the marginalization of the human measurer seems to be a condition of a valid measurement – apparently minimizing contamination originating from individual persons and cultures, and excluding effects of personal caprice or incompetence. With the who, the when, or the why of the relevant observations being placed off-stage, measurement seems closer to belonging to a view from no-one.

The apparent erasure of the conscious subject from measurement is reflected in this passage from Rovelli: "In quantum physics parlance, an 'observer' can be a detector, a screen, or even a stone. Anything affected by a process. It does not need to be conscious or human or living, or anything of the sort".[47] In case we haven't got the message, he spells this out in *Helgoland*: "Any interaction between two physical objects can be seen as an observation. We must be able to treat any object as an 'observer' when we consider the manifestation of other objects to it. Quantum theory describes the manifestation of objects to one another".[48]

It is easy, it seems, to overlook the connection between measurement and consciousness – a view that, as we shall see, leads to the mistaken idea that measurements are natural phenomena and to the complementary idea that natural phenomena count as measurements, that as Hossenfelder quoted above, claimed, "a measurement in Quantum Mechanics does not require a conscious observer".[49] Given that measurement results in numbers, we are en route to the view that a world picture that mathematizes nature is the truest portrait of, or the very substance of, what-is. En route we pass through the paninformationalism that we discussed in Chapter 3: all the "its" of the material world are "bits".

Contra Rovelli, a measurement is clearly more, much more, than "anything affected by a process". If a falling stone breaks a glass, which is the observer and which is the observed? Neither of course. Without conscious subjects and their questions that prompt and shape the physical events that constitute the measurement, the items that are used to make the measurement do not have the status of instruments. Nothing that comes out of the process is "a manifestation" – never mind "a result", "a datum" – if it is not registered by a conscious subject making sense of it. A measurement is an answer to a question formulated in a certain way, posed within a certain frame of reference. Events in the material world do not constitute the asking of questions nor do subsequent events amount to answers. Measurements are not merely the last step in a series of physical events but the outcome of a goal-informed succession of events requisitioned to deliver a particular end. Indeed, in the absence of an explicit goal entertained by a subject, there is no beginning of the process or an outcome. What-is does not deliver itself as that-it-is in the form of how much it is. Somebody must turn up to transform the slumbering apparatus into a source of data.

Measurement, in short, takes place in the extra-natural human world and the data that result also belong there. It is of course in this realm that instruments are devised and calibrated against standards, and quality controls are introduced into observations. *Pace* Rovelli, a stone would seem to be underqualified. His view that all interactions between physical objects are observations is an especially brave version of paninformationalism.

Measurement always takes place at a certain scale. There are, as we have mentioned, no scales in nature.[50] Scales are ultimately rooted in scales of attention, though the extra-natural nature of attention – like other aspects of the consciousness of the human subject as scientific observer – is hidden by the formalization and sophistication and technologization of observation. It is equally valid to describe a stretch of the material world as a single beach or as a trillion grains of sand or as a trillion-trillion-trillion atoms. Global differences in the behaviour of entities in the macroscopic world of classical physics and the microscopic or nanoscopic world of quantum mechanics should therefore be looked at in the light of the fact that the distinction between these realms belongs to the science of physics and the realm of physicists and not to the physical world. There is no natural border between the macroscopic and the microscopic realms because the realms are not naturally defined.[51]

To suggest that the distinction is any more than a question of the grain of attention would seem to imply that the realms could and, indeed do, exist side by side, which clearly they do not, any more than Eddington's two tables are next to one another. It is this that makes the very idea of a passage from one to the other, or the interaction between a nanoscopic entity and a macroscopic environment difficult to accommodate within physics.[52]

Another manifestation of the error of taking measurement for granted as a mere series of physical events, as a manifestation of the natural world, is that of seeing the human organism, as opposed to the human person, as a measuring device. This is exemplified in Bas van Fraassen's claim that "the human organism is, from the point of view of physics, a kind of measuring apparatus".[53] This overlooks the vast distance between any organism as a material object interacting with laboratory kit in accordance with the laws of nature, human or otherwise, and *Homo scientificus.* The physical world as instantiated in the material of the human organism does not stand outside of itself, and extract and quantify parameters. What performs the measurement is not (just) the human organism with its physiological (and hence ultimately physical) properties but the human person, the conscious subject drawing, as we have said, on the community of minds to which he or she belongs. It is the *am*bodied subject, the citizen, who finds inches in her hands, cubits in her arms, and feet in her feet. The measurer and the measured are and are not in the same space. Thinking that they belong to entirely the same space is a manifestation of how science takes itself for granted, how it overlooks itself – which is a supreme example of the tendency to overlook the process of making what-is explicit.

While measurement is a physical process, to see it in purely physical terms is to overlook the subject-object distance, the contrasts between agent and substrate and instrument and datum, necessary for series of events to qualify as a measurement – and indeed for those events to be brought together to happen in the coordinated sequence necessary for a measurement to be accomplished. All of these contrasted pairs mark the distance between the physical elements of the act of measurement and the datum that results; and between a datum and the physical event that corresponds to it.

Nothing counts as a measurement that does not involve the measuring *intention* of an individual conscious subject or a community of such subjects or (typically) an individual subject informed by the community. Measurement goes beyond physical substrates, apparatus, and the material bodies of physicists pressing the buttons, switching the switches, and noting the pointer readings. Which is why, as Einstein pointed out in the passage quoted earlier, measuring rods and clocks, which make space and time explicit, do not fit into the physicists' picture of nature at the fundamental level. This obvious, but often under-appreciated, point will prove to be central to putting some of the paradoxes of quantum mechanics in perspective.

The story of the relationship between measurement and the conscious subject is, however, a conflicted one. On the one hand, we have seen how consciousness lies at the heart of measurement. On the other, marginalizing the individual consciousnesses of subjects or even the collective consciousness of the culture to which they belong, also seems to be central to measurement, the basis of the agreement between observers and successive observations, and the grounds for the claim that measurement is true to what is measured. Implicit in scientific measurement is commitment to a form of objectivity, in which we are on the side of the object, that we get ever closer to seeing what-is as it is in itself. That is why the community of scientists transcends individual cultures and why technology that is born in a laboratory in a particular place works throughout the world, indeed the accessible universe, such that truths discovered on earth seem to apply to events on stars billions of light years away.

The fact that the result, if the measurement is carried out properly, will be the same no matter who makes it, encourages the mistaken belief that it doesn't matter if *no-one* makes it; that measurement reveals intrinsic properties of the physical world; that a table which does not own its appearance does own its length. Measurement seems aperspectival, de-situated, independent of persons, and hence of the variables that influence what things "look like":

the accuracy of a measurement is reflected in the fact that, to an important degree of approximation, it does not matter who makes it, when the measurement is made, and from what angle it is measured. It seems that the table still has the size we may translate into "6 by 4" even when no-one is looking at it. This is an approximation of course that does not take account of the finite speed of light and the dependency of measured length on the relative velocities of the frames of reference of the observer and the observed. But even that relativity can be defined quantitatively and so allow for correction. Relativity theory cancels relativism by defining the conditions under which observations would be invariant.

The price of decontaminating observation-as-measurement of the viewpoints of particular observers, so that what is observed appears to be intrinsic to the object of measurement, is the reduction of what-is to how much, to pure quantities. This is bought at a heavy price, as we have seen. The more that observation confines itself to quantities, the less of the world is captured in it. This loss becomes evident at the most basic level. If I tell you that there is a solid item that is 6′ by 3′ and that there is an aperture that is 6′ by 6′, you will be confident that the item will be able to pass through the aperture, even though you have no idea whether the item is a table or a baby elephant and the aperture is a hole in a wall, or the entrance to a burrow. In short, the abstract portrait of what-is, composed entirely of the results of measurements, is emptied of any of the characteristics that define its character. Adding further measurements would never be sufficient to reconstitute the type of object it is.

There is another source of the distance between the quantitative portrait of what-is and anything particular to that which is portrayed: units, which we have seen are the necessary mediator of quantification. The table is not "3" but "3 feet"; it does not weigh "20" but "20 kilograms"; and its temperature (at present) is not "25" but "25° Centigrade". Thus, as Max Born reminded us in the passage we quoted previously, we are "dealing with ratios, relative data concerning units which are themselves to a high degree arbitrary". The last toehold on reality – which may give us a hint as to the kind of item we are dealing with, connecting numbers with something definite – is chopped off. As we have already noted, nature has no units, any more than it has scales. A table may have weight, but it does not have kilograms. There are no kilograms in the world; or not without the assistance of the conventions of conscious humans for whom there are kilogram weights which encompass tables, boulders, and bodies. One part of nature does not quantify another part against standardized measures. The arbitrariness of units to which Born referred is highlighted by their progressive refinement with the advance of science – such that, as we noted, a kilogram is no longer the weight of a particular entity in Paris but is defined with respect to Planck's constant or 6.626×10^{-34} joule seconds. In applying units to objects, we drain them of their individual characteristics.

Quantification reduces place to space and space is in turn reduced to a featureless matrix of points borrowing any specification they may have from a frame of reference defined by co-ordinates imposed upon it. Place is no longer something that is inhabited by subjects or even occupied by objects. Parting the veil of appearance in pursuit of the object-in-itself thus involves an exsanguinating dephenomenalization. It therefore comes as an affront that subjective consciousness – which is necessary for but hidden in macroscopic measurement – seems to return in the most powerful and universally applicable of all scientific theories – namely quantum mechanics. We shall discuss this in the next section.

The problem with units – or unitized quantities as a portrait of what-is – is exacerbated when we consider the most prized product of the scientific method: *equations* expressing laws connecting aspects of the physical world teased out as variables. Consider an unfashionably simple example: Boyle's law.

According to this law, the pressure and volume of a fixed quantity of gas are inversely proportional, if other relevant conditions such as temperature are kept constant:

$$\text{Pressure} \times \text{Volume} = k \text{ (constant)}$$

Pressure is measured in pascals and volume in cubic centimetres. The equation connects these fundamentally different units, highlighting that the separation of aspects of the gas into parameters which permit quantification, is an artefact of the scientific gaze. No gas is or has a pressure separated from a volume. The constant k is consequently something that has no existence – even less a stand-alone existence – in nature.

It is worth exploring this further with perhaps the most famous of all equations: $E = mc^2$. If a portion of matter were converted entirely into energy, this would result in a generation of energy equal to the mass of the converted matter multiplied by the velocity of light in a vacuum multiplied by itself. Energy is measured in joules, mass in kilograms, and the velocity of light in metres per second. The equation that asserts a quantitative equivalence between what is on the left-hand side of the equation and what is on the right marginalizes what is on each side – the toehold on any feature of what-is. This distancing from any grip on reality is amplified by having light represented solely by its velocity – so that it has parted company with its essential feature, its luminosity – and is thus reduced to a number. The strangeness of light being represented in this way such that its (all and every) spatial transitions are divided by time – itself reduced to a denominator and then to a pure quantity – is compounded by its subjection to the mathematical operation of being squared. In the end, all we have are numbers either side of an equals sign, even though energy in all its manifestations, stuff, and force do not seem to be reducible to pure quantities. Nobody can be dazzled by light as pure numbers, burned by energy as pure numbers, or squashed by mass as pure numbers.

So much for the rightly celebrated equation that, along with the field equations of general relativity and those of quantum field theory, offers the most general, all-encompassing scientific account of what happens in the totality of things. Or, given the conservation of mass-energy, what does *not* happen. The left- and right-hand sides of the equation are connected by an equals sign, rather than an arrow marking successive states of the system observed. For this reason, the equation is temporally symmetrical and could have been presented as:

$$Mc^2 = e$$

Or equidirectionally:

$$E \leftrightarrow mc^2$$

This equal bidirectionality is true of any equation. For example, Boyle's law:

$$P \times V = k$$

could be unpacked either as

> Increased pressure on a fixed mass of gas causes reduced volume.

Or as

> Increasing the volume available to a fixed mass of gas causes reduced pressure.

Which equation is applicable depends on what counts as the reference change, most obviously defined where the change is brought about by an agent defining what we might call the "do" operator.

The central point is that what-is, seen through the fundamental equations of physics, is entirely featureless: everything boils down to numbers of featureless stuffs. There are no secondary qualities (colours, smells, tastes) – of course; but nor are there primary qualities. Shape, size, etc., are reduced to pure magnitudes. Anything beyond magnitude is tainted with "secondariness". As Stephen Shapin put it: "In the wake of Newton, all natural processes were now conceived to take place on a fabric of abstract time and space, self-contained, and without reference to local and bounded human experience".[54]

The quantitative gaze ultimately looks straight through the particulars of what-is to its most general features. Mathematical what-is is effectively featureless. As I have (not entirely seriously) suggested elsewhere, if $E = mc^2$ were a true representation of the universe, it would not seem very enticing as a tourist destination. It is tempting to modify Galileo's much-cited claim and consider that, if the book of nature really were written in the language of mathematics, it would be written in invisible ink. This should be evident in the full quotation from Galileo: "[Nature] is written in the language of mathematics and its characters are the triangle, the circle and other geometrical figures without which it is humanly impossible to understand a single word of it".[55] Not quite invisible ink, perhaps, but mere hints of ghostly forms in emptiness. After all, triangles, circles, and other geometric figures are without content, and their boundaries are without thickness. Pure geometric figures are nothing in themselves. They are abstractions, ideas.

That the quantification of what-is terminates in featurelessness is hardly surprising. Number is not a property of anything since it can be attached to anything. Threeness does not become the property of individual entities – for example sheep – when we note that there are three of those entities. Threesomes – three sheep, three planets, three grains of sand, three political parties, three thoughts, three numbers, and three wrong ideas – have nothing in common in virtue of their being equinumerous. The number that they share is not a property of anything. After all, an individual sheep could be seen with equal validity as one (beast), a trillion (molecules), or a millionth (of the sheep population of the earth). Given that a trillion, 1, and 1/1,000,000 are not equal, the sheep could not have these magnitudes as intrinsic properties.[56] The quantitized world picture – according to which what-is boils down to different quantities of how-much – is homogeneous. Hence the attraction, discussed in Chapter 2, of homogenizing atomic accounts of the fundamental stuff of the world.

The featureless world of fundamental science as the endpoint of the quantification of what-is may seem consistent with the suspicion that, at the fundamental level, nothing happens – or there is no net change – because everything is fundamentally the same. Equations describe "conservation-despite-surface-change": nothing is lost or gained. Conservation can be described structurally as in Noether's first theorem for the conservation of energy.[57] And Einstein acknowledged that his Block Universe, the spatio-temporal continuum was intrinsically unchanging. He accepted the soubriquet "the Parmenidean" from Karl Popper.[58]

In summary, there is a tension at the heart of the idea of measurement. As the project of quantification progresses, the portrait of the world is increasingly drained of anything corresponding to sense experience – it becomes a mathematical representation of a view from no-one, no-when, no-where. At the same time, that portrait could not have written itself. It is ultimately based on, and checkable against, observation, notwithstanding that those observations are remote from the ones we make as we look about us and enjoy and suffer the lived reality of our daily life. As science progresses, observations are increasingly mediated – by instrumentation, concepts, communities of inquirers. The relationship between the experiences of the observers in the lab and what counts as the data added to the empirical base of advancing science becomes more distant, so that observations are ever more remote from what is delivered to our sense organs in everyday life. It is consequently easier to overlook the role of the conscious subject in generating what aspires to be an all-inclusive account of what-is or what-really-is. This – the problematic nature of measurement – is the true "measurement problem", though it is not what that phrase is usually taken to mean.

4.7 THE MEASUREMENT PROBLEM

When, at the cutting edge of fundamental physics, the role of the conscious agent in measurement recovers prominence, it is seen as something between a mystery and a scandal. The scandal is not that measurement involves a conscious subject; after all, it seems unlikely that unconscious nature would measure itself and, with a view to discovering its own properties, carry out all the remarkable things measurers have to do. The measurement problem highlighted in quantum mechanics, therefore, seems like the return of the repressed, though there is more work to be done to establish this

For some the problem that (for example) an elementary particle does not have a definite location or momentum until the appropriate measurement has been made seems to undermine the claim of physics to be the ultimate authority on metaphysics. It seems to throw into question the status of measurement as a revelation of objective reality. For others, it undermines the very idea of objective reality independent of observation. The concern was expressed by Heisenberg when he claimed that "What we observe is not reality itself, but reality exposed to our method of questioning".[59] The problem is also encountered – though in a different, less dramatic, form – in general relativity where, as we noted earlier, the status of measuring devices – clocks, measuring rods – and their failure to fit into Einstein's account of the physical world, was of profound concern to him.

The measurement problem is connected with the problem of how, or indeed whether, the wave function described by Schrödinger, which is a linear superposition and evolution of different possible states, gives rise, in the context of measurement, to a definite state of the physical system. The wave function is a probability wave: it depicts the relative frequencies of different outcomes of measurement. The problem has been summarized by Tim Maudlin as the incompatibility of three claims: (1) the wave function of a physical system is a complete description of the system; (2) the wave function always evolves in accordance with linear dynamical equations, such as the Schrödinger equation; and (3) a measurement always yields a single definite result.[60] If the wave function of an isolated physical (quantum) system is both a probability wave and a complete description of that system, it seems to be incompatible with a definite result from measurement. Michel Bitbol characterizes this as "a gap between the quantum domain of superpositions and the classical domain of sharp properties".[61] Somehow, a superposition of a range of possible values randomly "collapses" to a single measured value.

This goes against the most fundamental assumption, or faith, of natural science: that what is observed, not to speak of the quantitative values that it instantiates, exist prior to observation. It is clearly at odds with what is seen in the macroscopic world that we all, including physicists, occupy. If superposition were scaled up to the level of macroscopic objects, then we would have the possibility of such objects co-existing, until they are observed, with themselves in incompatible states, a possibility dramatized in Schrödinger's thought experiment in which a cat is both dead and alive.[62] Superposition in the macroscopic world would be a problem because the states that are superposed are rival occupants of the same state space variable. There is no shortage of explanations for how quantum superposition does not translate to the macroscopic world but none of them is satisfactory.[63]

Equally disturbing is the Heisenberg Uncertainty Principle if it is taken to say something fundamental about reality rather than merely to specify the limits of measurement. The uncertainty in question is the impossibility of, for example, predicting the position and momentum of a particle based on initial conditions and the equal impossibility of determining the precise position and momentum of a particle at the same time. One or the other must be indeterminate. There are states of particles in which they have a definite location but no definite momentum; and there are states of particles in which they have definite momentum and no definite location.

The uncertainty is not, so we are told, merely a feature of the limitations of *measurement* methodology – of, for example, the fact that measuring two variables requires different, incompatible experimental set-ups and apparatus – for there is a *precise* degree of uncertainty. The particle may have a superposed small range of positions and a superposed small range of momenta. As measurement cones down on position, so the range of momenta widens; and as measurement cones down on momenta, so the range of positions widens. At no stage are both a precise position and a precise momentum co-present in the same particle. Which parameter arrives at precision depends on the kind of measurement that is made.[64]

Schrodingerian superposition and Heisenbergian uncertainty together suggest that there is indeterminacy inherent in the states of the entities at the microphysical level. This seems to be compatible with the traditional understanding of an overall deterministic world. And also seems incompatible with a realm of actuality rather than possibility. After all, if an electron does not have a definite position, it does not have a position period. And if it does not occupy a definite position, does it have a definite size? And if it does not have a position or a size, in what sense does it exist?

There have been many responses to the measurement problem. The boldest – some would say ontologically the most expensive, perhaps the silliest – is the many worlds interpretation, according to which, every time a measurement is made, the universe splits to accommodate the actualization of all possibilities: all the superimposed states continue to exist, but they are located in separate, non-communicating universes. If there is a 50 per cent chance of an atom decaying, until a measurement has been made on it, there is no way of telling whether it has decayed or not. Prior to measurement, you must treat the atom as if it were in a superposition of the two states. After measurement, there must be two universes – one in which the atom is decayed and the one in which it is not.

It is difficult to know how an initial budget of universes is added to at the time that the measurement is made and what is the distinguishing nature of a measurement that confers upon it such creative power. It is difficult, also, to know what is meant by a "universe" and how (according to some interpretations) there could be an infinite number of them.[65]

To describe the multiverse solution to the measurement problem as being to take a sledgehammer to crack a nut would be the understatement of the century. Moreover, it is not amenable to confirmation or falsification because we are not able to leave our own parish to inspect parallel universes.[66] Nevertheless, as a metaphysical or ontological interpretation of quantum mechanics, it has a significant following among theoretical physicists, ranking just behind the Copenhagen interpretation (see below) and decoherence.[67]

I shall not argue its particular merits – notably that, for some, it seems to provide an explanation of how it happens that we live in a universe fine-tuned to the conditions of life.[68] An alternative response to the multiverse hypothesis is to see it as the *reductio ad absurdum* of conflating what is revealed by measurement with what-is, confusing the mathematical map with the terrain that is mapped. It is arguable that it is a projection to see them as inherent in the material world beyond any gaze. The fact is that, in the case of something so vast and detailed as a universe, any actual state it has is highly improbable.

For many decades the most widely accepted response to the measurement problem was the Copenhagen interpretation most closely associated with Neils Bohr. For Bohr, a microphysical object such as an electron does not exist anywhere until an observation or measurement is made. As he put it, "an independent reality in the ordinary physical sense can neither be ascribed to the phenomenon nor to the agencies of observation".[69] John Wheeler expressed it succinctly when he asserted that "no phenomenon is a real phenomenon until it is an observed phenomenon".[70] For Max Jammer, "quantum states are nothing more than formal devices for encapsulating the probabilities of observation"[71] – a mere algorithm for connecting observations.

To address the measurement problem, it is necessary, but not sufficient, to draw on what we have already discussed: that measurement – and more generally scientific (and pre-scientific) observation – is the product of an interaction between a piece of the material world and a conscious subject. The ghost of contamination haunts all observations, even in the most scrupulously quality-controlled measurements, because they necessarily involve interactions between an observer and what is observed. As Heisenberg expressed it, "no observation of atomic phenomena is possible without their essential disturbance".[72] (The disturbance, it is to be noted, is "essential".)

As we have already noted, it is an illusion that observations can be made in the absence of a conscious subject: an event in an instrument counts as a datum only when it is registered by a conscious subject and is the product of a succession of events chosen by subject. It is this that takes us beyond the mere matter of physical interaction necessary for an observation. The ghost of the observer remains, notwithstanding all the steps taken to exorcize variations arising out of subjectivity: the deployment of units, the mobilization of rigorously calibrated instruments, the sterilization of the field in which observation takes place (keeping "other things" equal), the checking and re-checking of findings and, most importantly, pushing pretty well all of the experience of the subject to the margins so that nothing seems to remain except pure quantities or the generalized relationships between quantities. The phenomenal character of the readings on a meter, of the location of a pointer, of the position of hands on a clock face, of the graphs drawn across a screen, are looked past or through to pure numbers. Nevertheless, the number remains the product of an interaction between a conscious subject – or a community of subjects – and a physical set-up shaped by human intentions connected with the purpose of an experiment defined within the discourse of a particular science.

This may not itself seem particularly disturbing – even point-missing. Does it not simply affirm that what we find in the world to observe will depend on the direction and scale of our attention and more broadly on the intelligence, discursive structures, and assumptions that shape our awareness? And that this will shape the act of measurement? The nature of the interaction between the measuring instruments and the object of measurement will be expressed in the choice of the experiment to be performed, which will determine the kind of interaction, and hence disturbance, that measurement will involve.

Why, however, does this matter so much in quantum physics? Why is it *here* that the problematic nature of measurement becomes foregrounded as "the measurement problem"? Does it matter less in macroscopic science because the "essential disturbance" associated with the interaction of measurement will be negligible: when I distort a table, and thus alter its measured length, by applying a tape measure to it, the pressure on the table has a negligible effect on its length. If, however, we are near the limit of the granulation of what-is, the physical impact of the measurement is proportionately higher. The choice of experiment can determine which parameter has to bear the burden of the interaction and the disturbance. As Kumar has expressed it, "Heisenberg identified the act of measurement to determine, for example, the exact position of an electron as the origin of a disturbance that ruled out a simultaneously precise measurement of its momentum".[73]

The measurement problem, however, is not simply a matter of the limits to precision when what we are measuring is close to the Planck limit. After all, Bohr's interpretation goes as far as denying that elementary particles such as electrons have position or momentum *before* they are measured. The problem isn't, therefore, merely the result of a nudge that disturbs that which is measured, blurring the result – or of some kind of exotic interaction between that which is measured and the consciousness of the measurer. We are not obliged, that is, to subscribe to the more radical implication that the properties of a quantum entity uniquely lack definite values until they are measured, whereupon the wave function which describes the distribution of the values of properties, collapses. This would license a form of idealism – quantum idealism.

Idealism is a surprising conclusion for a tough-minded discipline such as physics. Moreover, it seems more profoundly undermining of our usual way of seeing things. What can it mean to assert that an entity – even if it is a theoretical construct such as an electron - does not have a definite position or a definite momentum until it is observed; and which of these it has depends on the kind of experiment defining the kind of observation being made? It is, however, a logical consequence of the radical empiricism that was so influential in early twentieth-century science, according to which it makes no sense to ascribe existence to entities or values of variables that are inaccessible to observation or have no observational consequences.

Do we need to draw such radical conclusions? Perhaps we should reflect on how, even at the macroscopic level, there is a mutual exclusiveness of *measured* parameters – and they have independent existence only insofar as they are measured – and hence of their having definite, precise values.

Consider a ball thrown through the air. A measurement, or *a fortiori* an observation, over time (and it will always be over time because it takes time) can determine its velocity but not its location; more precisely, a thrown ball does not have a definite location so long as it is in motion – rather it has a *trajectory*. Location is at a moment and the more quickly it is measured the less it will be spread out or smeared. It could be measured by means of instantaneous photography, but this would cut off access to velocity. The concept of velocity, on the other hand, requires time – typically, it is averaged over an interval of time. Admittedly, the idea of "instantaneous velocity" is a perfectly respectable concept in natural science. It is, however, an abstraction because when t as the denominator is reduced to zero, to an unextended *point* in time, the value on the numerator has likewise to be reduced to zero, otherwise instantaneous velocity would be infinite: displacement that took no time would have to be infinitely fast.

There is, in other words, a profound sense in which (exact) position and (exact) momentum are incompatible. The position occupied by a moving object is occupied for a non-extended instant – zero interval; the momentum of a moving object can apply only to extended time – more-than-zero interval. Which is not to say a macroscopic object does not have both position and momentum and, indeed, have them simultaneously. I can say that the ball was continuously moving at 20 mph between position 1 and position 3 and between t_1 and t_3 and that it was at position 2 at t_2. The problem begins when they are *separated* – something that is routine in natural science but not possible in nature. Then an entity cannot have an exact value for its velocity and an exact value for its location as they are methodologically incompatible. Techniques to measure velocity and location will have to accept approximate results for one parameter if they generate exact results for the other.

So, what of the quantum level? Yes, of course, an experiment can seek and discover a definite value of either the momentum or the position of an electron but not both, even though in reality they are not separable. It is not possible simultaneously to carry out momentum-measuring and position-determining experiments on the same electron. But it does not follow from this that it does not have both parameters at once, if it is accepted that the separation of properties of an entity into parameters is *an artefact of measurement.* Indeed, it would be difficult to ascribe reality to an entity that lacks, for example, a position and had only momentum. The paradox arises because we are too used to taking literally purely mathematical conceptions such as "position" (defined by co-ordinates) or "instantaneous velocity" – velocity over no interval of time and no interval of space and the idea of the position of a moving object – a position of something that is not still and hence does not occupy a given spatial point for a period of time greater than zero.

The macroscopic example of the thrown ball shows how there is nothing special – or especially disturbing – about uncertainties, indeed indeterminacies, at the quantum level. The precision of Heisenberg's indeterminacy principle may suggest a difference between quantum and macroscopic observations. The relative imprecision of the uncertainty of velocity when position is privileged, and of position when velocity is privileged, in the case of a macroscopic object such as a thrown ball is due to the fact that measurements are contaminated with an irreducible level of noise. At the quantum level, where the magnitudes in question are close to that of the quantum interval, the scale of contamination is equal to that of that which is measured. At both levels, there is an incompatibility of parameters: it is not possible to ascribe a position to an entity when its velocity is being measured, nor a velocity when it is ascribed a position. The challenge that nanophysics presents to the metaphysics of macroscopic everyday life, given that macroscopic objects are made of nanoscopic ones, does not, after all, seem so profound. Bohr's assertion, quoted earlier, that "Everything we call real is made up of entities we cannot call real" seems a little panic-stricken. If anything is unreal, it is not entities but the parameters that are separated prior to measurement.

If we accept that position (the point in space the entity is at) and momentum (how quickly it is moving from point to point) of moving objects are not available as objects of simultaneous measurement – not the least because position is *at* a time and momentum *over* time – then we can understand how there is an uncertainty linking these parameters. When the size of what is being measured is at the Planck limit, that uncertainty is 100 per cent. I think this is similar to the claim that the quantum to classical transition is due to "the coarsening of measurement"; that it is explained by the fuzziness of measurement references.[74]

So much for a deflation of Heisenbergian uncertainty. It rests on something that is central to this chapter, regarding the unnatural nature of measurement. There is a sense in which what-is becomes definite, in the sense of being bounded, individuated, or defined, only when it is made explicit and becomes quantitatively definite only when it is measured. The measurement problem, when definite results are replaced by probabilities, becomes evident only at the microphysical level, when what is measured are the fundamental constituents of matter. This is the level at which, so we are told, the continuum breaks into discrete units – quanta – that are not amenable to further division; the level at which we are measuring the size of houses using the house as a tape measure.

Another difference between the quantum and classical world may also be relevant to the obtrusiveness of the measurement problem at the quantum level. Quantum entities lie – and will always lie – beyond the reach of direct observation: at best they are inferred from measurements. Macroscopic objects are directly observed: measurement is downstream of ordinary perception. There is no "perception problem", analogous to the measurement problem, though what is seen will depend on the location and state of the sense organs of the perceiver – it is not, however, accessed primarily via measurements with all the uncertainties they bring.

Thus, one attempt to make (common) sense of the measurement problem. Another, all too understandable, response is to lose one's temper with it. It is easy to sympathize with the physicist John Bell's famous, enjoyable tirade "Against Measurement". "Surely", he begins, "after 62 years, we should have an exact formulation of some serious part of quantum mechanics".[75] Alas we do not. The reason for this, he argues, is the non-mathematical language we use to interpret quantum mechanics: "Here are some words which however legitimate and necessary in applications, have no place in *formulations* with any pretence to physical precision: *system, apparatus, environment, microscopic, macroscopic, reversible, irreversible, observable, information, measurement*".[76]

The range of terms placed on Bell's index – climaxing with "measurement" – is striking. The cast of banned words seems to suggest that the interpretation of physical theories should not utilize terms that (accurately) describe the practice of physics that has led up to the theories. Nevertheless, Bell was not happy with those who were indifferent to what quantum mechanics meant about the real nature of things: about what he called "beables" and those elements of a theory "which are to be taken seriously corresponding to something real". As Baggott summarizes Bell's discontent (one which will be shared with many philosophically inclined readers of this text) "what is the purpose of a scientific theory if not to aid our understanding of the physical world?"[77] The central notion of quantum mechanics, the wave function – "a mathematical description of the quantum state of a system, which contains all its measurable information"[78] – remains enigmatic. Nobody knows what it is. We are not only unable to solve the measurement problem; we don't understand the question. It seems odd that we should not seek an answer that takes account of, and tries to understand the nature of, the processes by which we arrived at the quantum story. In short, the problematic nature of measurement.

One valid reason for placing the term "measurement" on the index is because it is ill-defined. The other terms that Bell wishes to exclude from quantum theory – *system, apparatus, environment, microscopic, macroscopic, reversible, irreversible, observable, information* – simply highlight what a tangled notion "measurement" is. Moreover, as Jonte Hance and Sabine Hossenfelder have expressed it, "if quantum mechanics was fundamental then the behavior of macroscopic objects (like measurement devices) should be derivable from it. *The theory should explain what a measurement is, rather than require it in one of its axioms*" (emphasis added).[79]

As we have discussed, there are at least three players in the measurement problem drama: the measuring subject (with her internalized hinterland of the science she is pursuing); the measuring apparatus; and the entity or process or event that is being measured. The borders between them are ill-defined, as also are the borders between the microscopic processes that are being measured and the macroscopic entities being wielded to make the measurement.[80] And when it comes to something seemingly as simple as an observation – uniting the observer and the observable – this is in fact far from simple. We have (for example) a pointer position, its translation by a dial into a reading, the energy transfer from the pointer-plus-dial to the sensory system of the experimenter, the harvesting of the measurement as something sought and understood, and all those things the experimenter does to incorporate a particular measurement into a body of data, to interpret it, and to transmit it to someone who is going to use it to feed into a calculation to support an hypothesis.

The further transmission and sharing of measurement – its incorporation into a shared body of knowledge that belongs to an evolving, ill-defined community of educated consciousnesses – makes the question as to the nature of measurement even more entangled. When is a measurement made? What are the boundaries of the measurement process, given that it is subject to collective regulation and agreement as to methods and standards of measurement? Imagine a photosensitive plate being affected by a shower of photons. The resultant image is transformed by a computer into an array of numbers that are then automatically printed out and placed in a file by someone who may have no idea what they mean and sent by email to another individual who has requested them and does know what they mean. Does the completion of the measurement have to wait until the collaborator opens the email and looks at the relevant page?

Such questions become more pressing when we think of the measurements that take place at the cutting edge of physics. Measurements are made by teams – present teams building on geographically and historically absent ones. Even when an embodied individual is attached to a measurement, we cannot think of the act of measurement as "a physical event". Certainly not a common or garden one, analogous to the physical events that are measured. A measurement is above all a purposive observation based on a solicited experience, triggered by an intervention, whose purpose defines a beginning and an end.

There is nothing in the physical world corresponding to the contrast between the assembling of the appropriate conditions of measurement and the observations that extract the result; between applying the tape measure and reading what the tape measure says. Even less is there anything corresponding to the agonizingly slow progression of the accumulation of data to reach the 5-sigma level for the peak at 125 GeV necessary to justify confidence in the reality of the Higgs boson.[81]

In short, measurements are unnatural. No wonder they present a challenge to the self-styled "natural sciences". The problematic nature of measurement, if not in the form of "the measurement problem", is present throughout science.

Bell's ban on certain terms seems justified. However we try to define measurement, it is evident that the boundary between observing an event and recording it as a measurement is impossible to define in terms that belong to any quantitative science never mind fundamental physics. To define the boundaries of measurement in terms of the operation that is delivered by it is, of course, circular.[82] And all these uncertainties are present before we move on to consider the fact that measurements, or data, are requisitioned by, and made available to, the shared consciousness of a research community.

The appeal to measurement as the fundamental means by which the wave function collapses has extraordinary implications. As Bell (somewhat rhetorically) asks, "Was the wavefunction of the world waiting to jump for thousands of millions of years until a single-celled living creature appeared? Or did it have to wait a little longer for some better qualified system … with a PhD?"[83] The question reminds us that there is almost a contradiction in ascribing the results of measurement, such as the position or momentum (as isolated aspects) of an electron, to an entity in itself – before it has been measured. While the mountains on the moon were a certain height before they were known to human observers, those heights were not "so many feet" or "so many angstrom units". Nor were those heights stand-alone entities. It was only when humans with tape measures approached it that the boulder as the base of a mountain became 6 feet by 6 feet.

Bell's point about the wave function of the world "waiting to jump" until it is observed raises a more profound issue about the fundamental nature of the world. Ward Struyve made the point that "if the wave function is all that there is, then the physical arena does not seem to be physical space (or space-time) but rather configuration space of perhaps some Hilbert space ... [T]his means that one cannot meaningfully discuss the possibility of locality".[84]

Bell's hostility to the use of "measurement" in interpretations of quantum mechanics is echoed by Peter Lewis in his recent essay "Against 'Experience'" where he argues that the list of banned words should be extended to include "experience", and cognates such as "awareness", "perception", "observation", and "consciousness". As Bell argued, while it is fine for such words to be used in the applications of quantum theory, they should not be used in its formulation, as primitives in the theory. And this applies not only to measurement and to experience but to a term that, in virtue of occupying a half-way position, is seemingly innocent: "observation".

In view of all this conceptual confusion, it is hardly surprising that there is a body of informed opinion, both within and beyond the physics community, that shares the view advanced in this chapter that quantum mechanics is not a reliable guide to the fundamental nature of reality. There are grounds for suspecting that measurement is not the best, even less the only, path to unpeeling fundamental reality. When Bell deplores the fact that after 62 years we don't have "an exact formulation of some serious part of quantum mechanics", he explains that "by 'exact' I do not of course mean 'exactly true'. I mean only that theory should be fully formulated in mathematical terms".[85] We should resist – indeed reject – an extreme scientism which insists that philosophy should wait in the anteroom of physics for instructions as to the fundamental nature of what-is.

As for the measurement problem, it is ultimately rooted in the problematic nature of measurement – in its stubborn refusal to fit into the physicist's portrait of the world; in the entirely understandable refusal of the physicists' world picture to accommodate the physicist; the inescapable failure of the picture to depict how it is built up out of pixels while still remaining a picture.

Let us look at another reason for rejecting the claim that physics is the royal road to metaphysics. In doing so we shall anticipate our discussion in Chapter 9: the nature of possibility and probability as the most explicit manifestations of explicitness.

4.8 PROBABILITY AND POSSIBILITY IN QUANTUM MECHANICS

At the heart of quantum physics is the "wave function". This central item is neither a wave nor a particle. This much is generally agreed but its actual nature is fiercely contested. For Bohr, "The wave function has no physical reality; it exists in the mysterious ghost-like realm of the possible".[86] And there is no way out of this ghost-like realm: until an observation is made, microphysical objects such as electrons that might seem to be directly or indirectly represented by wave functions do not exist anywhere. There are only abstract wave probabilities, and these are merely mathematical entities, a representation of the possible outcomes of an observation. Consequently, according to Christopher Fuchs, "a quantum state has no 'ontic hold' on the world".[87]

The Born rule which defines the probability that a measurement on a quantum system will yield a certain result is not "a descriptive law of nature in the usual sense".[88] Rather, it is a normative statement. By this it is meant that the rule is "an empirically motivated norm of rationality a wise agent should follow in addition to those whose violation would render the agent's degrees of belief incoherent. As usually formulated, the Born Rule specifies probabilities for various measurement outcomes given a quantum state".[89] Contrary to what is often suggested, the passage from future probabilities to present actualities captured in the Born rule is not a transition from a multitude of superimposed states to a single state. There is, according to this "Quantum-Bayesian" view, no quantum world, only an abstract quantum description. It embraces a subjectivist understanding of probabilities being in the minds of observers and rejects an objectivist one of their being properties of nature.

That, of course, is the beginning not the end of the story. The bitter arguments between those who wish to reify the wave function and others who want to reduce it to probabilities, have not abated. Eddy Chen[90] has attempted, honourably but inconclusively, to adjudicate between a range of interpretations of the wave function: (1) ontological interpretations – such as that it is a field in high-dimensional space, or a vector in Hilbert space;[91] (2) nomological interpretations that places the wave function on a par with a law of nature; and (3) *sui generis* interpretations, according to which it is a new kind of entity. The failure to settle on the nature of the wave function might have been foreseen, given (so we are told) that it is a function defined in high-dimensional space, with values in complex numbers. Since these numbers include the square root of -1, it is no wonder that it is such a slippery character.

The shortest route, then, to ending the hunt for the wave function may be to accept that it is not a kind of stuff, a thing, an event, a state, or a process; rather it is a mathematical tool for capturing the *probabilities* that define how what-is unfolds at a microphysical level. It is a way of summarizing, or portraying, the distribution of those probabilities, a calculational recipe for prediction. To speak of it as a "wave" is to take its graphical portrait rather too literally; and to speak of it as a "function" (though it is a perfectly respectable mathematical term) may tempt some to believe that it is something that causes things to happen.

At the heart of these misunderstandings is the misreading of the nature of probability. In a short, devastating paper, David Mermin who has embraced Bayesian interpretations of quantum theory, argues that "the idea that the collapse of the quantum state is a physical process stems from a misunderstanding of probability and the role it plays in quantum mechanics".[92] Mermin begins by defining a quantum state as "a compendium of probabilities of all possible answers to all possible questions one can ask of the system". This makes especially clear sense when we remind ourselves that making a measurement – which interrogates the composite of a physical system being measured with a physical system making a measurement (the apparatus) – is an intervention that interrupts the continuous, deterministic unfolding of the physical system. There is a reference point – a "now" as the investigator closes in on the system to be investigated – and a future opened up by the possible value of an outcome of measurement. Hence the origin of probabilities: they "correlate the possible answers given by the state assigned to the original system with states assigned to the apparatus that indicate those possible answers".[93]

The important point, according to Mermin, is that probabilities are not objective features of the physical world. They are the correlatives of bounded, incomplete knowledge. There is no statistical probability without ignorance – of what has not yet happened but what might happen – and explicit not-knowing, like knowing, is a state of a conscious subject. Mermin cites probability theorist Bruno de Finetti's assertion that probability "if regarded as something endowed with some kind of objective existence ... is an illusory attempt to exteriorize or materialize actual probabilistic beliefs". Finetti compares ascribing objective existence to probability with "superstitious beliefs in the existence of Phlogiston, the Cosmic Ether, Absolute Space and Time ... or Fairies and Witches".[94] The key word here is "beliefs". As we shall discuss in more detail in Chapter 9, possibilities exist only insofar as they are *entertained* – by conscious subjects. They may range in explicitness from the unexpressed assumptions which fill in the spaces in our present knowledge (and set the scene for fulfilled or unfulfilled expectations, and nasty or nice surprises) to fully articulated claims that have varying degrees of precision and are embraced with varying degrees of confidence in the face of inner doubt or the scepticism of others.

The key point is this: irrespective of any phenomenological content, *probabilities are inseparable from conscious subjects*. Contrary to what is often said in accounts of quantum theory, therefore, there are no such things out there in the physical world called "probability waves", propagating through uninhabited space. They owe their existence to possibilities which in turn owe their existence to the conscious subjects who entertain them. A wave function, therefore, is not a picture of something that is the case but of the reasonable expectations of any observer regarding what might be the case. The Born rule, which assigns probabilities to rival outcomes, does not have to deal with the co-present, superimposed, reality of all these outcomes.

This is highlighted in the macroscopic world by the fact that the definition of the range within which possibilities fall, and the number of outcomes within that range that are considered as possible, depends on the interests of conscious subjects and how they divide up or define what counts as an outcome. As we shall discuss in Chapter 9, the possible outcomes of tossing a coin may be classified as "heads" or "tails" with the probability of each outcome in the long run being 50/50 so long as the coin is a valid one and the coin-tosser is not manipulating the result. But there are many other possible outcomes - such as the distance from the coin-tosser's feet, the numbers of wobbles before settling for heads or tails, the pattern of light on the coin when it has come to rest, the depression it leaves on the pile of the carpet, etc. Each of these is equally valid as a physical outcome.

This highlights how possibilities and the probabilities assigned to them are not intrinsic to the material world. Without a definition of the outcomes of interest, with their connected uncertainties, there is no fixed range of possibilities that have probabilities assigned to each of them.

To say this is not to imply that what is possible and what is not possible, the pattern of happenings in the natural world, are at the behest of conscious subjects. That is magic thinking. While I may have limited, or (as when I am watching cricket on television) no, control over the outcome of a coin-tossing, what *counts* as an outcome will be defined by the direction and scale of my, or someone's, interested attention. More broadly, what counts as the outcome of an event will depend on how that event is defined, what its borders are. Those borders are defined by a conscious subject – something that applies equally to the much-contested question of what counts as a measurement.

There are deeper objections to seeing probabilities as being part of nature. Nature is what-is, not what-might-be: there is no "might be" in the physical world any more than there is negation. Indeed, there is no modality in nature. It follows from the fact that I am in Prague that I am not in London, Chester, etc. But my not being in London or Chester are not properties of the material world. Possibility – what might be – is also caught up in tensed time which is not part of the natural world. Built into any possibility that I envisage is the sense of a *future* in which the possibility might be realized or in which I make the discovery that it has or has not been realized. A measurement does not cause a range of existing possibilities or superimposed states to "collapse"; rather it is the transition from a range of future possible states to an actual present state now.

The immaterial nature of possibilities gathered up into probabilities is further highlighted by the fact that serious scientists are willing to deploy *negative* – less than zero – probabilities in their endeavours to model the quantum world. Richard Feynman defended the idea of a more than 100 per cent impossibility – given that zero per cent probability is 100 per cent impossibility – on the grounds that we accept the idea of negative numbers applied to the real world.[95] To compare probabilities to numbers is, unless one is a Platonist, to deny them status as stand-alone entities. Paul Dirac also argued that negative probabilities should be taken seriously, comparing them to negative money.[96] This seems an even more desperate ploy, given that negative money makes sense only in the context of human institutions such as currency and equally human notions such as that of indebtedness.

There is an irony in the replacement of causation in classical physics by probabilities in quantum physics. David Hume described our tendency to ascribe causal connectedness, *de re* necessity, to the connections between types of events that reliably follow one another as a manifestation of the mind's "great propensity to spread itself on external objects".[97] The transformation of quantum indeterminacy into waves of probability existing in the material world seems to be another manifestation of the mind's propensity to spread itself on external objects. Our uncertainty as to the outcome of measurement is translated into the co-existence of probabilities of different outcomes. This in turn becomes a superposition of states in the material world. In this case, it is not the transformation criticized by Hume of our *certainty* of the sequence of events, based on perceived regularities, into material necessity. Rather it is the projection of our *uncertainty* into a spread of probabilities that are projected into the material world.

The key point is that, in the physical world, there is no passage from what-might-be (located at time t_1) to what-is (at time t_2), from possibility to actuality. What actually happens is not the end-stage of a transition beginning with what-might-be. What-might-be is no more part of nature than what-is-not – that which is expected to happen and then disappoints us. Possibilities, but not nature, include that which does not, as well as that which does, happen, as do probabilities. Even the highly probable may not happen and that which does not happen is nothing: it is only in the eyes of a deceived subject that unrealized probabilities are real things, expected happenings that fail to happen.

It is obvious from this that so-called probability waves are not parts of nature, even though they may be central elements of a powerful *theory* of nature which enable us to predict what would happen under certain circumstances with a staggering degree of precision. Such entities are located in the minds of conscious subjects – or a collective of conscious subjects – reflecting on, and anticipating, nature from a reference point that counts in those minds as an initial state. Happening in the natural world is not the emergence of a definite actuality from a cluster of competing possibilities.

The projection of quantum probabilities into nature, the realist interpretation of Schrödinger wave functions, their reification as entities in the physical world, is the result of the fact that they are not properties of individual minds. They belong to communities of minds who access and understand physical theories, realized in individuals who are not defined by any determinate viewpoints, who contribute to and draw on the rich and complex story that is the advance of physical science.

It is worth revisiting another aspect of possibilities and probabilities, already referred to, that denies them a place in the physical world: that they are related to the *future*, to the not-yet, to the might-be. There are three things to be said about the future. The first and most obvious is that it must be entertained in order to exist. The second is that its status as future depends on its being related to a present, which will provide the necessary reference point. "Now" – like other aspects of tensed time – does not exist in the natural world. For this, we have Einstein's word: "Once Einstein said that the problem of Now worried him. He explained that the experience of the Now means something special for man, something essentially different from the past and the future *but that this difference does not and cannot occur within physics*. That this experience cannot be grasped by science seemed to him a matter for painful but inevitable resignation"[98] (emphasis added). An instant of time in the spatio-temporal continuum – or in the quantum spatio-temporal "discontinuum" – does not count as "now" without the frame of reference provided by a conscious subject. And thirdly, as already noted, the anticipated future may not happen: possibilities are not "actuals" though they may be actualized when they will cease to be possibilities.

As for quantum probabilities, they are as defined by the interests of the physicist inquiring into the system as much as are the probabilities associated with coin-tossing; more particularly by the choice of quantitative questions seeking quantitative answers. In the absence of such questions, then there isn't a superposition of possible states – just as an untossed coin does not have superposed states of heads and tails or all the other possible outcomes of interacting with it that fall under the heading of "tossing". The many answers there may be to a question about a physical system are not co-existing rival states of the physical world.

My aligning the macroscopic and the microphysical level in this way may still seem controversial. It cannot, however, be denied that what counts as a physical system – at any level – is significantly determined by a frame of reference constructed by particular interests, even if those interests are the highly standardized and regulated interests of natural science. Mermin expresses this as follows: "What one chooses to regard as the physical system and what state one chooses to assign to it depend on the judgement of the particular physicist who questions the system and who uses quantum mechanics to calculate the probabilities of the answers".[99] The updating of probability in the light of measurement is the abrupt and discontinuous part of any classical process and, as Mermin says, "Nobody has ever worried about a classical measurement problem" – notwithstanding the problems discussed earlier surrounding the idea of the instantaneous velocity of a moving object or the position of an object that is not at rest .

At the beginning of this chapter, I argued that the predictive and practical power of quantum mechanics has little to do with any claims it has to metaphysical truth. We cannot therefore invoke those powers as evidence of its metaphysical authority. This same point is made very lucidly by Mermin:

> We still lack any consensus about what one is actually talking about as one uses quantum mechanics. There is a gap between the abstract terms in which the theory is couched and the phenomena the theory enables us to account for so well. Because it has no practical consequences for how we each use quantum mechanics to deal with physical problems, this cognitive dissonance has managed to coexist with the quantum theory from the very beginning.[100]

We may set aside the idea that the wave function is a real thing and yet still be left uncertain as to how to understand what quantum science is telling us about the world. This is echoed by William Seagar:

> But a fundamental metaphysical problem remains: What is the nature of the world itself? The mundane intelligibility of the mechanical world view simply took at face value the quotidian perception of matter as massy, impenetrable, moveable *stuff* and extended that into

> a mathematical elaboration. But the nature of matter now seems quite unintelligible. Instead, we have an extremely successful and precise recipe for determining what we are likely to observe (experience) in any situation to which we can apply a viable quantum description.[101]

For the present we note that denying objective, physical existence to the probabilities depicted in the wave function is sufficient to address the measurement problem.

The limits to knowledge should not be interpreted as ontological truths. There is, however, a tendency to do so. Hence quantum idealism – a projection of the role of consciousness in the measurements that create the quantitative model of the physical world into the physical world itself.

4.9 TOWARDS QUANTUM IDEALISM

It is difficult to escape asking a challenging question. Is the entirety of existence, rather than being built on particles or fields of force or multidimensional geometry, built upon billions and billions of elementary quantum phenomena, those elementary acts of "observer-participancy", those most ethereal of all the entities that have been forced upon us by the progress of science?[102]

According to the most radical version of quantum idealism, entities such as electrons are not small things, localized in space and time, and having definite dimensions. They exist only insofar as they are observed. We earlier cited Bohr's belief that an unobserved electron does not exist. This position is arrived at by a drift from: (a) an acknowledgement of the indeterminacy of measured variables – such as position, momentum, and spin – ascribed to a microphysical object; to (b) questioning the existence of those variables in the absence of measurement; and to (c) finally denying the existence of the entity that would instantiate the variables, having them as their properties.

The place from which this drift begins – the discovery that it is not possible simultaneously to *measure* the position and momentum of an electron – does not, as we have seen, license the conclusion that the electron does not have these properties simultaneously. Once this has been conceded, and the actual is conflated with, or confined to, the measurable – what-is with how-much – and the criterion of measurability is a measurement with a definite result, the path is opened to the conclusion that microphysical entities do not exist until they are observed or, indeed, measured. After all, an entity that has no location or momentum or other physical attributes until they are measured – but only the probability of having such attributes within a certain range – seems to fall short of being anything in particular; and hence, of being anything.

This may explain the drift from "External reality does not exist in a definite state" – where "definite" is itself defined as realizing certain parameters that have a definite measured value – independently of our measurements to "External reality does not exist independently of our measurements". And it is no mitigation to say that, while the particle has no definite location, it does have a rough one and a tightly defined range within which that location falls. A well-defined degree of indefiniteness of (say) the location of an electron that depends on the precision with which another property (momentum) is measured seems contradictory. Either an electron is somewhere in particular – over there – or nowhere in particular, and hence nowhere at all. It cannot *in itself* be "roughly over there": if it is anywhere, it is precisely where it is. If, on the other hand, electrons do not exist until they are observed, what would give their probable positions and/or momenta a defined range within which the results of measurements made on them would fall? It is obvious that if a thing does not exist, these parameters have no definite value; but it should be equally obvious that the parameters do not have an *in*definite value either, if that value is envisaged as an intrinsic property of the entity in question.

The idealist suggestion that an entity does not exist until it is observed – in the special sense of "observation" as it is used in the context of an advanced science such as quantum mechanics – seems outrageous. After all, if quantum idealism is applicable to microphysical entities and the macroscopic entities that populate our world are made of microphysical entities, quantum idealism would apply to the macroscopic world: it would exist only when it was observed. Arthur Eddington's second (scientific) table would evaporate, like a Berkeleian object, when it was not observed. This raises the question as to what sort of physical reality worms and bacteria inhabit. Did electrons exist in the billions of years of the universe prior to the invention of the devices used in fundamental physics? Or (to requote John Bell's tongue-in-cheek suggestion) "Was the wavefunction of the world waiting to jump for thousands of millions of years until a single-celled living creature appeared? Or did it have to wait a little longer for some better qualified system ... with a PhD?" And, if sub-atomic particles did not pre-exist single-celled organism or physicists, are they therefore hallucinations of the community of scientists? If this is the case, how has invoking electrons as playing a central part in a vast range of physical phenomena proved so useful as evidenced by the electronic devices that fill the landscape of our lives?

Shall we conclude, therefore, that electrons (and the other inhabitants of the menagerie of particles, fields, forces, that makes up the Standard Model of the atom) are just *theoretical constructs*? As Davies and Brown have expressed it, "What Bohr's philosophy suggests is that words like electron, photon, or atom should be regarded ... as useful models that consolidate what is actually only a set of mathematical relations connecting observations".[103] Electrons, after all, are unobservable, only being inferred. Likewise, quarks, spin states of sub-atomic particles, and photons.

Whether entities being unobservable in principle means they are less real – so that photons are less real than light – remains uncertain, particularly as the notion of "real" is so complex, as we shall discuss in the final chapter. As the children of mathematical models of what is observed, they clearly would not exist when not observed – preferably by someone with a PhD, as John Bell pointed out.

The ontological cloud under which they sit is sufficient to make many "particles" secondary to fields. According to John Dupré and Daniel Nicholson

> Quantum fields, which are dynamic organizations of energy distributed in space-time, appear to have purged classical notions of elementary particles from the ontological picture ... Although contemporary physicists still routinely speak of 'particles', these no longer refer to solid micro-entities or tiny impenetrable granules, but to quantized excitations of particular fields. Quantum fields, in other words, are primary, and the various kinds of particles that physicists refer to are derivative entities, appearing only after quantification.[104]

This leaves conspicuously unexplained the origin of the discrete objects – such as cups and trees and human bodies – that seem to be sufficiently distinct from fields to be acted upon them as from without, as in the case of a pebble falling through a gravitational field or iron filings drawn by a magnetic field. Nevertheless, it is not clear how much of the applied success of physics depends on the reality of the menagerie of particles of the Standard Model. They are explanatory hubs that help to organize what is known about the way things unfold – serving, to quote Bohr again, "to extend the range of our experience and reduce it to order"[105] and thence to direct the ways in which that knowledge might be put to use in practical applications. If the particles were constructs out of measurement, it would not be surprising if the postulation of these constructs were efficacious in the ways specified given that explanation, prediction, and application seem to be usefully defined in quantitative terms. We could express it like this: even if quantum mechanics merely gets the maths right, this would be sufficient to account for its practical use because its applications are guided by mathematics as a pattern of the unfolding of material things.

Thus, it is possible for a theory that swims in a sea of uncertainty – as to whether fundamental entities are waves or particles or both (depending on how they are measured) – not only to have astonishing predictive power but also to guide the development of technologies that have transformed our lives. From this we may conclude that the predictive and practical power of fundamental physics does not prove that it has unique access to the true nature of things, thereby making traditional metaphysics redundant. The efficacy of the application of quantum mechanics evident in computing, optics, AI, all the electronic gadgets that populate our world – does not arise from, and is therefore not proof of, anything that may give it metaphysical authority. We can – and should be – impressed by science, but what impresses us has nothing to do with any claim it may have to pronounce on the fundamental nature of reality. After all, humanity had vastly increased its collective powers relying on pre-quantum, classical physics.

Hence the validity of Gilbert Harman's eloquent protest: "*[W]hy* exactly is it the mission of philosophy to limp along after the science of its time? It is not clear why philosophers must prematurely unify their own speculations on space, time, and substance with those of a quantum theory and relativity that are not yet even unified with each other".[106]

The propensity to overstate the metaphysical implications of discoveries in fundamental physics is illustrated by the excitement around a recent experiment which, it is claimed, demonstrates that there is no such thing as external reality and that this supports quantum idealism.[107] The claim is based on a study aimed to test a thought experiment first proposed by Eugene Wigner over 50 years ago.[108]

In the thought experiment, one experimenter – Wigner – measures the polarization of a photon. Before measurement it can be both vertically and horizontally polarized. After measurement it can have either vertical or horizontal polarization but not both. Wigner does not know whether his friend in a distant lab has made a measurement on the photon. It is possible therefore that the photon could be in a pre-measurement state so that for his friend both polarizations are superimposed. The state of the photon would be observer-dependent: pre-measurement, it has superimposed polarizations; and post-measurement, it is either horizontally or vertically polarized. There is no objective fact of the matter. This conclusion was confirmed experimentally by Massimiliano Proietti and colleagues.[109]

There are in fact several different interpretations of the findings but the question for us is whether, on any interpretation, it could justify the conclusion that there is no such thing as external reality and that quantum theory justifies an idealistic account of what-is. Surely not. After all, at the microscopic scale the experiment presupposes that there are entities such as photons, with modes of polarization, and that there is such a thing as a re-identifiable, self-individuating photon. The experiment design refers to individual photons. More obviously to the point, it is assumed that there is a real world that includes a laboratory where the experiment took place, and that that experiment (really) took place at a particular time, and that it was carried out by the individuals named in the paper who are parts of a community of experts. In short, any uncertainty as to the objectivity or otherwise of reality, had to be entertained and tested in a large chunk of reality situated in a vast hinterland of further reality, leading ultimately to a real scientific paper printed in a real journal and read by real people, including myself, situated in a real world.

A more persuasive interpretation of the experiment is not that there is no such thing as objective reality (which is necessary, after all, for the experiment to take place) but that quantum theory has lost touch with reality or (more generously) that quantum mechanics is the wrong instrument to reveal it to us.

The tendency to exaggerate the actual-world implications of the extraordinary observations in the fundamental physics laboratory (other than their spectacular practical applications) may account for the lack of interest that many serious players in the discipline of quantum mechanics have as to how it might be interpreted. This highlights the dissociation between the efficacy of quantum mechanics and what it has to say about the world at the deepest level. Paul Dirac, for example, whose theories united quantum mechanics and special relativity and predicted the existence of the positron that was discovered shortly afterward, was indifferent to any non-mathematical interpretation of his theory: interpretation "seemed to be a pointless preoccupation that led to no new equations".[110] Moreover, an absence of consensus as to the status of the wave function at the centre of quantum mechanics has not impeded the latter's progress and its practical application. Given this fundamental lack of clarity as to the physical status of the entities that populate the millions of pages of discussion about quantum mechanics, their metaphysical implications will be even less clear – if that were possible.

By the time scientists are postulating virtual particles or completing the Standard Model of the atom with the confirmation of the existence of the Higgs boson, the science has moved a long way away from anything that might, even in principle, cast light on the fundamental nature of the reality in which we live and have our being. It is difficult to feel that transient, unobservable-in-principle *virtual* particles that play such a part not only in the Standard Model of the atom but also in the physicist's version of the creation story are in any sense "real".[111] The fact that they are also invoked to sew together a story of the emergence of the universe out of nothing, given that something-out-of-nothing breaks all the conservation laws that are (or were) central to physics, it is difficult to think of them as anything other than a way of balancing the books;[112] of the creation story as creative accounting, as we alluded to in the Preface. The claim that "the vacuum contains a ferment of transient virtual particles of matter and anti-matter fluttering in and out of existence"[113] sounds as weird as any theological claim about the Creation. The uncertainty principle, we are told, allows brief violations of energy conservation in spacetime. This seems a daunting job description for something as ontologically dubious as probabilities.

One could be forgiven for thinking that every time mathematical physics runs into trouble – when, for example, infinities rear their ugly heads – physicists invent a new particle, a new stuff (dark matter, dark energy), or a new mode of being of existing entities.

The "no miracles" argument of Hilary Putnam in favour of a realistic rather than a purely instrumentalist interpretation of science now seems less decisive in supporting the claim of science not only to be true of certain aspects of the material world but *the* truth of the intrinsic nature of things. It is possible to embrace an instrumentalist vision of science rather than a literal-minded realist one without having to account for "a miracle" – that science works. The realm in which science works is not that in which fundamental reality is revealed.[114]

A mystery remains, however. If, as we have discussed, measurement leaves so much out – as it most certainly does – and there is an ever-widening distance between the realm of measurement and that of the lived experiences and meanings of the everyday world where we benefit from science, how is mad, mad science of such practical use in our common-sense world? There must be something fundamental connecting the two realms: fundamental science, however strange, must have got a lot right – more than at any time in the history of human consciousness – about the realm in which we live and have our being.

Now, there's a job for philosophy. And this thought seems to command wide agreement between physicists who want to look beyond the efficacy of equations and philosophers who don't feel obliged to grovel before science.

4.10 FROM QUANTUM IDEALISM TO MATHEMATICAL IDEALISM

This brings us to the heart of our inquiry into measurement and the consequences of the belief embraced by some thinkers – notwithstanding all that we have discussed – that what is revealed by measurement is ultimate reality; that what-is is how much; that everything boils down to quantity. If this were true, so the argument or the mind-slither continues, given that measurements are quantitative, it follows that the order of things is the order of the magnitude of things, with the things being merely derivative of magnitudes, perhaps shorn even of units. In short, given that the best portrait of what-is is mathematical, what-is is itself mathematical.

There is a short answer to this. It is possible that science enhances our agency only because what it gets right about the physical world – quantities and their relationships – is sufficient to enable us to manipulate it more effectively because, in turn, the goals of our agency are defined quantitatively. We need to know only that an object is x feet wide, and an aperture is 2x feet wide, to be confident that we can pass the former through the latter without knowing anything else about the object or the aperture – that, for example, the object is a table, and the aperture is a door.

Mathematized reality, however, seems too ethereal: the (abstracted) *size* of a table does not occupy space in the way that a table does. Even less does it support cups and saucers. It lacks particularity: 2×2 has no features. And no amount of mathematics can capture all those many properties that are dismissed as secondary qualities, though they are essential to, even constitute, the presence of what-is. In short, mathematical idealism cannot account for the distinctive features of anything that is: for the entities that physicists deal with in pursuing their science, and for all the modes of presence which fill their lives and on which they depend. Even a classical equation such as $E = mc^2$ would be unable to accommodate the *presence* of anything to anything else. Nothing in a purely mathematical universe could bump into, interact with, anything else. Particularities are subsumed in featureless generalities. They lose their presence.

The power of mathematical accounts of reality does not, therefore, justify embracing mathematical realism; even less, an industrial-strength Pythagoreanism according to which what-is is composed of mathematical forms.[115] For mathematical physics to be as powerful as it is, it is necessary only for it to get the maths right and it is possible to get the maths right without getting much else right. This does not mean getting other things *wrong* but bypassing them. We need to distinguish the claim that reality (seen in a certain way, through a certain lens, at a certain depth) has a mathematical structure from the claim that it *is* a mathematical structure. Even in $E = mc^2$ some non-mathematical heterogeneity of the world is retained. E, m, and c are not purely mathematical entities. They say much more than $x = yz^2$ where x, y, and z are just numerical variables.

Given the pedigree of the minds that have embraced mathematical idealism, it deserves more detailed engagement. The case for the superiority of a mathematical account of what-is is discussed by Mauro Dorato: it is founded in the fact that the laws of nature are overwhelmingly mathematical.[116] They are typically expressed in equations – for example $P/V = k$; $E = mc^2$ – flagging up a mathematical equivalence. The case for inferring that nature is mathematical from the fact that its most general laws are mathematical is, however, less compelling than some seem to think.

First, it is not the laws of *nature* but the laws of *science* that are mathematical. Given that science is, and will probably remain, a work in progress, there will always be a distance between what are referred to as the laws of nature but are, strictly, the most fundamental and universal habits of nature, and the laws of science. The latter are teased out of nature under very special conditions, and they can then be employed in an unnatural way to guide behaviour directed towards specific ends. I know that, if I want precisely to double the pressure of a body of gas, I need to halve the volume into which it is squeezed – so long as other variables such as temperature are kept constant. As we noted earlier, gases in nature do not, however, have separate independent variables (in this case, volume), dependent variables (in this case, pressure) and uncontrolled variables that need to be controlled to get the desired result (for example, temperature). The laws of science, that is to say, are artefacts extracted from the uniformity of nature.[117]

Secondly, those relationships that are flagged up as "the laws of nature" *must* be mathematical to qualify as laws of science. There are many aspects of scientific inquiry that foreordain that its laws should be mathematical, most importantly that it is based on quantitative measurement. Science sets aside that which is not reducible to quantities, thence to patterns of quantities, and ultimately to relationships between patterns of quantities; in short it excludes that which does not fit with increasingly simple mathematical patterns.[118] If there were not a predictable mathematical relationship between the pressure and the volume of a gas, Boyle would not have had a law named after him. Scientists would have looked elsewhere for laws.

Connected with these two points is something mentioned earlier in the chapter: much of what is happening in the observed world is classified in experimental science as "noise" to be controlled (or eliminated by averaging and statistical analysis) in order to extract what count as signals; filtered out as the *ceteris* that is excluded as an uncontrolled source of variation.

It is this that secures the path to the necessary homogenization that makes what happens "the same" such that what is observed in a particular experiment can be seen as representative of what is happening elsewhere in the universe.

This privileging of that which is constant and the reduction of what-is to "how much" is also secured by the use of units that are underwritten by a principle of sameness of stuff, enabling us to get our heads round what is out there. When, for example, I say that pebbles, potatoes, and people weigh so many kilograms, I am implicitly asserting that the difference between these kinds of items does not go all the way down; that, at some fundamental level, they are all the same; they are different quantities of the same substance – something called "matter". The generalizations that this predicts underpin applications that "work" in everyday life. And as we have discussed, their practical use gives them a standing that is not entirely deserved: the generalizations are seen as ultimate truths about the intrinsic properties of the natural world that are otherwise hidden from the consciousness of those that live in it.

We could approach the question of the mathematical nature of the laws of science from a slightly different direction. Laws are, by definition, general; and the more general they are, the more respected they are. The route to generality is via the filtering out of particularity and, as we have seen, the most powerful filter of particularity is the reduction of whatever is being observed to a quantity. It is this that rounds up pebbles, potatoes, and human bodies, into a single pen – that of quantities of matter. "All these different things boil down to similar things, forces, fields, etc." is the guiding idea of physical science so that the gain in the generality of its laws will be an acceptable price for losing differences, for the de-differentiation of reality. Indeed, this loss of difference is presented as a gain; that generalization, as a result of defocusing on, indeed filtering out, particularity, enables the gaze to penetrate closer to fundamental reality, where this is understood as what everything has in common. Treating pebbles, potatoes, and human bodies as being composed of "matter" enables all sorts of useful predictions to be made about them on the basis of laws of motion.

One of the most spectacular examples of loss understood as a gain is what Einstein described as his "happiest thought".[119] In a thought experiment, the equivalence of gravitational mass (weight) and inertial mass (resistance to change of velocity) was revealed through imagining circumstances in which they would be indistinguishable. Einstein envisaged the occupants of an elevator in empty outer space deprived of all clues as to the nature of their movement with respect to any other object. Under such circumstances, they would be unable to distinguish between entering a gravitational field or undergoing accelerated motion, as both would (for example) result in increased pressure under their feet. Thus, the principle of equivalence of mass and weight.

The thought experiment licensed the transition from special relativity – where accelerated motion was distinguished from a state of rest or uniform motion in a straight line – to general relativity whose equations also encompassed accelerated motion or movement along a curved path. While this brought an immense gain in generality, there was loss in particularity. That inertial and gravitational mass are identical as expressions of the curvature of space leaves unexplained the different ways they are expressed in the world in which we pass our lives.

We can see the wisdom behind Bertrand Russell's response to the question as to why the laws of nature are mathematical. "Physics is mathematical not because we know so much about the world but because we know so little. It is only its mathematical properties that we can discover."[120] It is echoed by quantum physicist David Bohm's observation that "Quantum mechanics says that nature is unintelligible except as a calculus, that all you can do is compute with the equations and operate your apparatus and compare".[121] By "nature" he means nature seen as a whole. There are, after all, other, local modes of intelligibility, evident when we make ordinary, moment by moment, sense of the world around us. All of these disappear as nature is gathered up into quantities and quantities are gathered up into patterns of quantities and those patterns are connected in a mathematical structure.

The royal road to generalization – and hence to general laws – is through draining the infinite variety out of what-is so that it is amenable to being reduced to (almost) pure numbers. We should adjust what Russell said: Physics is mathematical not because we know so much but because we need to set aside so much of what we know, in particular experience, to bring the world to an order we can get our head round in order to amplify our capacity to manipulate it.

Surely, it may be argued, even if we were to accept that increased generality, with a consequent homogenization of the totality of things, brought science ever closer to the hidden reality of nature, this would not result in pure mathematization of our portrait of what is. Measurement does not result in the complete loss of differentiated actuality and arrive at pure quantity – "2" or "1,000,000" – or mathematical structures created out of pure quantities. The answer to "How much?" is (for example), "2 feet" or "1,000,000 kilograms", not a naked number. Any measurement that is – as all measurements must be – expressed in units is mathematically impure. The extraction of parameters may homogenize things that we see as different – pebbles, potatoes, and human bodies all have a weight expressible in grams – but these items also host other parameters. The material of which they are made remains as the keeper of differences: hard and inedible in the case of a pebble, soft and edible in that of a potato, and conscious in the case of a waking human body.

Against this, we can point out that some of the impurity imported by units will be to some extent cleansed by the fact that, as we discussed earlier, units are themselves mathematical ratios – so that "1,000,000 kilograms" will be a number: the multiple of the mass of a, far off, standardized entity, the standard kilogram. Moreover, as we noted, this has recently been replaced by something even further off from any material object with specific properties – namely 6.626×10^{-34} joule second, a designation that conjures up no image. The passage from what-is, to the quantification of what-is, and thence to what-is portrayed as generalized quantities, is further propelled by the relationship between quantitative laws either at the same level of generality or at an ever higher level of generality until the totality of what-is is defined as a mathematical reality.

Among the many reasons for resisting the conviction that what-is proves ultimately to evaporate to pure quantity, to mere number or mathematical structure, is that it overlooks the difference between counting and measuring. The equations of chemistry and physics are not essentially the same as those of arithmetic and algebra. $2H_2 + 0_2 = 2H_20$ is fundamentally different from $2 + 2 = 4$; and $E = mc^2$ is fundamentally different from $3x - 15 = 9$, where "x" is 8. The equations of physics and chemistry, what is more, cannot be proved mathematically: they require (often messy) observation. $E = mc^2$ is a discovery, a climax at the end of a long chain of discoveries. By contrast, the equations of arithmetic and algebra cannot be upheld (or indeed disproved) by experiment.

Even the barest equation in natural science is more than mathematical equivalence: it has an irreducible linguistic element. Without this – without say the difference between "matter" and "energy" or between "matter", "energy" and "the speed of light" – Einstein's portentous equation would be empty of content. While measurement paves the road from the natural world to mathematics, where there is content, as in an equation, there is more than mathematics – though this content is reduced as we approach the most fundamental level: content is traded off for generality.

(Two-sided) equations strip measurements even of the non-mathematical contaminant of units. Consider Boyle's law: $PxV = k$. As pressure (P) applied to a fixed quantity of gas rises, its volume (V) falls. The *constant* which captures this relationship is unitless: it is a pure number. It is not a number of anything. By the time we have reached $E = mc^2$, there has been an homogenization of all modes of energy to "E" and of all things that have mass to "m". Light, stripped of its luminosity is reduced to its velocity – a mere number, which is amenable to the indignity of being squared, something that could not happen to light itself. The what-is of light is present simply as the "how much" of one of its parameters – velocity – and that how much is further pushed to being a pure quantity when it is multiplied by itself. It is colourless, lacking luminosity, sunshine, or moonshine, and the velocity to which it is reduced lacks movement or whizz. In short, it imports only a sterile number nailed to its emptiness by being multiplied by itself. An equation which equates energy and mass in this way – everything that happens and the substrate in which it happens – reduces both to numbers. In being thus mutually translated, they shed even the units that made them quantifiable. The equals sign erases the differences either side of it. There is blankness.

The emptying of content is not only homogenizing but it prepares entities to be dissolved into a structure of relations without *relata*. All measurements, after all, are relational and all portraits of what-is executed in measurement paint will be relational.

Of course, it is difficult to conceive of definite, distinct relationships without *relata* to instantiate them. If there are only relations, it is not possible to justify locating an object "over there" as opposed to "over here", something that can engage in some relations and not others, and be present rather than absent to a conscious subject located in a particular place. It is difficult to think even of fields – their location, intensity, and changes over space and time – without objects located in space and time to situate them. A purely mathematical world will be without locations, and, in the absence of location, it is difficult to imagine entities being individuated – a conclusion that many mathematical idealists are willing to accept precisely because the removal of concrete, individual entities as *relata* reduces the physical world to a mathematical structure.[122]

This notwithstanding, the notion that reality is fundamentally mathematical, and that quantification-through-measurement is the royal road to revealing the ultimate nature of what-is, dies hard. Hence the willingness among some quantum physicists to accept that microphysical entities – presumably the fundamental components of stuff – do not have definite properties until they are measured or the more radical idea that they do not exist except when they are observed, even though, as we have seen, this is beset with problems. Its attraction may lie in its combining a tough-minded commitment to measurement as the path to the fundamental truth of things and the tender-minded suspicion that things may be closer to ideas than we customarily believe. The path from hard-nosed materialism to tender-minded idealism seems shorter than we might think.

At any rate, there is no place in the mathematized reality portrayed in fundamental physics for the concrete, discrete, solid objects that populate everyday life. At best there are structures. The case against this mathematical structuralism – and more broadly against a relationism that denies *relata* or concrete objects – is lucidly expressed by Sebastian Briceno and Stephen Mumford:

> Simplification and systematization aid our understanding of the world. But there are also associated risks. We should not of course then assume that the abstracted aspects of the world are the only ones that are real. It would be foolish to assume that the aspects we ignore, for the purposes of simple theory construction, thereby do not exist. Reality is all-embracing. We seek general theories; but actual concrete reality is always the point of departure.[123]

Even if mathematics really were the most faithful and complete *portrait* of what-is (which manifestly it is not) it would not follow that mathematics is the stuff of what-is. To assume that it is not only confuses the map with the territory but the map with the material of which the territory is made.[124] It is perhaps not surprising that if our scientific image of the world is an expression of our endeavour to get our mind around the totality of things, that the world according to science looks mind-like. Or, rather, looks like what the mind focuses on when it is trying to encompass the universe as a whole, bleached of individuating qualities.

If reality were in fact mathematical, it would not be clear what kind of mathematics would make it up. Plato, we are told, was particularly impressed by geometry – to the point where he denied entry to his Academy to those who were unfamiliar with this discipline – for, as he said in the *Republic*, geometry was "the knowledge of the eternally existent".[125] As Wittgenstein pointed out, however, mathematics is "a MEDLEY"[126] (of all sorts of methods, arguments, entities), and it is not possible to place them in a definitive hierarchy where one aspect of mathematics is more fundamental than another. Even less is there a point of convergence where mathematics can render itself coherent and transparent. (I shall resist the temptation to harvest the low-hanging fruit of Godel's theorem demonstrating that there is no such point of convergence.)[127] Even the mathematics of physics – that has for some justified the notion that what-is is composed of mathematical structures – is irreducibly a medley, where arithmetic, algebra, geometry, topology, calculus, tensors, Hamiltonians, Lagrangians, vectors in Hilbert space jostle for position with any number of other mathematical methods, equations, entities that most of us have not heard of, never mind understood.

If the universe were ultimately composed of mathematical entities, it is difficult to imagine how they could give birth to new entities as in the endless multiplication of axioms, disciplines, and methods that is the history of the subject; or how mathematical types can generate mathematical tokens – thought, spoken, written, and discussed by individual conscious subjects at particular times in particular places; or how a mathematical universe can accommodate living, breathing mathematicians or indeed the local expressions that is the mathematics as we know it – as an object of study, as something that is used. It is even more difficult to imagine how mathematical entities could give birth to the rocks, trees, beasts, and human activities that surround us. What would propel the genesis of seemingly non-mathematical things? More mathematics? Answering that question is not made any easier by the suggestion that rocks, trees, beasts, and human beings, the realms of seemingly non-mathematical stuff as we know it, are illusions. What mathematical processes would create the illusions of non-mathematical processes and the apparently non-mathematical entities that entertain them?

The (exaggerated) respect for mathematics, and the tenacity of the belief that reality is mathematical, are exposed when the mathematics of fundamental physics runs into difficulty – when there are internal problems with infinities or incompatibilities between different theories. New particles, fields, or forces are mobilized to save the mathematics. The endeavour to perfect the mathematical portrait of the universe seems to grant physicists the license to print particles or dimensions or to invoke "effective field theory" frameworks. Hossenfelder has argued that the appeal to "naturalness", supersymmetry, and the multiverse are wrong turns motivated by the conviction that mathematical beauty is the sign of truth.[128]

4.11 CONCLUSION

Natural sciences (and most impressively fundamental physics) are successful because they reduce what-is to how-much and how-much is expressed in quantities of units. While we should be grateful for the way this has transformed our lives, our gratitude does not oblige us to believe that physico-mathematical science gives us privileged access to, or is the last word on, the metaphysics of reality, on the fundamental nature of what-is. While getting things mathematically right explains the predictive power of science and the empowerment we have enjoyed through its applications in novel technology, this does not license the conclusion that the essence of what-is is mathematical, that the world is a mathematical entity, or even that everything boils down to quantities. We should not confuse epistemology with ontology, particularly when quantum epistemology so manifestly excludes much of what makes up what-is. The fact physics has empowered us does not justify our conflating its epistemic truths with an ontology of what-is. We do not have to know the intrinsic nature of things to be able to manipulate them better. Yes, physics has given us recipes enabling us to act on the world in ways that were unimaginable to previous generations; but the recipe is not the meal and, what is more, the purely mathematical deliverances of fundamental physics do not even add up to recipes.

If, in accordance with Lord Kelvin, we see measurement as the beginning, middle, and end of knowledge worth having, and if furthermore, progress is to be judged by the extent to which qualities are reduced to, or displaced by, quantities, it is hardly surprising that science pushes towards the view that what-is is essentially mathematical. We should push back against this for two connected reasons.

First, the world-as-maths is homogeneous and (as the other aspect of this), empty of individuated content. While there is some truth in what Lord Kelvin said – that knowledge without measurement may be in some sense "of a meagre and unsatisfactory kind" – it is equally true that knowledge confined to measurement is also meagre. I can know a lot about an object such as Eddington's table in virtue of my innumerate interactions with it – by bumping into it, carrying it from place to place, sitting at it. All of that is lost as Eddington's second table displaces the first. While nothing in mathematical physics explains why this table is a lovely walnut-brown, this does not mean that it is not walnut brown. Secondly, the purely quantitative world of fundamental physics void of qualities is remote from the macroscopic world in which physicists do their physics and live the lives necessary for them to be able to be physicists.

The claim, underpinning mathematical idealism, that there is only quantity ultimately empties science of content. It might be argued that quantities are quantities *of* but what they are of is marginalized. When it is asserted that $E = mc^2$, the very equivalence between what is on either side of the equation minimizes the difference between them. E and m are at the fundamental level the same: what distinguishes them in a purely quantitative world is purely quantitative – which is why the equation could after all have read $M = e/c^2$.

The long and winding path to the erasure of the richness of the world in which we live begins with privileging what things have in common over their differences. Vastly different phenomena are gathered up into, say, the effects of magnetic force; magnetism is fused with electricity into electromagnetism; the conception of electromagnetic forces widens to include light; and so on until we reach the notion of undifferentiated energy. A similar path leads to "mass", as something common to the space-occupying entities in the universe. And finally mass and energy merge in mass-energy. We arrive at an account of what-is where that which is common to all things, reduced to pure quantities, is privileged and all distinctive features of actual things are brushed out of sight.

In short, the scientific project of burning off innumerate experience, in order to part the veil of appearance, results in – well, *dis*appearance. This may seem to be exemplified in Bohr's world picture in which the fundamental elements of what-is signified by "words like electron, photon, or atom should be regarded ... as useful models that consolidate what is actually only a set of mathematical relations connecting observations".[129] "Connecting observations"! There is the tripwire. Appearance is back with a vengeance.

The tension between the appearance necessary for measurement and the disappearance that is the final outcome, or the asymptote, of quantitative science, is noted by Michel Bitbol for whom the view from nowhere and the view from somewhere (which is the view from someone) are both aspects of the phenomenological realm of appearing. Paraphrasing John von Neumann, he argues that (for example) the switch from the superposition of the wave function to the definite, sharp eigenstate of observation is not some physical process such as a "collapse" triggered by consciousness but a switch from a neutral to a situated mode of description: "A quantum entangled superposition (involving the system and anything correlated with it) holds for anyone who would like to anticipate probabilistically a measurement outcome, whereas a sharp state holds for someone who has observed this outcome and wants to take it into account for anticipating the outcomes for future measurement".[130] Both modes of knowledge, Bitbol emphasizes, "hold for the same concrete situated person, but with two different stances: the stance of the anonymous predictor, and the stance of an individual observer becoming aware of the outcome of a particular measurement".[131]

It is difficult to resist the conclusion that the measurement problem of quantum mechanics is an expression of the problem of measurement, and a comeuppance for the apotheosis of measurement, expressed in Lord Kelvin's claim that without measurement, "knowledge is of a meagre and unsatisfactory kind". It is ironic that measurement that aims, and indeed claims, to uncover a reality independent of the mind – certainly as it is manifested to individual conscious subjects with their prejudices, preoccupations, and the vagaries of their intention – should have generated a science that has culminated in quantum mechanics that, according to many interpretations, makes even quantities mind-dependent. That there is a collapse of the distance between what-is and that-it-is; that the erasing of explicitness results in everything evaporating into explicitness; matter evaporating to thatter.

Why, however, should this comeuppance come up so late, so recently, in the history of physics and at the sub-atomic level? And why should the degree of uncertainty be so precise? It is because what is evident at that level is entirely mediated by measurement: it is largely a theoretical construct woven out of patterns of quantities, out of mathematical abstractions. We are remote from realms that can be accessed by experience as ordinarily understood.

When we appreciate how strange an action measurement is and what a peculiar stance to the world it expresses, we find it easier to not take it so seriously. Measurements are remote from the perceptions, qualia, the sense data from which they take their rise. Even when we are noting a result – for example a position on a clock face, a dot on a screen, a pointer position - much is excluded from the experience of doing so. Wilfrid Sellars' claim that "Science is the measure of all things, of what is that it is, of what is not that it is not"[132] looks less compelling. Whatever science is, it is *not* the measure of *all* things. Yes, "Measurement began our might" but it is an instrument – the instrument of all instruments, as the hand is the tool of tools – not the ultimate guide to what-is.

And what-is does not boil down to mathematical entities, if only because measurement presupposes a distinction between that which is measured, the measuring device, the conscious subject who performs the measurement (and the epistemic community to which she belongs), and the actions of that subject in performing it.

This is admitted at times even by the most radical Pythagoreans. Max Tegmark, for example, argues that whatever exists mathematically may *also* exist physically, as if physical existence were an add-on.[133] And a rather peculiar add-on. Tegmark believes that some of those mathematical structures that exist not only in a mathematical sense, but in a physical sense as well, include self-aware structures (such as human beings) that might inhabit some of these mathematical structures.[134]

If the asymptote of theory is the widest, deepest, least bounded account of what-is, it will also be the endpoint of a process of homogenization that, by claiming similarity under all differences, will eradicate (necessarily specific) content. As such, it will not allow actualization – understood as a descent from the abstract, general form of what-is to anything, any particular, that is. The outstanding task that remains when a Theory of Everything is arrived at will not be just to discover, as Stephen Hawking famously expressed it, "what will breathe fire into the equations",[135] but what will restore stuff to a universe that has been reduced to equations that have no substrate. As has often been said, a final theory may tell us nothing about the actual world; which is not in the slightest bit surprising. A Theory of Everything – of what everything has in common – will be a Theory of Nothing in Particular – which may not be readily distinguishable from a Theory of Nothing.

We can unpack this claim in two ways. First, looking inevitably involves some *over*looking. The wider, the deeper, the more universal, the gaze, the more overlooking is built into looking. Secondly, the path to generalization involves tidying up the field of inquiry, of separating signals from noise.

Eventually, when universal homogeneity is sought, all difference, all particularity, is set aside. Behind both of these features characterizing the pursuit of completed generality, in which what-is in all its manifestations is reduced to a few simple relationships, is overlooking the looking, the creature who looks; overlooking that in virtue of which (or whom) what-is is made explicit. While the great equations such as $E = mc^2$ or the Schrödinger wave equation are remote from any imaginable, that is to say particular, experience they are built on many layers of explicitness.

Pace Wilfred Sellars we cited earlier, "In the dimension of describing and explaining the world, science is the measure of all things, of what is that it is, of what is not that it is not", it tells us nothing of the transition from what-is to that-it-is, even less about the transition to what-is-not and the judgement that-it-is-not. Measurement arises out of the realm of that which is measured and seemingly reduced to what measurement reveals. Einstein's acknowledgement (referred to earlier) that the world picture of the physicist cannot accommodate clocks and measuring rods – material objects transformed by conscious, informed agents into scientific instruments – highlights what necessarily lies outside of the physical world to make physics possible. The physicist's time as "little t" cannot accommodate the act of timing – of making time explicit.

For all of these reasons, we should reject the notion that the last word on what-is is a number or a numerical structure. Such a portrait of what-is – and the claim that it is a sketch of a Theory of Everything – is even less able to accommodate the supreme expression of explicitness that is science itself.

It seems appropriate to end this chapter – devoted to challenging the belief that physics has superseded philosophy as the guide to metaphysics – with an observation from a volume devoted to one of the most spectacular achievements in recent fundamental physics: the confirmation of the existence of the Higgs boson. Frank Close, himself an eminent physicist, reflects that while this discovery completed the Standard Model of the atom, "internal completeness is a mathematical requirement, whereas describing the world around us is the demand of natural philosophy".[136] Philosophy is not dead yet; nor should it declare itself redundant. This has been well expressed by William Seager, "[T]he existence of the large number of interpretations of quantum mechanics (Wikipedia currently lists some eighteen) testifies to the desire for some way to make quantum mechanics intelligible. Obviously, mundane intelligibility is not going to be forthcoming so it looks like metaphysical intelligibility will have to do the job".[137] Metaphysical intelligibility will not, however, be delivered by a world picture woven out of measurement. It will not therefore begin from, or be constrained by, or be subservient to, fundamental physics, wonderful though it is.

And it is indeed wonderful. It is difficult not to be awestruck by the detection of the Higgs boson, given that its presence is signalled by the difference between two numbers that, written in decimal notation, are the same in the first 14 units, differing only in the fifteenth.[138] Such achievements and countless others make frequently repeated claim that "physics is in crisis" implausible – and would be so even if modern lives were not immersed in a landscape of artefacts made possible by quantum mechanics.[139]

Yes, there are problems. Only 5 per cent of matter – once seen as the basic stuff of the universe – is covered by the Standard Model. Twenty-five per cent of matter is dark matter and nearly 70 per cent of energy is dark energy. The nature of dark matter is, as it says on the tin, dark. Beyond this there is no consensus as to whether it is a new kind of stuff or merely a manifestation of variations in the laws of gravity – the so-called modified Newtonian dynamics or MOND explanation that proposes a modification of Newton's second law of motion to account for the observed behaviour of galaxies. In addition, there is persistent uncertainty as to whether the universe is or is not expanding. Discrete objects, space, and time are difficult to account for in an entangled quantum universe. Infinities continue to haunt the sub-atomic world, and, given the regulation of nature by constants fine-tuned to endless decimal points, nature seems, well, unnatural. We need to tune dark energy to 123 decimal places to make habitable galaxies.[140] The appeal to the multiverse to account for this is always in danger of overdelivering universes. The suggestion that they might be infinite in number breaks the rule that actualities must always fall short of an infinity: it is always possible to add to any number, however large, and still not arrive at infinity. And then there is the small question of the unintelligibility of quantum mechanics: "Quantum mechanics says that nature is unintelligible except as a calculus, that all you can do is compute with the equations and operate with your apparatus and compare the results".[141]

Judging by its spectacular practical and theoretical achievements, however, other sciences might wish for such a crisis. No, physics is not in crisis. But physics-as-metaphysics, as the last word on what-is, most certainly is.

PART II

First-person explicitness: circling round the self

CHAPTER 5
Ambodiment: the marriage of (that) it is and (that) I am

5.1 MY BODY AND I: THE NECESSITY FOR EMBODIMENT

The more closely we inspect brains, or ionic currents in nervous tissue, the less we understand what role, if any, they play in the generation of consciousness. They fail to account for the distinctive nature of conscious subjects – entities to whom a subset of what-is is *present* to them as a world or as a situation that is *their* situation, a reality they must engage and cope with, exploit, enjoy, or suffer. In short, nothing in the brain explains how certain portions of matter come to matter to themselves and, as a consequence, other portions of matter come to matter to them.

The body beyond the brain seems even less likely to deliver this transition. While the extra-cerebral body is essential to ensure ongoing brain function, what happens in and under the skin, the function of organs such as the kidney, or of circulating fluids such as blood and lymph, acting either singly or in concert, does not seem to account for the transition from an "It" to an "I", from the realm of what-is to that of that-it-is and thence to "that-I-am". Indeed, those elements remain stubbornly "it".

Granted that some of them, by means that remain utterly mysterious, may move closer towards the status of "I" when they are sore or give pleasure. Nevertheless, they still fall short of fully-fledged first-person being. The toothache that demands our attention seems to be in a littoral zone between "it" and "I", between something that we have and something that has us; between an it-happening and an I-happening. The suffering we experience is not identical to the decaying flesh or neural activity triggered by decay that we believe explains it. Our body presents itself as the innermost layer of a sometimes hostile environment.

While the properties of the body inside and outside of the brain do not explain why or how that body is a person who matters to herself, subjects are inseparable from bodies. We call those bodies "their" bodies as if they were possessions but they are, of course, closer than possessions: we are incarnate, corporeal, *em*bodied – or, as I shall characterize it, *am*bodied.

The idea of the person as "embodied subjectivity" is particularly associated with Maurice Merleau-Ponty and it has proved an illuminating way of characterizing, if not understanding, human being.[1] I shall touch on this relatively briefly because it is a territory I have explored extensively elsewhere, where some of the key concepts in what follows are developed in more depth.[2]

Much of the recent emphasis on the body beyond the brain in the philosophy of the self has been presented as a corrective to approaches to consciousness or the mind that have centred on the brain-as-computer or as "information processor" and an excessively intellectualist approach to the active life of the human agent, who is, after all, caught up in, rather than being at a distance from, the world. Being-in-the-world, living a life, is not a spectator sport. This is also acknowledged by the so-called enactivist turn in the philosophy of mind, which brings cognition closer to the physical interaction between the active organism and its environment.

Notwithstanding the failure of even the most promising candidate organ – the brain – to deliver the beginnings of a satisfactory account of subjectivity, our subjectivity is clearly inseparable from our body. There are several reasons why it is appropriate to characterize people as embodied subjects. The death of the body seems to spell the end of the subject. Not everyone, of course, accepts this as an incontrovertible, indeed brutal, demonstration that we do not merely inhabit our bodies as spirits renting flesh for the duration of an earthly lifespan before moving on elsewhere. For many, discarnate existence after death is the central idea of their lives and a profound source of hope and consolation. All the evidence we have – the kind of evidence that we draw upon in daily life – seems, however, to suggest that posthumous life, existence as a disembodied spirit, is not available. It is not unreasonable, however, to keep an open mind on this matter and settle only for the belief that *pre-mortem* conscious subjects are inseparable from their bodies.

The resistance to identifying ourselves with our body is understandable – identifying the Bloke with the Beast or the Blob. My body is largely hidden from me as I gaze at myself – whether I am looking at my image in the mirror, examining my feet, or palpating my abdomen. There are many layers of opacity. The things I can see or otherwise sense as happening in my body are vastly outnumbered by the events that happen – and must happen – in order that the body can continue to live and to be the primary agent of my agency. If my breathing – increasing and decreasing the size of the greatest hollow in my body – is visible in the mirror then it is only as a slight, regular shrugging of my shoulders. The sensations I have in my chest and the sound and feeling of the air passing through my airways tell me little about the vast sponges of lung tissue that are expanding and contracting and even less of the transactions involving oxygen and carbon dioxide. Of my beating heart, I can feel only muffled footfalls – directly in my chest and indirectly as pulsing in my neck and my feet. I can palpate it in my wrist, where it sullenly declares itself as fast or slow, regular or irregular. I cannot sense the fact that my heart has four chambers – any more than I could experience the fact that I have two kidneys or a spleen not far from my liver. The circulatory system that unifies my various pulses is largely hidden from me. The pilgrimage of faeces down my colon, of urine passing from kidneys to bladder, the secretion of hormones that will have such exquisite effects on my well-being, and the nano-pipetting of sweat before it gathers into experienced damp, drops, or trickles, are mainly hidden, not only from my gaze but from my other senses. In sum, my body lies largely beyond my hearing, touch, sense of smell and taste, and even from the interceptors and proprioceptors that register the state and posture of my body. My carnal being presents itself to me only on a need-to-know basis, and then patchily.

There is also the question of scale. My experience of my body is largely macroscopic – certainly well above the cellular and molecular level. At the microscopic level, my body is unhaunted. The privileging of certain scales is, as we discussed in Chapter 2, not something that we see in the physical world beyond sentient organisms. My "amming" of my body takes place at a human scale. There is a kind of existential tautology at work here: I am at the level at which I *am* that which I am.

The lack of sensory transparency is compounded by epistemic opacity. The cardiovascular and respiratory system deliver many things, most importantly oxygen acquired from the surrounding air, and this is delivered to 37 trillion cells of 200 different kinds, few of which most body owners could name. Even less could they describe what those cells get to do, to ensure their own growth, replication, and survival, and to contribute to the work of the organs and systems that ensure the mirror-gazer's moment-to-moment, day-to-day, and year-to-year continuation.

My gaze, which anyway usually stops at my skin, could not directly uncover the facts biological science has accumulated about my body – understood as an instance of *any* human body. One of the most poignant marks of its beginning as anyone's body is the navel that is a relic of the strange time when, lacking subjectivity, we grew to the possibility of ourselves, the self-building flesh making possible the particular life and the basis of the particular person we would become.[3] The navel is a poignant relic of a time when we lacked any kind of knowledge of the body we were growing into or, indeed, of the world into which we were heading.

In sum, for most of the time, there is no "what it is like to be" most of our body – either its parts or its totality. There is no what-it-is-like to be a nephron, a kidney, or an entire urogenital tract. Even when hidden parts of our body remind us of their existence through sensations – such as an itch in a foot, the feeling of warmth in the sunlight, or something as intimate as tiredness or nausea – they reveal little of the structures that are affected, little of what is going on. Looking at one's eye in the mirror betrays nothing of the processes in virtue of which there is seeing or, indeed, of the eye it is seeing. Even less, is there a "what it is like" to be a potassium level in the blood of 4 millimoles per litre (normal adult level). While a low potassium may be associated with fatigue, weakness, and muscle cramps, this is not what it is like to be a low potassium but for a conscious subject to *have* a low potassium. And there is no time at which there is a what-it-is-like to be our body as a whole

Our ignorance of an entity which, of all the objects in the world, is the one most identified with our self – such that when we point to our bodies we point to ourselves and where our bodies are is where we are – is a reminder of the vast distances between what it is like to be Raymond Tallis – which includes moment-to-moment experience of his body – and what his body gets up to. There is a wide gap between the story of my life and the happenings in my body; between my C and its V. This distance may be dramatized in a video of me dreamlessly asleep, when for my body it is largely business as usual, while my life is on pause

The complex relationship of identity and non-identity between the "I" and the body is further complicated by the possibility of experiencing our bodies in the modes of disability and suffering, which may make bodily parts more obtrusive but no more transparent. Disability transforms parts of our bodies into objects that have to be struggled with. Lifting my weak legs into bed presents them as obstacles rather than primary agents of my agency. In walking with effort, we encounter our lower limbs both as agents of our agency and as barriers to that agency; as explicit tools with which we "do" walking as opposed to merely walking.

As for suffering, its primary manifestation, pain, is alien, howsoever wearisomely familiar it may become. It arises from the hidden interior of the body that is itself part of a strangely intimate outside, a reminder of how the stuff of which we are made is other than the person we are. The painful (or nauseated) body is an intruder in our lives. Distant from and yet not separable from us, suffering is engulfing and inescapable. It is something I want to get rid of and yet also something I am sometimes forced to *be.* I put it "out there" as a happening other than me when I give it a name, seek a diagnosis, look for treatment; but it sticks to me, invades me, overwhelms me. As do hunger, thirst, and cold. And so, too, with lesser intensity, do tiredness, sleepiness, malaise, nausea, fullness, which have no clearcut or precise localization within the body.

Through these experiences I am forced to be my body – but in an incomplete sense. While the obtrusiveness of my body clouds my openness to the world I inhabit as a subject and is itself for the most part opaque sensorially and, more broadly, epistemically, I am still open to that world.

It is worth emphasizing that the opacity of the body is not reduced by biological and medical knowledge, most obviously because such knowledge is general. Learning how my kidney functions tells me little or nothing about myself, not the least because the kidney of the textbook is anyone's kidney and indeed is in important respects the kidney of any mammal. Although as a doctor I know that my serum potassium should fall within a certain range, I do not know precisely where it is falling within that range at any particular time. Moreover, knowing that my serum potassium is, say, 4.0 mmol/litre is not a form of self-knowledge. Like so much else, it belongs to the "it" of the body rather than the "me" of the subject; to an itself rather than a myself.

For this reason, most of what lies beneath our skin would not be recognizable as our own, even less as "me". Indeed, some of us will not be aware that we have ureters, spleens, and various other occupants of the subdermal zone. We would not be able to pick out our own brains – the neurophilosophers' seat of selfhood – from an identity parade of pictures of brains. Much of what is in us – nails, faeces, hair, saliva, blood – has only a distant relationship with ourselves. Nail-trimming, defaecation, haircuts, spitting, non-fatal blood loss, do not diminish the self.

While there seems to be a hierarchy whereby our faces – animated by a variety of conscious and less conscious expressions – may seem more first-person than our buttocks or our inner organs, even our facial muscles have an impersonal heart, something hidden from us. The 11 muscles it takes to make a smile or the 12 to make a frown seem, for the most part, to mobilize themselves. The intuition of externality (of outerness, of otherness, of not-me-ness) haunts all of our carnal being – such that what I am is also what I am condemned to be. The capacity of the I that chooses to do this rather than that does not extend to the body knitting itself together *in utero* and much of its subsequent biological development.

"Amming" something does not deliver "it-knowledge" of that something – not the least because such knowledge places events and objects under categories. And while "amming" delivers more to the ambodied subject of what it is, the "am" of the "I" has, as we have discussed, only limited penetration, and at a certain level, of the body from which the "I" is inseparable. The gap between objective knowledge of the body and bodily experience is in most cases no narrower for an individual whose body it is than it is for others. So much for panpsychists who, as discussed in Chapter 3, believe that being, or "amming", a brain gives us fundamental insight, generalizable to the whole of the material world, into its intrinsic properties.

The limitations of our third-person knowledge of our bodies is an aspect of the fact that our relationship with our bodies is not an external, contingent one: bodies do not, for example, merely *house* our subjectivities such that we are lodgers, temporary inhabitants. As Descartes said, "I am not merely present in my body as a sailor is present in a ship, but that I am very closely conjoined and, as it were, intermingled, with it, so that I and the body form a unit".[4] This fact is highlighted by the co-development of our bodies and our subjectivity, our being a viewpoint on a world, The process began with the moment when two cells fused to form a gamete. Without this co-development, there would be no reason for individual subjects as it were to "alight" on particular bodies at a particular time; nor, indeed, would there be any basis for the ability of a subject – conceived as neither located in space nor spatially extended – to accomplish such a landing.

The incomplete fusion or co-penetration of the am of the "I" and the "is" of the body in ambodiment is beset with complexities, some of which are reflected in the distance, we have touched on, between direct experience and mediated knowledge – feeling awful because my serum potassium is low and knowing that my serum potassium is low. I do not need to be told that I am in pain but I may need to be told what is causing it and certainly of the details of the processes that constitute it. Bodily experience is confined to privileged scales whereas knowledge of the body can be of the organism as a whole, or at a cellular, or indeed molecular, level.

The gap between the "am" of experience and the "is" that is the intentional object of knowledge raises questions as to how that gap is to be crossed. The two most pressing questions relate to the standing of is-objects, given that they have to be revealed by am-experiences. The question dealt with in the following sections concerns the construction of objects out of experience.

5.2 OUR MODE OF BEING IN SPACE

The primordial "here" of the conscious subject and that part of the material universe that is disclosed as her "there" whose boundaries define the border of "elsewhere" – the border zone between presence and absence – is centred on her body. While memory and thought – which we shall discuss in later chapters – permit the conscious subject a margin of phenomenological disembodiment, this does not alter the fact that the realm of our actual exposure to and engagement with what-is is fundamentally shaped by the location and state of our body, though the nature of that engagement is influenced by actual past exposure and envisaged future exposure of ourselves as individuals and as members of a community of minds. The here-and-now, which is the theatre of my agency, is ultimately defined by where my body is – what I am exposed to and what lies within my literal reach. However mediated my actions, I still have to press a literal button with my literal index finger.

That I am where and when my body is – notwithstanding woolgathering thoughts, temporally deep memories, and all-consuming preoccupations – is not altered by technology that permits us virtual occupancy of an e-elsewhere. Yes, I can talk to someone in Australia, and we can hear each other's voices, though my body is in England. And, yes, I can "see" what happened in the First World War courtesy of the television. But neither of these things would be possible did I not have a phone in my hand, or my body were not parked in front of a television screen. The boundless elsewhere opened up by speech, or mediated by page and screen, still has to be made available through a body that hears speech, looks at a page, or stares at a screen. Even the metaverse has to be accessed through goggles, helmets, etc., placed on my actual head. My body is the place where all the heres, theres, and elsewheres are harvested as immediate or mediated presences.

It would hardly seem worth spelling out these rather obvious truths were it not that the embodiment of the subject has been overlooked or at least marginalized in some traditions in the history of philosophy according to which objects are constructed out of percepts.

The most interesting – and certainly the most subtle, complex, baffling, and widely discussed – variant of this view of objects is Kant's transcendental idealism.[5] In *The Critique of Pure Reason*, Kant argues that it is courtesy of the synthetic activity of our minds that we gather up sense experiences into objects located in space and time. Things as they are in themselves – whatever this might mean – are not thus located, since space and time are modes of representation originating from the transcendental subject who imposes these forms on sense experience.

If this were true, it would be difficult to understand what it is that determines or constrains what we experience. The spatio-temporal location of the body and of other objects could play no part in determining the contents of the mind because the body and other items would have spatio-temporal location only courtesy of the mind. To put this another way, if my mind were not the subjective aspect of an embodied subject, itself an object localized in space and time, then neither it nor its (my) physical surroundings would have any say over what I am aware of. Without embodiment supplying a location to the subject, there would be no ground for some entities qualifying as being present and others as absent.

Without being embodied in an entity that came with its own location, I would have no physical surroundings defined by spatio-temporal relationships between extra-corporeal objects (including the bodies of others) and my (physical) body or, more specifically, the sense organs embedded in my body. Nor would there be a theatre of, or guide to, actions given that they are, at the most basic level, *inter*actions between the material of the body and an array of material objects. Nor, again, would there be grounds for the subject sensing that she is moving towards or away from an object; or is in a room in England as opposed to on a boat in the Atlantic Ocean.

It is, in short, not possible to see how a subject disconnected from a living body, or how a transcendental subject – whose activity is, according to Kant, required to synthesize perceptions into bodies located in space and time – could relate to a particular world as the theatre of its life, a life that is in part had in common with other embodied subjects. The very idea of something as simple as a handshake or an embrace would not be possible if the space in which subjects were co-present was separately constructed by the minds of each.[6] And more complex interactions such as those mediated by speech or writing would also be difficult to conceive. Shared space is a fundamental precondition and manifestation of an *inter*subjectivity which, ultimately, depends on intercorporeality. This is one of the reasons why I shall argue that it is not the transcendental subject but the embodied subject who makes objects.

Furthermore, there are the *intra*corporeal spatial distances – made available by the body's extension in space – necessary for most actions. A grip depends on the length of fingers, walking involves spatial separation of legs, depth perception requires the interval between the two eyes, vision the crossing of a spatial gap between the light bouncing off a seen object and the retina of the subject. The functions of the eye, the heart, the lungs, the kidneys, etc., depend on their being arrayed as a smorgasbord of spatially distinct structures.

The way we relate to the space that surrounds us and to the objects in that space, while depending on embodiment, is not, of course, a purely physical relationship between material objects and a material body. Yes, there is some similarity between how I am related to a cup next to me and how the cup is related to the table on which it stands: there is a physical interval common to both modes of "being next to". But that is only a small part of the story of the relationship between conscious subjects and the objects of which they are conscious.

The difference is highlighted by the distinction, emphasized by Merleau-Ponty,[7] between being (physically) located in space – as is the case with the cup – and *inhabiting* space as in the case of conscious subjects. Inhabited or lived space is the space of conscious relations and of opportunities, and mandated interactions, of the connectedness between moments of the life of the embodied subject and the things by which she is aware of being surrounded. "Surroundings" are spaces of meanings that, among other things, provide the proximate agenda for agency. The embodied subject, that is, has a "situational spatiality" orientated towards actual or possible tasks and includes the location of foreground goals and the background that makes them possible and indeed meaningful. In this space, the cup is "handy" or out of reach, relevant or irrelevant, an "affordance" – an entity you can do something with or requires something of you – rather than a mere material object. In none of these cases is the relationship between the embodied subject and the cup reducible to a mere distance understood as the interval between two physical points, even though that distance may be a crucial determinant of how subject and object interact or, indeed, whether they could or should interact.

The difference between my relationship to the cup and the cup's relationship to me goes much deeper than the contrast – associated with Martin Heidegger – between the ready-to-hand and the (merely) present-at-hand.[8] For Heidegger, this difference is captured by that between a hammer when we are busy using it and a hammer when we have laid it aside. Beneath this difference, however, is something fundamental they have in common: that they are made explicit as presences to a conscious subject.

We are on the edge of territory that has been explored at great length and in many thousands of scholarly papers and volumes by phenomenologists (in particular those influenced by Merleau-Ponty) and enactivists. Given that, as J. L. Austin said, one must be a particularly egregious kind of fool to rush in where so many angels have trodden already,[9] I shall not attempt to engage with the massive corrective represented by Merleau-Ponty's *oeuvre* or, in recent decades, the enactivist movement – except to offer a brief engagement with those correctives.

Merleau-Ponty's assertion that we are not located in space but *inhabit* it should not be taken to justify overlooking the fact that we must be located in space in order to inhabit it and, indeed, in order to engage with entities that are located in space as our mode of inhabiting it. Without spatial location there is nothing to inhabit; nothing, that is to say, that distinguishes correctly knowing where I am and dreaming that I am somewhere where, in fact, I am not. I am not "in this room" in the way the desk is "in the room" – I inhabit it in the way that the desk does not – but both my body and the desk have to be spatially in the room.

Which brings us back to Kant. There can be no basis for my being in a particular parish of reality, with particular experiences, and having particular needs and duties and goals and hopes and relationships, without my also being at some level an object among objects, a locus of physical events and processes connected with a world of physical processes and events. If, as we have discussed, space and time were impositions by our minds on our experiences – they were mere "forms of sensible intuition" – there would be no justification for experiencing ourselves as located in one place rather than another or, indeed, no justification for our having certain experiences rather than others.

The relevance of this issue to our fundamental theme of explicitness is that it illustrates an error opposite to that which we explored in Chapter 3. The claim that neural activity in the brain can deliver the transition from what-is to that-it-is is in some respects the opposite of idealism, according to which what-is is the product of, or a construct of, experience, of that-it-is. Kant's more subtle – and indeed enigmatic – position is that, while the origin of sense experience is not clear, its transformation into the presence of entities that are located in space and time depends on the top-down, spontaneous, activity of the mind organizing experience. The fundamental features of what-is – namely being laid out space and time – are, according to Kant, imposed on the deliverances of the senses. In short, far from being features of what-is, spatio-temporal location and order are aspects of higher-level explicitness. Imposed by the mind, they are as "secondary" as colours and sounds. While Kantian idealism, unlike Berkeleian idealism, does not dissolve what-is entirely into that-it-is, it dissolves enough of what-is to render it (understood as the thing in itself, the noumenon) as irremediably problematic and beyond the range of experience.

Our sense of the nature of the objects "out there" – in particular our intuition that they exist independently of anyone's awareness of them – is, I shall argue in the next section, connected with our (seemingly justified) sense of ourselves as embodied subjects.

5.3 MY BODY AS THE PRIMORDIAL "OBJECT IN THE WEIGHTY SENSE"

In what follows I want to rehearse, and then develop, a claim I have advanced elsewhere.[10] It is that our experience of ourselves as embodied subjects licenses the intuition, at the heart of our sensory experiences, that they are of physical objects that exist in themselves and, as such, are more than those experiences; and that we consequently also intuit a world populated by objects that are extra-mental in the sense of being more than that in virtue of which we get to know them, than that which is made available to us. Unlike episodic and patchy sense experiences, these objects have permanence and spatio-temporal continuity; and they can (and, indeed, overwhelmingly do) exist unperceived. In short, the intentional objects of transitive consciousness we perceive as being "out there" are what P. F. Strawson characterized as "objects in the weighty sense".[11] When I see a cup, what I see is something that transcends visual or, indeed, other sensory experiences. It is mind-independent.[12] More to the point, it is me-independent. Or, to connect with the fundamental preoccupation of this book, they are experienced as elements of what-is that is not reducible to the that-it-is in which they are made explicit.

The path from a sense experience to a full-blown intuition of an object existing in itself, continuing to exist when it is out of sight or otherwise unsensed, may seem to be laid down by the accumulation of sense experiences. I acknowledge that, when I see a cup, I see merely a part of the surface of the cup. Looking at it from different angles, picking it up and feeling its smooth surface, its weight and resistance to my grip, would supplement my intuition that I am seeing only part of the surface of the cup and that it has depths of its own. Much of its surface and all its depths are hidden from my gaze, as are its feel, its weight, and many other aspects and properties of it. I can correct this deficiency by further, multi-modal inspection. Seeing may be believing but touching and handling are more secure grounds for belief.

Would, however, multiplying sense experiences in this way take us beyond sense experience to the justified idea that such experience is of an object that is independent of those sense experiences? Could our minds by this means come to intuit a realm of entities that are mind-independent? It is not clear that it would. The ontological postulate that the object is more than all that is revealed in sense experiences, indeed, that it is more than a construct out of sense experiences, seems to lack justification. It remains a matter of faith. This is where what Jan Patocka called "the lived corporeality" of the body,[13] and what we may designate as "amming the flesh", provides the basis for the jump between experiences and the ontological intuition of an object in itself underwriting those experiences. Our bodies, I shall argue, are the primordial objects in the weighty sense.

This is a far from original thought. It is lucidly explicated by Huei Ling Cheng.[14] But how does the body deliver this sense? He quotes Quassim Cassam: "For one to be in a position to conceptualize one's perceptions as perceptions of objects in the weighty sense, one must be intuitively aware of one's self as a physical object".[15] Both Cheng and Cassam consider whether this awareness of one's self as a physical object is an intuition or a (potentially linguistic) conception.[16] My own position is that, through being lived, it goes deeper than merely being conceived or intuited.

At the most superficial level the sense of any objects being more than what we experience of them at any given time – so that we intuit them as existing independently of us – is exposed by what we might describe as "cross-modal stereopsis" of our own body: cross-modal inasmuch as it involves more than one sense; and stereopsis as implying ontological depth perception.

Consider my perception of my hand. I see it from without as an entity out there. At the same time I feel it from within as warm, heavy, perhaps aching and, in effortful grip, as the primary mediator of my agency. By this means I am directly apprised that there is more to my hand, experienced from within, than there is evident as I experience if from (visual) without. The contrast between "within" and "without" is not reducible to that between the visible or tangible external surface and an invisible interior hidden from sight and touch. Importantly, there are two viewpoints directed at the same entity: an allocentric viewpoint on an object that is seen as "out there" and thus as other than myself as a perceiving subject; and an egocentric viewpoint of something directly perceived, as part of my sense of being an embodied self. Thus is a space opened up within perception between the perception and a perceived object.

Surely, it might be objected, this kind of cross-modal stereopsis would deliver only a sense of epistemic depth – betraying that there is more yet to be known of an object than what is delivered to a particular set of senses – but not ontological depth whereby the object is sensed to be more than anything that could be accessed by any senses: that experience is *of* something other than experience; that its intentional object is a physical object. What is it, therefore, that makes the body of the embodied subject the primordial object such that it is the ontological paradigm of material items out there? It is that the body is not only living but is *lived*.

An aspect of its being lived is something we discussed in the previous section: the sense of our body having a dark, private interior and an illuminated public exterior. This contrast is both literal (as in the contrast between the inside of my abdominal cavity and my external abdominal wall) and something more profound. The rumours of what is "going on" in my body are experienced as sensations – the feeling of saliva in my mouth, the humming in my ears, aching in joints and muscles, the throbbing of the hidden heart, the rush of air into the lungs, the silent waves of peristalsis, and so on. The signals are intermittent, but they (literally) flesh out the sense of an entity that is more than is revealed to my senses at any particular time.

It is important not to place too much emphasis on sensations inside the body to avoid the risk of seeming to confuse a literal interior with an ontological reality that transcends sense experience, with an intrinsicality providing a paradigm for the in-itself of an object in the weighty sense that amounts to more than any experiences of it. Nevertheless, the quasi-interiority of sensations arising from the body deepens the sense had by a lived body that it is more than is experienced of it at any one time.

A particularly forceful expression of the status of our bodies as objects in the weighty sense is our kinesthetic awareness of the coordinated movement of our limbs and trunk and head. This presupposes a continuity between parts that are not currently being experienced, an implicit body schema that underpins actions and, more profoundly, the agent's sense of what makes those actions possible. The intuition of the body as a whole is reinforced as the latter changes its form when the body engages in activity; and it is placed in italics by effort, by a sense of the lived weight of the body, and by the cooperation of different parts. If, for example, I am typing, my fingers will be in the foreground and other parts of the body will be located in a receding background that will form the necessary infrastructure to support the action, a *presupposition* set out in space as that which will make it possible. There is a carnal logic at work in the connection between fingers and the arms to which they are attached, thence to the trunk that supports the hands, the buttocks that support the trunk, and the legs and feet that join the trunk in contributing to core stability. Much of our body, though not the object of explicit attention, with associated sense experiences, is thus the implicit condition of our physical engagement with the world upon which we are acting.

The stereopsis that arises out of the union of the body that is lived – as what I am – and the body that is observed, used, and judged, thus has many layers, building a sense of ontological depth, of an in-itself beyond what is experienced of it at any time. There is a constant fluctuation within our body of the status of its parts as being lived by me and as identified as something that is there, even "out there". The hands I may inspect for cleanliness will be being experienced, and lived, in the aftermath of my vigorously washing them. In addition, while experience will flicker around, highlighting different parts of the body, parts in the sensory shade will be implicitly present. I may not be aware of my legs when I am standing talking to someone. Nevertheless, they must be tacitly present in my standing as I talk; otherwise, I could not make sense of the position of my head.

The key point is this: the "amming" highlighted in agency – which also underpins the continuity of the body through time and change – brings together parts of the body not being experienced with those that are being currently experienced in a continuum that is implicit in the necessary coherence of the moving parts of the body in action. Behind the distinct elements that are mobilized in my action, there is a wider whole, a unity underpinning the actions of what is mobilized: a *body schema* implicit in coordinated movement.

The concept of body schema has attracted a huge literature. The thimbleful of exposition I have extracted from the ocean of scholarship is sufficient to highlight what the term draws attention to: the unconscious carnal logic that unites large stretches of flesh during the course of action, making it an implicit presence in the absence of sense experience; or making a continuous presence out of patchy, discontinuous experience. It reminds us of how much of the body can be lived, or "ammed", even when it is not being sensed.

There is an analogous unity-across-difference, a collapsing of distance, when we for example experience a sensation in a part of our body that is sensed as distant from the capital of our flesh – the head. Consider an itch on my ankle. I have to reach out to scratch it and I see the place that I scratch as "out there" or "over there". Nevertheless, it is not entirely "out there" or "over there". It engages, is part of, "me here". But it is also necessarily part of a continuum in virtue of which my looking head, my scratching hand, and the itching ankle are one. (It is the faking of this continuum that makes the pain in a phantom limb doubly distressing.) The unsensed but implicit continuity connecting areas of my body that are sensed provides the otherwise seemingly paradoxical example of an object's being intuited as existing in the absence of its being sensed: that it is present and yet is not epistemically transparent.

The totality of my body experienced as its summed weight, typically experienced in my feet and legs, adds to my intuition of its spatially continuous existence-in-itself transcending my patchy awareness of it. In short, the status of my body as the primordial "object in the weighty sense" is made even more secure because my experience of its weight is not localized to any perceived part of it.

Thus, does living our body give us epistemically privileged access to the ontology of objects.[17] The implicit continuum may be underscored by clothes. Most obviously they may put individual parts into italics through dialogue with tactile surfaces – as socks do feet, gloves do hands, tightly-fitting trousers the legs, a scarf the neck, or a hat the head – but, more importantly, they may generate an enhanced awareness of the unity of these parts out of the coherent sense by which they are connected. Hat, scarf, gloves, pullover, leggings, socks, are clearly part of an integral whole – of *an* outfit. And this is true even before the real or imagined gaze of others is felt to synthesize and judge what is worn as ill-fitting or smart, appropriate or inappropriate.

More fundamentally, the assumption of the material continuity of the body across sensorially silent areas is embedded in an assumption of the permanent availability of its parts for agency – most notably for actions other than those, associated with experiences, in which it is presently engaged. The chewing mouth will be confident of its availability for kissing, reciting poetry, or carrying items that cannot be accommodated when both hands are full.

It will be evident that I have shifted the emphasis from the lived body as the site of largely passive experience to something that generates and shapes its own experiences (present and future) through activity. I have given reasons for thinking that this gets us closer to the intuition of an object that has an existence irreducible to that which is sensed of it. Passive experiences of our body – such as the visible surface and the hidden warmth of the hand, opaque to one sense and experienced from within its explicitly hidden depth – take us beyond successive experiences in relation to which it remains a leap of faith that they are experiences of something that is not reducible to experience, but perhaps not far enough. In the case of agency-in-progress, bearers of the succession of experiences will be present in some form throughout the action, even when they are not currently engaged: in the form of an unfolding bodily schema, bodily parts that are not headlined will still be implicitly present; their existence will be intimated in the absence of, and hence independent of, actual sensation, as part of a largely implicit, largely silent continuum of flesh.

Thus, we arrive, in the case of our own body, at the presence of an object that is experienced as being independent of, more than, the sensations by which it is revealed to us. By living my body, I synthesize parts that are separated by sensory silence into an object that transcends my experiences of it. The fluctuating blush of "am" into a spatially extended "is" illuminates an object in the weighty sense. If this lived, enacted reality of an object in a weighty sense, is not good enough demonstration of the body as having an in-itself, then it is difficult to know what could be.

5.4 ONTOLOGICAL DEMOCRACY: MY BODY AND OTHER OBJECTS

So much for my intuition of my body as a primordial object in the weighty sense. I have existential or lived grounds for judging my body to be an entity that amounts to more than my experience of it. What grounds do I have, however, for extrapolating my sense of my own body, an entity that uniquely fuses "I am" and "It is", as having an existence beyond whatever is experienced of it, to other entities that are encountered only through the mediation of exteroceptive experience, known only from without? How can we justify generalizing from one (admittedly rather special) body to all bodies to conclude that they are weighty in Strawson's sense?

There are many pointers to a fundamental similarity between the body that I (incompletely) am and the extracorporeal objects with which I interact, either through sensing them or, more intimately, acting upon them or being acted upon by them. The cup I take in my hand and the hand that takes the cup seem in many respects to be comparable entities. First, the stereopsis of senses – vision and touch – highlighted in relation to our own bodies is also seen with extracorporeal objects. Secondly, in both cases there are coherent, repeatable, predictable changes in appearance as the object – hand or cup – is moved and/or the angle and/or distance from which it is observed is altered. Third, when I switch on a light, my hand and the cup are both illuminated, and both cast shadows. Fourth, when I take hold of it, the cup, like my hand, has weight and the interaction between the two has a certain equivalence, a reciprocity: as the hand presses against the cup, the cup presses against the hand. This reciprocity is seen when, for example, I sit down on a chair. My bottom presses down on the seat and the seat presses up on my bottom. If I sit long enough, I leave a mark on the cushion and the cushion, if of inferior quality, may return the compliment. Finally, when I lift my leg out of an awkward position it declares its weight as does any other entity I may lift.

There are further, more compelling, similarities. If, carrying a cup, I stumble on the stairs, the cup and I both fall under the influence of the same forces, may leave marks on the place where they have landed, and that landing place may leave its mark on both of us: I break my leg, and the cup breaks its handle. More fundamentally, my body and the objects that it interacts with both occupy finite spaces and their positions, like that of my body, relate to one another, as we are forcibly reminded when we squeeze through a gap. We have also relative positions in the same space as is revealed by motion: when I move forward, the objects around me seem to move backwards. There is a similar continuity in their trajectories and mine. And the same applies to other changes. I can cause ice to melt by leaving it in the sun or, less efficiently, by gripping it in my hand.

The democracy of the interaction between my body and other bodies in the material world is rehearsed within my body, as when my crossed-over legs leave marks on one another. Or when parts of my body warm one another. In the case of one hand gripping another, there is an hierarchical relationship – with (say) the left hand as agent gripping the right hand as patient. The fact that the roles can be reversed – so that the gripped hand is like any other object that is gripped, pushing against the gripper – highlights the status of the embodied subject as having at any time parts that are objects, even if they are sentient objects.

In sum, my democratic interactions with other objects – body-on-body activity – reveals our ontological kinship and our common subjection to the same natural laws. But is this enough to justify concluding that there is an ontological equivalence such that being-in-itself that we ascribe to our own bodies can also be attributed to the objects with which our bodies interact?

Well, it seems unlikely, indeed inconceivable, that the interactions we have described would be possible if the extracorporeal participants were merely constructs built by the embodied subject out of its sense experiences but did not actually have any independent reality. The *inter*actions seem to require that both bodies – the cup and the hand, the shadow-casting body, and the carpet on which it leaves its marks – should belong to the same ontological realm of physical entities set out in physical space. It strongly suggests that extracorporeal objects could not be reducible to that in virtue of which they are known; their what-is to that-it-is; their matter is thatter. It would be difficult to understand the facts of our interaction with the physical world, the object-on-object interaction that is the basis of our moment-to-moment existence, if one of the partners were a solid piece of stuff and the other were merely constructs out of the experiences of that one piece of stuff.

To anticipate the subject of the next section – intercorporeality – we can also highlight the fact that our world has more than one embodied subject. If you and I are walking side by side and interacting with the same objects – reaching for them, pointing them out to each other, avoiding them – it is difficult to think of those objects as being constructed – presumably separately – out of our separate flows of sense experiences. It would be even more difficult to imagine what is going on when we cross the intervening territory, say a carpeted floor, to shake hands – not the least because there would be an implicit solipsism in the claim that each of us constructs the objects that we both see. In the absence of a Berkeleian god to make our experiences compatible it would be difficult to see how our worlds could be orchestrated into the great overlapping that is the shared reality in which we pass our lives.

So, there is an overwhelming plausibility in the belief that the objects we perceive are ontologically of the same kind as our body that perceives them, notwithstanding the asymmetry according to which objects "out there" are revealed to the embodied subject while the latter are not revealed to the former.

It may seem that I am being inconsistent, given my harsh criticism in Chapter 3 of the defenders of panpsychism who extrapolate, from the fact that the brain, which is the one object we know from within – in virtue of being or, more strictly, "amming" – is conscious, that consciousness is an intrinsic feature of the entire material world or even the universe. The present argument does not, however, simply assert that what I know from amming one object can be extrapolated to all objects. Rather I adduce reasons for doing so from the democratic way the body and extracorporeal objects interact with each other and add up to a world inhabited by the body or, more tellingly, co-inhabited by any number of embodied human subjects. If the embodied subject is pulling her weight as an agent in virtue of her body being an object in the weighty sense, then it must be necessary that the world in which she acts should also be populated by objects in the weighty sense.

There is another reason, beyond my experience of the in-itself of the body, of myself as a conscious subject, for ascribing weightiness to the objects around us. If extracorporeal objects had no intrinsic properties, no being, when they were not being perceived, there would be no grounds for experiencing individual objects as a such-and-such (as a cup or as a cloud) and in a certain way (from close up or from far away, from in front or from behind).

We are back to the reasons for challenging Kant's claim that the objects of perception, insofar as they are located in space and time, are generated by the operation of the spontaneous activity of the mind on experiences. What grounds would there be for locating an object in one place rather than another? And what would be the status of the human body as an entity located in space?

In the absence of answers to these questions, how could I explain that I am aware of a particular cup in my study but not another cup that is in New York? I have discussed this question elsewhere[18] and I raise it here only to reinforce the argument that the objects that we embodied subjects perceive must belong to the same ontological realm, most particularly the spatio-temporal realm, as the body of the subject that perceives them.

So that's it, then: our body, as a primordial object which we experience from within, vouches for the weightiness of extracorporeal items with which it interacts democratically. Job done. Or is it? The belief that there is a cup-in-itself, continuing to exist when it is not observed, is as well-founded as my sense of my body as being in-itself could be turned upside down. We could say my ascription of stand-alone reality to my body and to cups is equally *ill*-founded.

At any rate, in presupposing that securing the ontological status of material objects points only one way – from our body to extra-corporeal entities – I may have oversold the standing of the lived body as the primordial object. The interaction between the material item that is the body that I am and other items in the world is, after all, two-way.

It seems reasonable to conceive of an *iterative* process whereby objects that are given intrinsicality by projection from the body return the compliment. The intuition of one's own body as something that transcends sensation feeds outwards to the everyday metaphysics of a world of objects that are also more than is sensed of them, of self-individuating entities that seem to exist in the absence of sensation, and this reinforces – and justifies – the sense of one's own body as likewise being an object that is more than what is experienced of it by the subject it embodies. In the end, however, the ultimate underwriting comes from the "amming" subject. After all, it is difficult to imagine how I could be mistaken as to the fact that my body has a being that is more than my experience of it at any time. At the very least, it has to have being-in-itself in order to house and locate and integrate experiences that count as experiences *of it*.

Thus, the existential revelation, courtesy of "amming" our body, of a what-is that exceeds that which is perceptually revealed as that-it-is. It is in or through the body that self-revealing what-is does not evaporate into that-it-is. The passage to explicitness does not leave behind that which is made explicit. The story is not entirely straightforward, as we shall see, especially in Chapter 10 when we see how "am-being" makes "it-being" into what counts as "reality". But the fundamental principle seems sound.[19]

5.5 A SIDEWAYS GLANCE AT INTERCORPOREALITY

The notion of "intercorporeality" – particularly associated with Merleau-Ponty – highlights "the interaction between embodied subjects".[20] It reflects the social nature of the body – as when I am aware that aspects of my body are being sensed, interpreted, and judged by others – and the bodily nature of social relationships. Sartre identified aspects of this as "the-body-for-itself-for-others".[21]

Without social interaction, the embodied subject may seem something of an island bounded by the skin. Nothing, of course, could be further from the truth. While the body is sharply delineated from other bodies, following its own trajectory through space and time, the subject embodied in it is dissolved to a greater or lesser degree in an ocean of shared meaning. Although the human race is divided into portions of flesh, embodied subjects encounter themselves – get a sense of who and what they are, and what they should be about – through the medium of the countless languages, dialects, customs, and practices, that define many groups large and small to which each one of us belongs. "I am" encounters him or herself in a sea of "we are" or (where we is present only through discourse) "they are". As Jensen and Moran put it, subjects are not only embodied, but they are also "embedded in social and historical life-worlds, and essentially involved in with other embodied subjects and in an intersubjective cultural world … Human beings are embodied intentional agents – expressive, meaning-construing and meaning-intending beings embedded in a world that is loaded with significance, overlain with fantasy, imagination, memory and all kinds of projection".[22]

Even before being gathered up into the seas of language, the embodied subject's sense of herself is caught up in networks of significance woven out of the direct and mediated presence of other embodied subjects. We are, after all, manufactured in another's body, at birth are gathered into our mother's arms, and are fed directly from her body. In short, our sense-mediated presence to ourselves is extended, expanded, reinforced by our presence to others.

My status as an object (as well as a subject) is ceaselessly affirmed by my visibility, audibility, and so on, to others. The fact that you can see and hear me – that I am an object of another's perceptions just as cups and saucers are – adds validity to the intuition of ontological democracy discussed in the previous section and my sense of being an object in a weighty sense. Those objects are affirmed as real through being actually or potentially present to others. My presence to you as an object of your awareness and your presence to me as an object of my awareness reinforce my sense of my body being an entity that transcends my experiences of it. We stand surety for each other's objective reality.

This is elaborated in many modes of co-presence. Even at the most elementary level of joined attention – as when a toddler points out something to its parent – there is a good deal of complexity. The toddler's pointing is an affirmation that the object it is pointing to and its pointing finger are visible to the parent with whom it is sharing its experience. If the pointer and the pointed-at are assumed to be objects of another's perception, this will support the belief that the objects of perception are more than anything any individual perceives. This is consistent with the belief in objects having an ontological standing as things in themselves seems unlikely to be the product of an epistemological conspiracy – if only because such a conspiracy would require to be delivered by persons in a weighty sense who must, anyway, be necessary to explain the desire to share experiences that appears so early in life.

Of course, the body of Raymond Tallis has a special status for him. So long as he is awake, it is continuously present to himself. The bodies of others – like all other material objects – come and go, popping up and disappearing at intervals that may extend over years. What is more, my body is the physical and existential centre of the space I inhabit and is a continuous presence in that space, because this space tracks the body. As already noted, the fleshly contents of that centre will be either all-engrossing or marginalized, foregrounded or backgrounded, depending on how much it insists on that attention.

We have discussed how many and various are the modes in which parts of my body are spatialized. There is the literal spatial expansion of reaching, standing up, stepping out, and the literal contraction of clenching one's fist, pursing one's lips, and curling up in a ball. But there are equally many ways in which movement in physical space realizes a location in *semantic* space – speech, facial expressions, gestures, and inscription, are obvious examples. Sometimes the movement is very simple, though the message is complex – as when one nods one's head in support of a particular position on an abstract issue.

Merleau-Ponty's seeming straightforward assertion, already cited, that we are not located in space but inhabit it connects with intercorporeality. Inhabiting space is inhabiting, and relating, to a nexus of objects, processes, and events of actual or possible significance, much of which will refer to items that are not physically present. That nexus will include the bodies of others whose presence will transform the shared space, most importantly through my judgements about their consciousness – of myself and of other aspects of our shared world.

At any given time embodied subjects are located in many kinds of non-physical space, corresponding to the many modes of their direct and mediated consciousness of themselves. As I hurry along the pavement to a meeting with you for which I am late, I am in some sense defined by my shortness of breath but also by my anxiety, fed by my sense of your possible impatience or anxiety. My location and my distance from my destination could be described by physical co-ordinates but this would miss much that is true of my hurry. I fear being late because of the importance of the appointment which in turn draws upon many sources of meaning, of my sense of what I am, at least in part defined by obligations – what I owe to others, and why.

While the most prominent aspect of my hurry, therefore, is a physical effort that ties me closer to this body, it has other aspects woven out of the complex networks of meanings that link me to the goal of my hurry, and the person in it, and what is to be achieved when I arrive. Other threads connecting my hurry with its target will be more general, structuring the kinds of obligations individuals in a certain culture might owe to each other. These may be legal, or institutional, as when, for example, the person I am meeting is a "client", with all that that term implies. Just before my arrival, I might attend to my body – mopping my brow and taking a few deep breaths – to conceal the fact that I have narrowly avoided being late with all the implications of incompetence or of not prioritizing the meeting. This will be modulated by previous encounters when I have or have not been on time.

The sense in which, as embodied subjects, we inhabit space as well as being located in it (as a condition of inhabiting it), can therefore be construed as different modes of cohabitation in a multitude of spaces, defined in many ways that go beyond the physical space in which we are co-located. The cohabitees may be individuals, groups, categories of individuals, or implicit or explicit communities. The shared, overlapping, and interacting consciousness that we experience when we live with someone is only to a small degree defined by the spatial relationship that makes it possible. And of course, a great part of many relationships is mediated through verbal communication that does not map on to, or link locations in, space.

Thus, the inseparable corporeality and intercorporeality of the embodied subject. Intercorporeality opens dimensions that – unlike its most direct manifestations in shared smiles and handshakes – may not be visible to third parties. Of course, there is more to other embodied subjects than I can experience; namely, their own experiences, though I may infer these from their behaviour. The example of a stranger observing me walking fast as I hurry to an appointment, or joined to the same queue as myself, shows how little we typically guess of the occupants of the intercorporeal realm we inhabit.

I have scarcely entered the rock pools on the edge of the ocean of the collective, social consciousness in which we individually participate. The sharing of consciousness through elementary gestures such as pointing discussed earlier is, if course, only the beginning. It is the anteroom of a vast echo chamber in which human consciousnesses past and present are shared and add up to a collective realm of "thatter". The mediated experiences and the epistemological promises with which they are laden – so that things I have been led by others to expect to encounter are indeed encountered – are a confirmation of the reality of a shared world, of a world that lies outside of our individual experiences, a world that is the weighty object of such experiences.

5.6 THE MIRACLE OF THE EMBODIED SUBJECT: THE TWOFOLD REVELATION OF A WORLD AND ITSELF

We come to one of the most challenging aspects of the mystery of the transition from what-is to that-it-is, of the enigma of our embodied being-in-a-world, of the body as that in virtue of which subjects are able to make what-is something that is present to them. The mystery thickens when we think of our bodies as the necessary condition for both transitive and intransitive experiences: sufficiently opaque to be self-present, to underpin a solid am; sufficiently transparent to be a window on the world; and able to connect the two so that experiences are "my" experiences, experiences of someone "here" referring to something "out there", to a world.

The archetypal example of transitive perception – of something that is experienced as being "out there", as having an intentional object that is more than its content – is vision. Vision is transparent or diaphanous. As I look and see a cup, I face an "it", an object revealed to my subjectivity. By contrast, intransitive perceptions seem to stop at the perceiver. "What it is like to be Raymond Tallis" is closer to the intransitive experiences of being itchy, tired, pleasured, than the transitive experiences of a landscape. RT's seeing a black object with a yellow tip is closer to seeing what a blackbird is like than experiencing what it is like to be RT. The less transitive an experience, the more it is a part of what it is like to be RT. RT can be itchy but not "landscapy" or "cuppy" – though he can experience being surrounded, or located in a large, open space.

The opposition between transitive and intransitive elements of consciousness is excessively simplifying especially as, for example, a standard example of an intransitive perception such as the itch I have just alluded to, may not be entirely intransitive. There still seems to be a gap, or a potential gap, between the experience and what it is an experience *of*; or rather between the experience and a something towards which it points. Even those sensations that seem closest to "what (at any particular time) it is like to be RT" have an otherness, or the possibility of it, built into them. Sensations associated with illness or malfunction, such as itches, pains, and tingles, seem to point to processes, to have an implicit intentional object, that have a potential significance – meaning something unwelcome going on in a body part which has too much to say for itself. While an itch in my foot is not a state of my foot (understood as a material object) in the way that the redness of a strawberry seems to be a property of the strawberry (notwithstanding its status as a secondary quality), it seems to refer beyond itself as a revelation of the foot. Even the most immediate non-pathological intimations of our body still potentially have intentional objects distinct from them. The feeling of bodily warmth, the blush of pleasure after a glass of wine ("I drink therefore I am" in italics), fullness after a meal, or the posture of one's trunk and limbs, have an actual or potential or inchoate intentional object. And, to refer to a previous example, seeing a blackbird may prompt a pleasure that is certainly part of "what it is like to be Raymond Tallis". As for emotions, there are intransitive elements

– such as the dry mouth of fear – interwoven with transitive elements – such as the numerous propositional attitudes (hopes, fears, thoughts) that might precipitate out of them.

No experiences, therefore, are entirely intransitive or lack the possibility of having, or seeming to have, an intentional object explicitly other than themselves. Itches, pains, tingles, pleasurable sensations, feelings of fatigue, and such-like can be judged with the aid of knowledge or guesswork to point beyond themselves. The locus of the pointee is the body, or part of it. Even a proprioceptive sensation has a pointee; for example, the position or positions of parts of the body. Nevertheless, the distinction between exteroceptive sensations such as seeings and hearings which are fully transitive, and tingles, etc., that have only potential definite intentional objects and are thus (relatively) intransitive, seems to be upheld. I can feel itchy but not (as a result of looking at a forest) feel that I am "tree-y".

Other contents of consciousness such as factual memories and thoughts, which remain private unless they are uttered, do not quite qualify as belonging to "what it is like to be" the conscious subject because they are, with some exceptions, rather flavourless. A thought should theoretically be even more transparent than, say, a visual experience, given that it is relatively aperspectival, not being dependent on the thinker's position or state. However, the token occurrence of a thought has a present context which may encroach on the thought a little. Nevertheless, thoughts seem to lie at the far end of the spectrum encompassing transitive and intransitive experiences. (We shall return to this in Chapter 8).

The reader may wonder where this discussion of the blurred line between transitivity and intransitivity in the contents of consciousness is going. My purpose is to draw attention to something which is often overlooked and, being a miracle, should be brought centre stage. While the border between transitivity and intransitivity may be ill-defined, and some contents of consciousness may be predominantly transitive or predominantly intransitive depending on the kind of attention paid to them by the subject, the distinction highlights the extraordinary achievement of the embodied subject in its double revelation of what-is as (a) that-it-is and (b) that-I-am, as it transforms being into presence. The miracle is the capacity to distinguish, within those items that are made present, what counts as parts of itself, of "I", and those that belong to the world other than the self, to "it". The embodied subject has both to reveal itself and to reveal the world, the situation, of the subject.

As if that were not complex enough, self-revelation is in turn a twofold task: revealing the body to the subject; and revealing the subject to herself not only as a body but as a set of relations to a world which are not translatable into mere physical relations.[23] There is an overlap between these two. What I am is to some extent the self-revealed body – which is the centre of *where* (physically) I am, though it is not just that. As one who is not merely located in the world but inhabits it, my status in the world as a localized being is transformed from a set of physical relations to something quite different.

That difference is evident at the most basic level when spatial relations are transformed into either proximity or handiness, or at a distance, being out of range. It is developed in different ways, as (for example) between public and private space, between the scene of my activities and what-there-is beyond the ever-changing horizon of that scene, between a world as revealed to my senses and a world making sense mediated through memory, knowledge and thought. The many layers of my situation – unfolding continuously or jumpily, varying in itself or on account of changes in the grain or direction of my attention, or switching between what is present before me and what is present only within me – instantiate the different modes of transitivity of the infinitely varied contents of my consciousness. My ever-changing sense of what is going on, of what is happening to me, of what is happening in the world, and of what is expected of me, and of others, is put together out of multi-hued torrents of transitive and intransitive experience in which, miraculously, the torrents are kept apart. Only thus can I allocate feeling itchy and being touched, blurred vision and a rainy landscape, to their correct locations – respectively "in my body" and "out there".

Given my dependence on relatively intransitive bodily experience to support my sense of being here, of being located in, as the moving centre, of what is out there, it is astonishing that the body does not obstruct, or at least distort, its own view. After all, it cannot efface itself or shrink to a dimensionless viewpoint. It is necessarily a fat viewblob (or a viewpool, perhaps, since we are 60 per cent water), self-present, explicitly here, in order for me to be *here* and for what is around me to be *there*. We need bodily sensation to *be* where we are, to inhabit the space opened up by our senses. The sense of our head moving round as we peep at, peer at, scan, inspect the world around italicizes the presence of what is made present to us. The miracle is that my being here as an embodied subject in the room and the room being there as the surroundings of my body of which I am aware do not interfere with each other. It is the miracle of explicitness placed in italics.

The transparency of exteroception becomes even more surprising when we think of what is going on in the machinery of the sense organs that support it. And even when that machinery obtrudes – as in the case of eye disorders causing blurred vision – we can see *that* our vision is blurred. On most occasions, we are unaware visually or in any other way of that in virtue of which we see.

My experiences both unite me with my surroundings and separate me from them so that they are experienced as my surroundings, rather than being fused with me. The intentional connection, far from collapsing the otherness of the object of perception, asserts it. And it does this in spite of the fact that the proximal end of the connection is an embodied subject, aware not only of the object but also of its own body.

This is what is involved when, for example, I enter a room and switch on its presence, as the corridor behind me dims to an absence or an implicit expectation of something I will find when I leave the room. My attention transforms electromagnetic radiation into the visibility of a familiar desk, carpet, curtains, word processor. I illuminate the light with my sensorium and make this part of the world audible, tangible, possibly odorous, and even tasty. The sunlit room is not illuminated until it is I-lit (or eye-lit) by a conscious subject of whom it might be said that "he lit up the room". He enters the room and lights up the light that lights up the room and transforms its contents into things that are present. Not that the contents of the room are likely to say that to one another when he walks out. Not only does he make the light illuminating, he makes the absence of light visible darkness. As well as lighting up the room, he imports his self-lit body into the room so that he experiences himself as being-in-the-room, making of the room his "here" and a place where he might meet with others.

The story of the embodied subject is of course not entirely written in the language of the body or of the successive (physical) situations of the body. That this is true is in no small part due to intercorporeality, to the fact, discussed in the previous section, that from beginning to end of his life the embodied subject RT is engaged with the presence of other embodied subjects who together create worlds and whose implicit or explicit presence makes him present to himself in a different way. This remains true notwithstanding that his experience of his body – or his body's experience of itself – and its sense of being located, provide the ground floor of what it is like to be RT and that the revelation of the body to itself is inextricably caught up in the body's engagement, as an agent, with its future and the extra-corporeal world where it is to be enacted.

The key point here is that the role of the body as embodied subject is twofold: revealing the world that is not-me; and revealing the body as me located in and interacting with the world. How these are co-present and yet distinct, united and yet separated, is a profound mystery which is ultimately rooted in intentionality and its different degrees of transitivity but also requires that I should experience some, but not all, bodily events as being part of me. The distinction between what is me and what is not me is at least in part defined by aspects of agency. This is something to which we shall return in Chapter 7.

The twofold nature of experience – supporting self-presence and the presence of, and to, a world – presents another barrier to explaining explicitness by means of material events in a biological entity – irrespective of whether we are talking about a whole organism or its brain. There is nothing in the body as seen from without – even, or especially, through the gaze of science – to account for the transition from "is" to "am", from the "it" that merely is to the "I" that is bodily present to itself, makes present a world which in turn alters its presence to itself. It is not clear how the body makes the subject possible nor how that subject makes the body – and a world as its world – explicit. It is even less clear how the body can be made explicit, in order to locate the subject in a world, without blocking the subject's view of that world.

Embodiment is the primordial eruption of a subject transforming continuities and discontinuities in the universe into "here" and "there", presence and absence, gathered up into an ever-changing world whose changes are in part driven by the activity of the subject.

5.7 CONCLUDING THOUGHTS

The that-I-am and the that-it-is have to be kept separate and yet are rooted in each other. The mode of being that is "amming" – specifically, "amming" a body – delivers, as we have argued, the intuition of material objects existing when no-one is looking; of there being more to them than what is sensed of them. First-person being pitches its tents in a borderland between the subjectively experienced and the non-experienced body, creating the sense of being in a world. "Amming" my body underpins the transformation of my transitive experiences into experiences of enduring entities – objects in the weighty sense – that transcend those experiences. In the realm of explicitness, the "it" in "that-it-is" is construed as a full-blown, independent entity because the "I" that "am" is a body that is variably and only partially colonized by the "I" and because I am aware of the existence of parts of it even when they are not being made present to me through sense experience. I know that there is more to me – to my body – than meets that body's eyes and its other senses. There is no part of my body that is uninterruptedly "I", that cannot become "it".

A just criticism of this chapter would be that much of time it treats the issue of first-person being in a living body as if the "I" were primarily about the stand-alone existence of an individual person. The "sideways glance at intercorporeality" in Section 4.5 may seem insufficient recognition of the extent to which subjects are not only embodied but also "embedded in social and historical life worlds, and essentially involved with other embodied subjects and in an intersubjective cultural world".[24] To some extent, this deficiency will be corrected in our discussion of the self in the next chapter and of agency in Chapter 7. But perhaps, and alas, not entirely.

We end with a knot. There is nothing in the human body – not even the brain – that accounts for the transformation of part of the universe into someone's world or indeed makes the body explicit as someone's (my) body and, in a complex, slippery sense, "me". And yet, in the absence of the living body, there is no such transformation. It is not clear what it is about the body that it should illuminate part of the universe as "its or my situation" and the body itself as mine and not mine, as me and not-me, as me and an object in the world; that the mode of its being should be first-person "amming", notwithstanding the personless nature of what goes on within it, including those parts of the body whose function are necessary for me to be an "I". Nevertheless, it cannot be denied that it is the body's being made explicit as "amming" which makes the world with which it interacts explicit to it and gives content to its explicitness to itself. The knot remains impossible to untie.

CHAPTER 6
Selfhood

6.1 INTRODUCING THE SELF

Kant's extraordinary claim that "The fact that a human being can have the 'I' among his representations raises him infinitely above all other living beings on earth"[1] highlights something very special about the awareness that lies at the heart of the "I", the ego, the person, the conscious subject, or the self. What is special is highly relevant to our exploration, or recovery, of explicitness. "That I am", the amming of a body to create an embodied subject, is connected with all other modes of explicitness. The conscious "I" transforms the "it" into presences that, thus transformed, potentially have meaning. The explicitness switched on by the self faces two ways: to itself and to that which is other than itself; to the that-I-am – the "who" surrounded by the world, facing it, and engaging with it; and to the that-it-is – to the "what" that adds up to a surrounding world with which the self is inescapably engaged.

Notwithstanding its centrality to our endeavour to make explicitness explicit, the present investigation of the self will not do justice to the richness of the topic. Indeed, it will be consciously incomplete, not the least so that I shall avoid covering too much territory that I have covered elsewhere.[2]

The most hotly contested, and unavoidable, issue in the philosophy of the self is the extent to which it has substance and what that substance amounts to. We might expect that explicitness itself has no substance: that-it-is is not an additional mode of what-is. If it were, it would get in the way of whatever it is that it makes explicit, requiring itself to be made explicit. The self is not, however, pure explicitness, lacking any special association with a portion of what-is. It makes certain beings and not others explicit as itself, its world, and itself in that world; it illuminates one parish of being rather than another. Presence-to must be present to something or, more precisely, to someone. That something-someone is most obviously the body of the embodied subject but, as we discovered in the previous chapter, that is rather patchily and variably haunted by, or assimilated into, the subject. It would be rather an understatement to say that the relationship between the partners of the two-in-one of the embodied subject is complex and, when seen aright, enigmatic. The "I" of the self and the "it" of the body are not simply the *recto* and *verso* of a slice of being, of a portion of what-is.

All of which is by way of an introduction to the arguments between those who espouse a thin idea of the self – to the point almost of autocide – and those who embrace a more substantive notion. We shall deal with these separately. Let us begin with autocides.

6.2 DENYING THE SELF

In one of the most famous and widely quoted passages in all philosophy, David Hume argues that the self lacks substance because it is not available to introspection:

> There are some philosophers who imagine we are every moment intimately conscious of what we call our *self*... For my part, when I enter most intimately into what I call *myself*, I always stumble on some particular perception or other, of heat, or cold, light or shade, love or hatred, pain or pleasure. I can never catch *myself* at any time without a perception and can never observe but the perception.[3]

The killer phrase is that I "can never observe but the perception". When we turn our gaze upon ourselves, that self seems to evaporate. There is no self to be found in addition to the immediate contents of our consciousness. More specifically, it is not an additional content of consciousness. Not being a separate object of perception, additional to what we perceive, the self does not have any claim to existence. Or so the argument goes.

It is not necessary to look very far to find sufficient reason for challenging Hume's conclusion that the self is a construct to which nothing much corresponds. In the short space of a few sentences, Hume uses the word "I" four times. That "I" – let us call him David – does a lot of work and the work seems to be more than one might expect of a stream of perceptions, even less of those perceptions available to a moment of introspection.[4] When David "enter(s) intimately" into what he calls *myself*, he "stumbles on some particular perception or other". The acts of calling something "myself", entering "intimately" into that self, and the process of discovering and reporting that all that is to be found there is mere perceptions, are more than one might expect a non-existent self, or a bundle of perceptions, to be able to perform.

An inescapable conclusion is that there really is something corresponding to David which we may (perhaps) call a self but that David is looking for it in the wrong place. What David is looking at misses the David who is looking, and it is the looker which he should be looking for. At any rate, the David who is arguing for the non-existence of the self – arguing with other philosophers and perhaps an earlier version of himself – seems to have quite a bit of self.

It appears that there is a meta-self available to Hume the philosopher, able to reflect on his self and to arrive at a conclusion as to what it is – or what it is not: that it does not exist, except as a bundle of perceptions. How, otherwise, would a self that is but a series of perceptions be able to step outside of the series of perceptions and observe or conclude that it is (only) a series; somehow to generate a perception that is aware of its status as a member of a series of perceptions?

There is clearly a good deal to unpack in what it is that Hume is overlooking and the growing suspicion there is more to his self than meets his introspective eye (or indeed his outer eye should he inspect himself in a mirror). It is therefore worth reflecting on how he comes to miss it.

Supposing I were to note, courtesy of introspection, all the visual experiences I had. Absent among them would be something called eyesight. I could not see eyesight; even less the individual who is possessed of eyesight and who, courtesy of his eyesight, sees himself as located in a visible world. As we discussed earlier in Chapter 1, even if I were looking in a mirror, I would not be able to see my eyesight seeing my eyes looking out. I would see the eyes but not their sight. This would not, however, justify the conclusion that there is no such thing as eyesight; even less that there are only visual experiences and no such thing as a see-er, an individual possessed of eyesight – I-sight, perhaps.

6.3 AN AUTOCIDAL TENDENCY IN CONTEMPORARY PHILOSOPHY

Hume's analysis is not merely of historical interest. There are numerous contemporary deniers of the self.[5] For many of them, the self is to be denied because it does not fit into a physicalistic account of what it is to be a human being. It seems dangerously like a spirit or ghost haunting the machinery of the body. Most damningly, it cannot be found in the brain.

For Daniel Dennett, embracing the idea of the reality of the self is the result of taking abstractions too seriously. The self is a notion like the centre of gravity: "A centre of gravity is not an atom or a subatomic particle or any other physical item in the world. It has no mass; it has no colour; it has no physical properties at all, except for spatio-temporal location ... It is a purely abstract object. It is, if you like, a theorist's fiction. It is not one of the real things in the universe in addition to the atoms".[6] And the self is even more ontologically dubious than a centre of gravity because, unlike the latter, it does not have a spatial location. Any location it might borrow from the body would – in virtue of having a size and shape – be the wrong size and the wrong shape.

Thomas Metzinger is a yet more determined autocide. In *Being No-One: The Self-Model Theory of Subjectivity*, he argues that "no-one ever was or had a self. All that ever existed were conscious self-models that could not be recognized as models. The phenomenal self is not a thing but a process".[7] His basis for this argument is more respectable than Hume's introspection. It comes, we are to believe, from brain science. (He is contemptuous of philosophers who do not listen to what neuroscientists tell us.) What brain science tells us (or will tell us one fine day– this is not clear) is that the sense of self and indeed of the real world in which we live is constructed by impersonal processes which we cannot, and never shall, access. The contents of experience are fixed by the properties of my brain: the former supervene on the latter.

Metzinger's "neurophenomenological" approach claims that consciousness supervenes on brain states and that both consciousness and the objects it encounters are products of these subpersonal processes: "The brain constantly hallucinates at the world, vigorously dreaming of the world and thereby generating the content of phenomenal experience".[8] The naïve folk psychological belief that we have a self is the result of a trick played upon us by our brains. What we call the self is a cerebral process that generates a "transparent self-model". As for what is really going on, it is hidden from us by "autoepistemic closure" – an inbuilt blind spot that prevents us seeing the extent to which the self is constructed.

Metzinger's position is riddled with even more self-contradictions than Hume's. First, it puts into question the status of the brain as an object located in the body, housing the neural activity, notwithstanding that he believes it is the stuff out of which consciousness, including consciousness of objects, is created. If everything generated by the brain is an hallucination, so, too, must be our scientific encounter with the brain: the brain must be an hallucination generated by the brain. (In this respect Metzinger's view is similar to that of Donald Hoffman – which we shall discuss in Chapter 10. Hoffman, however, at least acknowledges that consistency demands that he shall deny the brain's existence as an object located in the skull.)

Secondly, even if the brain exists as a real object, "autoepistemic closure" would seem to make it impossible epistemically to stand outside of the brain sufficiently to see that it is the source of our view of a world containing objects and selves and to discover that consciousness and the world of which it is conscious does indeed supervene on neural activity. The specific assumption that there are close correlations – demonstrated by the effects of localized brain damage – between activity in particular areas of the brain and aspects of consciousness, however, presupposes a capacity to stand outside of the brain to see what they do and this is at odds with his assumption that we are epistemic neural prisoners. It is evident, then, that Metzinger's autocide is just part of a wider ontological massacre: "The book you are holding in your hand as experienced by you at this moment, is a dynamic low-dimensional shadow of the actual physical object in your hand, a dancing shadow in your central nervous system".[9] His conclusion that "we are systems which generate the intentional content of their overall representational state by pulsating into causal interaction space by, as it were, transgressing their physical boundaries, and, in doing so, extracting information from the environment"[10] is not one that we neural prisoners could arrive at, test, or confirm, even if we could understand it.

Metzinger's appeal to Darwinian evolution to support his interpretation of why the brain is inclined to create the illusion of objects and of the self is even more puzzling. We have, he says, the mental states we do have, because they are "functionally adequate from an evolutionary perspective".[11] Such an appeal to Darwinism implicitly embraces the ordinary world picture of beasts set out in space and time and interacting with each other – competing, preying and being preyed upon, and procreating – in the way that we usually think of them. It presupposes that this world picture is more than an adaptive illusion created out of the neural dynamics of the human brain.[12]

While Dennett's deconstruction of the self as a construct, a "narrative centre of gravity" is not as radical, or indeed as zany, as Metzinger's, it shares with it an important feature – namely that of sawing off the branch on which it sits. If the self were, indeed, the illusion Dennett says it is, it would be impossible to see how it could arrive at, or sustain, that illusion. For the meta-narrative of the self as a narrative centre of gravity to be possible we require an entity that (a) tells stories; and (b) adds up those stories to a story about itself. Narratives, and in particular narratives about the nature of the self, seem to require a self, just as do Hume's reports on the findings of introspection which, he claims, demonstrate that there is no self. For Raymond Tallis to tell such stories in which "I" figures in so many different and connected ways, and for such stories to be shared with other people – interlocutors understanding these narratives – would seem to require Raymond Tallis to have or to be an extremely complex self. A self worth having or, indeed, being.

6.4 WELCOMING BACK THE SELF

The denial of the self, with or without appeal to the authority of science, seems fated to result in a performative contradiction. What then is this self that those of us who are not autocides would claim to exist?

It is hardly necessary to point out that it is not a *thing*, in the sense of something that, over a period of time, occupies a portion of space and exists side by side with other things such as tables, clouds, and trees. Any thinghood it has is borrowed from the living body that is its necessary condition, which does indeed live side by side with tables, clouds, and trees. The self is the first-person dimension of an entity that is (also) a thing. That doesn't say much. We need to examine it further.

The self is fed by the succession of experiences, thoughts, and memories it has. It is, however, more than a mere Humean trickle of atoms of consciousness.

First and foremost, there is a unity at a given time that connects a multitude of contents of consciousness and gathers them up into a field of unfolding significance: perceptual experiences of the world outside the body and of the body itself are unified with each other and with thoughts and memories. As I write this, the surroundings of my study, the sounds of my own typing, of background music and noise from the next room, the slight ringing in my ears and the pressure of the floor on my feet, the thoughts connected with my writing and memories that feed into those thoughts, are gathered into a present moment in which they are one and yet (something that is rarely sufficiently emphasized) still retain their distinctness, so that they are available for separate attention. This synchronic unity draws upon a many-layered past and points towards a many-layered future that makes sense of its present. The successive moments of experience are transilluminated by the experienced past and the imagined future.

This does not add up to a thing, or a time-slice of a thing, nor does it need to do so. The body is in some respects sufficiently thing-like to underpin the continuity of the self with its own continuity, though (as I shall discuss) it is distinct from it. The body provides an audit trail marking the continuity of a self at and over time. But we should not conclude from this, as Metzinger does, that the self is a mere process, with the implication that it is a process similar to those seen elsewhere in nature. Nor is it a mere unfolding, analogous to material processes whose successive phases are *separate but connected.*

The synchronic unity of the self, therefore, is inseparable from diachronic unity over time. The latter is many-layered, densely and subtly interconnected, and particularly evident when (as is usually the case) we are engaged in projects – learning to speak French, refurbishing a house, or bringing up a child, and multi-tasking to allow each to make room for others. It is equally evident in the complex way we hold together in delivering on our responsibilities. And the sense of guilt and shame when we fall short reinforces the connectedness between past and present – and indeed the future, when we think of remediation or of "doing better next time".

A paradigm case is that of writing a book. Individual words make sense in the context of other words; likewise individual sentences, paragraphs, and chapters. That sense draws explicitly on previous thoughts, reading, and conversations. It can be interrupted by other projects that have proved more urgent, or mere events, but then resumed. Like many projects, it owes much of its sustained sense to its being part of larger projects – as in the case of learning the language of another country in order to be able to participate in its culture.

There are identifiable character traits that change only slowly, if at all, over time and they shape what others expect of us and what we wish others to expect of us. The self also makes sense of itself – and nails itself to itself – in duties, offices, professions, in being a such-and-such – a nurse, a social worker, a carpenter.

Other people – individuals, collectives, or represented by institutions – are, of course, involved in this kind of expression and reinforcement of the self. We are remote from the hopeless challenge that would be presented to a self-curating trickle of perceptions to build itself up into a self. Others' acknowledgement of what one is adds to the external scaffolding of recognition that properly reinforces the sense of oneself as that self. This is especially true if the source of acknowledgement is a significant other, as in an enduring relationship where the person to whom one matters in turn matters to oneself. No amount of misunderstanding can undermine this. The *re*cognition within our mutual awareness underlines the co-presence of past and present – a shared past and a shared present.

Others are not merely colluding in an illusion. They are part of a multitude of interlocking frames of reference in relation to which we make sense of ourselves. This is constantly reinforced in a multitude of ways when others re-cognize us – so that when I enter my home my wife does not call the police on account of there being an intruder. That recognition is mutual: I do not call the police when I find my wife in the home I enter. We each reinforce the sense of each other's identity in a way that has an existential reality that goes beyond any documentation: it is underpinned by a shared life, an overlapping world. And when, as the attending doctor, I visit a hospital ward, my identity is confirmed by the nurse who greets me, and asks me to see a patient for whose care I am responsible.

Documentation will also reinforce the objective reality of the self as a self-for-others, reflecting many aspects of our lives and, more broadly, our footprint in the world, effects that we refer back to ourselves and, for which, where necessary, we take (or consciously evade) responsibility.

In short, being acknowledged implicitly or explicitly by others whom we acknowledge in turn is central to the underwriting of the self. *I am* at least in part because *we are*, where the scope of "we" will vary from aspect to aspect of ourselves. When you recognize me, and I recognize you it is not an instance of collusion between two illusions. That which is between us, however many illusions and misunderstandings it may contain, has factual contents reinforcing the self that go beyond any evaporating trickle of perceptions.

A particularly striking expression of the cooperation of myself and others in the affirmation of myself and the embrace that gathers up so much into my identity is the use of my proper name. From baptism to burial, I and my many circles of loved ones, friends, acquaintances, and relative strangers, agree that I am Raymond Tallis. I introduce myself by that name, and I am called, summoned, greeted, abused, by it. It is the hub anchoring a thousand spokes of recognition and is the link between numerous aspects of myself. It is the key term on the countless papers that document who and what I am and make me responsible for the payment of far too many bills.

The portentous act of answering to the name that I have been given also embraces many aspects of myself evident to me directly or through the real or imaginary gaze of others. My name gathers up so many hours, days, years, roles, activities, modes of being alone and with others. The distance between "David Hume" and the trickle of perceptions he observes when he "enters most intimately into what I call *myself*" is in part sustained by the presence of the collective consciousness that fills out and validates the sense he makes of himself.

As will be evident even from this sketchiest sketch of the self, it is quite something. No, it is not a thing but that is not a deficiency: being a material object or something like it is not the only mode of being. A thing, after all, is an inert entity, a passive transmitter of impacts, acting – or seeming to act – only when it is acted upon. Nor, of course, is the self a ghostly spirit or an eternal soul. But it is certainly more than nothing. "That I am" makes itself and its world(s) present and, for this, it does not need to be a piece of what-is.

Let me end my defence of the self against the autocides with a return to my earlier charge that they are guilty of pragmatic self-refutation. Metzinger's book denying that neither he nor anyone else is, or has, a self seems like a mighty expression of a self. In order for the 643 pages of the book to be written, the professor would have had to unfold coherently over a long period of time in a way in which sustained interests, concerns, a body of knowledge, and a complex, coherent understanding of the world (expressed in, among other things, countless conversations about this topic and an audience for his views), his sense of his professional duties, and all the skills that enable him to negotiate the world in a particular way (using libraries, making the case for study leave to free up time for writing, attending conferences and seminars, pitching to publishers, ensuring an adequate income, looking after his health, dividing his time appropriately between professional and domestic duties, while he does so), would all add up to a continuous, interlocking sense of the world and his place in it that, by any reasonable judgement, amounts to a self.

My point is that it takes one mighty, highly developed, robust self to conceive, write, and complete a 643-page book denying the reality of the self. And then there are all the spin-off activities related to the book. The professor is invited to spread his word and so he will give lectures and, in order to do so, will draw upon a dense nexus of skills – making and keeping appointments, booking plane flights, writing, rehearsing, and timing lectures, preparing slides, travelling, and so on. His lectures will (or should) also draw on his sense of his audience – what they might know or believe and what they expect of him based upon past experience – and on his knowledge of the existing literature on the self against which he directs his views. So much, then, is involved in incurring and honouring commitments that have moved from a possibility located in a highly structured future that is his future to a present reality that is his, Thomas Metzinger's, present reality.

In short, denial of the reality of the self in the manner exemplified by Professor Metzinger is a sustained expression of a self that is more than a mere "process" – particularly if that process is seen as a Humean trickle of conscious experiences. It is consequently an egregious instance of a self pragmatically refuting itself.

One does not have to examine the self-expression of a professor of philosophy to discover how rich, profound, multi-layered all selves are. The lifelong experience of being someone is not an illusion, generated by the brain, to which nothing substantial corresponds. Indeed, if it were an illusion as Metzinger claims, but a lifelong one that somehow pervades, gathers up, informs, every aspect of how one engages with the world, and such a robust one that it is shared by others, it is difficult to see in what way it differs from the self as we conventionally think of it.

It is not a thing – of course. But there are other ways of being and enduring than that of being something like a pebble, a tree, or a bodily organ. Nor is it merely a "process", as Metzinger suggests, though this claim deserves a bit more examination.

A process, as it is generally construed, is the unfolding of the state of an aspect or region of what-is. It is typified by, say, the melting of a block of ice or the growth of a tree. While both processes and selves are characterized by internal connectedness, their mode of connection is fundamentally different. The successive phases of the processes such as the growth of a tree are mutually exclusive: the sapling does not have access to, co-exist with, the tree, nor the tree with the sapling. Likewise, the state of a block of ice at time t_1 is sealed off from its state at time t_2 and vice versa. In the case of the self, the states at different times have access to one another – most obviously through explicit memory, in particular autobiographical memory, where I recall something that happened to me, something I felt, or something I did at a time other than the present. That connectedness is evident when, in addition, I anticipate, dream of, attempt to bring about a future for myself, in the light of what I recall. It is present in the skills I deploy that I strove to acquire in the past with a view to this present that was its future.

The coherence of the self is not that of mere connection between successive states: rather it is a connectedness of sense-making in which, at any given time, that which is made sense of belongs to a personal world, and the sense reaches back and forth in time. It is actively maintained as I visit, talk about, my past and future, recalling the former and anticipating the latter. And, in a multitude of different ways, I reflect on myself: my behaviour, my track record, my prospects, my hopes and fears. All of these are features of the temporal depth that is central to the sense of who I am at any particular time

The reason many autocides deconstruct or debunk the self is because, as I have already discussed, it seems dangerously close to the discredited ideas of a spirit or an eternal soul haunting the body. Running a mile from anything that smacks of a religious soul or a Cartesian substantive spirit or mind still leaves the problem of knowing where to run to.

Others suspect that the notion of a self is the result of taking too seriously the "I" that is the subject of our life sentence; of thinking that grammar is a reliable guide to ontology; of mistaking the common tongue for an organ of revelation. That is why there is such resistance to acknowledging the rich, deeply connected, multiplicity of the perfectly ordinary self and wanting to reduce it to a succession of moments, instants, or a mere construct such as "a centre of narrative gravity"; why the self has to fight for due acknowledgment of its distinctive, un-thing-like reality.

6.5 THE SLIMMED-DOWN SELF: THE SUBJECT AS THE OWNER OF EXPERIENCES

Even those whose professional philosophical selves allow them to admit that they have – indeed are – a self are nevertheless concerned not to make too much of it. The slimmer the better. What are the options available to those who accept that the self is more than an illusion but do so only grudgingly?

The least ambitious way of allowing the self an existence in its own right is to make it an *owner* or *bearer* – of its experiences, body, life and life story, dispositions, habits, skills, accomplishments, powers, relationships, accountabilities, titles, and entitlements, and of course of possessions that may range from a fountain pen to an estate. All of these can uncontroversially be prefixed by "my". My ownership of disparate things – the itch in my arm, the arm itself, the experiences I had on holiday, my duties, responsibilities, preoccupations, status, or lack of it, what I call "my world", and of course my possessions in the literal sense – is a mark of the extent to which I make my history explicit in the way that no non-human entity (pebble, tree, a kidney) does. To own what I am – my experiences, status, offices, relationships, achievements, my situation with the kit and caboodle that prop up my days, and my life story – is to make myself explicit.

Some may feel reluctant to expand the notion of ownership in this way. Many of the "possessions" I have highlighted are hardly transferable: others could not lay claim to, even less steal, the itch on my leg, or my childhood. And they are not available to be passed around or sold on because when they are completed, they cease to be. There are intermediate cases. Parts of my body could be donated: my kidney could become your kidney. My body-as-a-whole is, however, mine and only mine and is so for its – and my – life. And while others can take over my duties, they cannot take over my experience of my discharging my duties. When I retired, someone else took over my professorial chair but they didn't take over my experience of occupying the chair as an outer layer of myself.

The sense in which my itch, my holiday experiences, my childhood, or my middle age, are mine is quite different from that in which my pullover, or my house, are mine. And there are profound differences between my ownership of an itch on my foot, of my foot, and of the sock I place on it. Indeed, the ownership of my foot seems to be half-way between my ownership of my itch and my ownership of my sock: the itch shares a location with my foot – indeed is necessarily co-located with it; and my sock shares with my foot the status of being a stable object. Even though it might be possible to induce an itch similar to the one on my ankle by applying to you the same species of mosquito as the one that bit me, your itch would not be the same as my itch because the itch that I possess, and you subsequently possess, are token itches that retain their unique relationship to me: its place on my body and its location in my life. Nevertheless, the first-person possessive pronoun seems appropriate as a way of capturing something real: the relationship of different kinds of mental entities to a subject, giving it (non-material, non-ghostly) substance. It seems reasonable to describe the itch from which I am suffering as "my itch" and the itch that is bothering you as "your itch".

Common ownership is one way of capturing the happenings and doings of the life of the self. The enduring subject who possesses them – along with acquired habits, skills, status, and so on – is invisible to the introspecting Humean philosopher who reported that "I can never catch *myself* [note: *my* self] at any time without a perception and can never observe but the perception". My itch is a perception but that it is mine is not an additional, or a modification of, a perception.

The case for the undeniability of this slimmed down self which is (unsurprisingly) invisible to the introspecting Humean being is well expressed by Ingmar Persson:

> Whatever else is necessarily true of a self, it is surely a *subject of experience* in the sense that it is something to which experiential states (and normally dispositional states, like desires) are attributable. It is plausible to think that experiential states, like perceiving and thinking, must have a subject – something doing the perceiving and thinking – just as much as they need an object – something being perceived and thought. If anything is to be a self, it must be such a subject. I call this *the owner aspect* of the notion of the self.[13]

This owner, this "I", is not added to, not merely accompanying, but implicit in, all my experiences – though I may make it explicit by reporting my experiences as experiences I am having, have had, or hope to have.

The matter of fact that my itch is indeed mine goes without being asserted. This has been expressed by Aristotle: "It seems that knowing, believing, perceiving, and thinking are always of something else but [also] of themselves on the side".[14] This passage also pinpoints something about the "selfness" of experience that it has a three-way explicitness. In making what-is explicit, it also makes itself explicit – itself the experience, and itself the experiencer. The last two elements, as we discussed in the previous chapter, are more evident in the case of intransitive experiences.

There is a case then for fattening the self at the very least as the owner of its experiences – riding above individual experiences in order to be the universal connection between, and the connector of, them. Ownership is a relationship. A relationship is between relata. What are the relata in the case of the kind of ownership we are talking about? One relatum will be an itch, a belief, an ambition, a life story. The other relatum will be the experiencing subject who has been built up out of the other experiences, stories, skills, that form the connected whole. The connection is possible because – to reiterate a point made earlier – while the material world in which we live is confined to the moment, the significance to an individual of what is experienced in that moment is not confined to that experience. The self is internally transilluminated across time by the sense or significance of what is being experienced.

The self-as-relatum is present even in the case of seemingly simple experiences. Take that itch on my foot. I feel it now and shortly afterwards affirm the unity of my eyes, my head, my arm, my hand, my trunk, and my foot, as I reach down to scratch it. As I do so, I mobilize memories of previous itches and bring to bear on it knowledge acquired at other times, from conversations or books, as to the nature, cause, significance, and treatment of the itch. The experienced itch, that is, is gathered into a nexus of knowledge and experience and embodiment that is unique to myself.

So much for the self as the owner of its experiences, its life. Is there more to the self than that? Dan Zahavi's assertion that "[A] theory of consciousness that wishes to take the subjective dimension of our experiential life seriously also needs to operate with a (minimal) notion of self [as a] built-in feature of experiential life"[15] seems unassailable. He then, however, speaks of seeking

> to identify a thin and minimalist notion of the self with the very subjectivity of experience. On this account, the self is present in experience, not as an additional experiential object or as an extra experiential ingredient (say as some kind of self-quale) but as the very first-personal givenness of experience. This notion of self is consequently not intended to refer to some persisting and abiding experience-transcendent self-entity but is rather meant to target and capture the ineliminable subjective and perspectival givenness of consciousness.[16]

This seems *too* minimal. My sense of being me is not merely present in the implicit truth that my experiences are subjective, first-personal, and had in, or from, a perspective. It is, as we discussed in the previous section, reinforced, indeed thickened in so many ways, not the least by fellow selves in the various communities to which we belong.

Other philosophers share this qualm. Marie Guillot examines certain features ascribed to our experiences as being intrinsic to them: *for-me-ness*, *me-ness*, and *mineness.*[17] Unfortunately, these prove extraordinarily slippery notions, not the least because it is difficult to know what they add to experience. The most tangible – and therefore perhaps vulnerable – is *mineness* in virtue of which I am aware that I am the owner of my experiences – in the very wide sense of their "belonging to" me that I have already discussed. My itch and my memory of a holiday belong to the same "me".

It is reasonable to question whether this "belonging to me" corresponds to a specific feature. Mineness does not seem even to be a *quality*, contrasted with, for example, "thineness". And no self would be in a position to make this contrast and hence to see it as a distinctive feature of their own, as opposed to others' experiences. I cannot experience your experiences in order to compare and contrast them with mine or see mine as a subset of the totality of human experiences – or not, at least, until they are articulated in a language we share, and even this does not capture the intrinsic, phenomenal aspect of either my experiences or yours. We both may agree that we have an itch in our foot and, may on the basis of our respective behaviours, conclude that your itch is worse than mine, and that it has lasted longer and kept you more awake, but this does not give us access to any distinctive "mineness" of my itch and "thineness" of your itch, even less a mineness common to all my experiences or a thineness common to all yours, a contrast that would also extend to our memories, thoughts, and emotions. After all, we might expect your experiences to be characterized by mineness, much as mine are. They would acquire the character of thineness only for another person who could not, of course, access them. I could never check the thineness of your experiences in order to compare them with the mineness of mine.

We would not expect belonging to a subject to be a distinctive – quasi-objective – feature of experiences. As we discussed in the previous chapter, it is part of the transparency of experience – it's being *of* something other than itself – that the experiencer should not get in the way, should not obtrude on it. The particular content of intransitive experiences such as itches and transitive ones such as sights experiences point away from the self. Besides, if the self were something that presents itself to me, through a universal quality of the mineness of my experiences, the self would be different from me in the sense of being an object of my experience. That is why Hume's observation that "I can never catch *myself* at any time without a perception and can never observe but the perception" is an entirely true report on the failure to catch the self in an inward-looking gaze, indeed any gaze, though this does not justify the conclusion that the self is a mere "bundle of perceptions", a conclusion, incidentally, that, as many philosophers have pointed out, would leave it unclear what, or who, would bundle the bundle. The conclusion is a consequence of looking for the self in the wrong place.

The very claim that mineness is *constitutive* of my experiences (and indeed of anyone's experiences) distinguishes them from things that are owned, given that being owned is not an intrinsic property of that which is owned. I have many possessions but there is no particular quality, even less phenomenal quality, associated with my possessing them. Yes, I may feel "possessive" in relation to them when I see my ownership threatened; generally, however, I am among them without relating to them in a specific way. The desk on which I am writing does not glow with mineness, notwithstanding that I own it. Indeed, most of my possessions are out of sight and out of mind until, say, I make an inventory for insurance purposes, or I am undertaking a decluttering exercise and deciding whether or not to keep them.

The other side of the fact that possession is not associated with a distinctive character or feel of a relationship to possessions in the usual sense is that, while I can discover that (say) a pair of glasses is mine after I have checked that they are not yours that happen to look identical, I cannot *discover* or even *be uncertain* as to whether an experience is mine. "Are those your glasses or mine?" is a reasonable question to ask; "Is this itch yours or mine?" most certainly is not. There would be no way of identifying the itch referred to – unless by pointing to a part of the body in which it is located, which would render the question superfluous. But it is also unreasonable because there is an important sense in which the experiences we possess possess us. The inescapable pain of toothache is not something that I can set aside or donate to others. By such possessions are we possessed.

Experiences may seem not to qualify as possessions on the grounds of their intangibility and, most importantly, their fleetingness: once they are had, they are gone, except, in the case of a minority: unreliable, quasi-narrative memories. Well, they can seem to acquire a standing character, to be solidified, and to be separable, indeed separated, from me, when they are reported and thus embalmed. It is in conversation with others that a pain becomes "my pain", "the pain in my ankle", "the pain I had yesterday in my ankle". Or a pain you experience becomes "your pain". But these are late developments, downstream of phenomenal experiences that, according to the ownership description, are to be directly experienced as "mine" – or as being "mine because experienced by me". If ownership were to rely on experiences being articulated, it would imply that they could be mine only after they have been distanced from me in the form of a report and this is clearly not the case. If there were such a thing as ownership of experiences, it comes packaged with them. On the other hand, it is not as explicit as it would be if it were possible that some of the experiences I had were not mine; that ownership applied only to a subset of my experiences. And this, of course, is not possible. All experiences, if they have owners, belong to the individuals who have them.

Some may feel that believing in the ownership of experiences is the consequence of taking the grammar of the language in which we describe our experiences too literally. Given that the grammatical form of reports of our experiences is highly variable, language is clearly not a reliable guide to the universal nature of the relationship between selves and experiences. In English, we say, "I am hungry" and, in French, "I have hunger".

Perhaps we should not dwell too much on the notion of possession. It might indeed seem too much of a hostage to grammar. What is important, it might be argued, is the relationship between the experience and the self and this may be expressed in a non-possessive form. It is, after all, equally valid for me to say, "I have an itch in my ankle" and "I am itchy in my ankle". In the former the experience is presented as a discrete item attached to myself via my body and in the latter as a less well-defined state of myself. When you and I contrast the ownership of your pain and my pain, we could have captured the difference as between, say, a pain in your ankle and a pain in my ankle. Ownership then passes from the experience to the body in which, and by which, it is experienced.

When I am *reporting* my experiences, there emerges a distinction between (relatively) intransitive bodily experiences – such as pains that seem ripe to be owned – and transparent, transitive, telereceptor experiences such as seeing a view. The contents of a visual field seem to belong to the outside world rather than to me, unlike a pain which can be spoken of as "my pain". Admittedly, visual experience typically acquires ownership when I look back on it as something experienced by me or I report it to others: "I have a lovely view of the landscape", "I had a ringside seat", etc. In some cases, the ownership can be shifted sideways to the organ of experience: "I saw it with *my own* eyes"; "I heard it with *my own* ears", etc. Or it can be mediated, as when I show you my experiences on my holiday through an interminable succession of photographs.

One aspect of the mineness of my experiences is that I have privileged access to them; I have a first-person authority as to their phenomenal content that cannot be over-ruled by others. I know what they are like. While you and I might argue whether a colour is bluish-green or greenish-blue, it is not an argument about what either of us is actually experiencing but how we should categorize that experience in a language which we share.[18] You may infer my experiences from my behaviour but not always and your inferences will often be inaccurate. I may feel a pain and indicate nothing about it or be irritable without telling you the reason for my irritability. And while I may be only intermittently aware, if at all, of your continuous pain, and choose to think about or not to think about it, my continuous pain does not allow me the luxury of intermittent awareness. And this is true of most experiences that are of less moment than pain or pleasure, such as the pressure I am currently experiencing from the seat beneath me or the sight of my hands moving over the keyboard. The privilege of unique access applies even more clearly in the case of past experiences. I can remember the pain I had yesterday, which has now passed, and appears to have no present significance, so I am not motivated to talk about it. To put this another way: I can hide my experiences, emotions, and thoughts from you but not from myself. Or not typically, anyway.

The mineness of my experiences, then, seems to be most apparent when I am reporting them to others who, of course, do not have direct access to them. This seems to make mineness a derivative rather than an immediate feature of experiences. This is no bad thing: we have in some sense to *encounter* our experiences – as if from without – to experience them as our own. As we have seen, mine is explicitly opposed to "your" (or "his" or "their"). That sharing my experiences highlights their mineness is therefore no surprise. Nor should we be surprised at – or, as defenders of the self, mourn – the extent to which our sense of our self is caught up with our sense of others as others and ourselves as other than those others. There is a dimension of the personal that is irreducibly interpersonal: we think about our experiences in the language we share with others, in our native tongue. Acknowledging, and being acknowledged by, others is necessary for our acknowledging ourselves – something we have touched on and shall develop. What is more, this deals with the apparent difficulty of my experiences being both mine – something that I "own" and hence is in relation to the self – and part of the fabric of the self. In the usual sense of the term, the owner is not built out of the things it owns. But we need to be careful that, by accepting the role of language in developing the sense of our ownership of our experiences, we do not allow the self to dissipate into discourse, thus giving credence to the suspicion that the self is a grammatical entity, the subject of sentences rather than the subject in the substantive sense that we have been defending. If we do, we shall find ourselves inadvertently siding with the autocides.[19]

Ownership of our experiences, then, is a difficult notion but it is not a lost cause. While ownership becomes fully explicit only when we talk about experiences to others – as when I refer to "my pain" – that explicitness seems to build on something that is implicitly present. If that is the case, we need a self, a someone, to own our experiences, enhancing their "me-ness". This will justify us in rejecting Humean autocide and, *a fortiori*, Dennettian or Metzingerian autocide.

Accepting the ownership of experiences, combined with acknowledging that the owner is the synchronically and diachronically fattened self we discussed in the previous section, also justifies rejecting the crash dieting of the self entertained by philosophers such as Galen Strawson.

For Strawson, selves are single, transient, episodes of experience. He echoes William James' view that each "perishing pulse of thought" is a self.[20] This doesn't allow space for ownership. Nor does it give a basis for the sense of ourselves existing over time; for the internal stitching that is most explicit in autobiographical memory, courtesy of which I am aware of something that happened to *me* yesterday or decades ago, in the facts of my life that I remind myself of, and in things I share with others as happenings to me. In the absence of a self that is extended over time, there is nothing to act as bearer of experiences, to fill the office of an "I" that experiences them and is aware of having had them as experiences. Most importantly, it does not account for something central to consciousness beyond mere sentience: the personal sense we make of what we are experiencing, and the broader significance we might attach to it, which draws not only upon what we are but what we have been, what we have come to understand through experience and knowledge.

One message we should take away from the arguments of autocides is that we should not exaggerate what is involved in ownership. It signifies that they are "my experiences", "what I am about", rather than possessions that I own in the way that I own my house or even my arm. We should anyway be suspicious of postulating too obtrusive a version of ownership – a taste of "mineness" – that is supposed to characterize *all* experiences. It would, as we have noted, have no distinctive content if it were common to all content. And while the owner of her experiences is not the sum total of all her experiences, she is more, much more, than successive moments of experience not adding up even to a succession. Having said this, it is not true that the ringing I have in my ears at the moment is owned by a combination of the individual who worried over an examination 50 years ago, laboured up Helvellyn mountain as a ten-year-old, and planned a trip to Prague last year, and everything else he did and experienced in between. There is, nonetheless, a deep connectedness between the successive moments, minutes, hours, days, months, and years of our life. That connectedness gives substance to the person who *has* the itch in an admittedly attenuated sense of ownership that makes it not in the slightest bit eccentric for him to report that he *had* an itch yesterday. Ownership seems less vulnerable than Kant's "thinkership", according to which every experience is claimed as something that an "I" thinks. Or that there is an "'I think' to accompany all our representations".[21]

There is something additional – that is at once more profound and yet lighter – to what we have tried to capture during our examination of the concept of ownership as applied to the self. When I have an experience, the experience points in two directions: towards that which is experienced; and towards the "I" that experiences it. It sometimes seems as if the inward-looking face of the experience is more evident in the case of intransitive experiences such as itches than transitive ones such as views. However, transitive experiences can be more freighted with me-ness when they are the fruit of active observation, as when I pan round in search of something and finally locate it. "Got it!" is an achievement. Be that as it may, in all cases, the it of the object of perception and the I of the subject support, give weight to, each other in mutual otherness.

My interactions with the world revealed by my consciousness implicitly reveal me as a being in that world. The self, however, while real, should not have too obtrusive properties – so that it does not, or would not, get too much in the way of revelation of what is out there. At the heart of the self that borrows substance from the body in virtue of which it is located in space and time, is a consciousness that points to itself. In making what is out there explicit, it implicitly makes itself explicit. That-it-is and that-I-am are inseparable.

The conclusion to be drawn from the discussion of ownership is that, while there is a sense in which we possess our experiences, the possessor is not to be found as either an entity, or an object of experience, or (*pace* Hume) an experience among those same experiences. While the self is seemingly obtrusive, as when we highlight "I" in reflection, its true nature lies in its being implicit in explicitness and making itself explicit indirectly as that in virtue of which other things are made explicit. It is not identical with whatever is made explicit; not even with the body as it is lit up in different ways as I experience, suffer, loiter in, act upon, the surrounding world. "Where are you?" you call and I respond, "Here I am" or "I am here in New York" and the "here" is the location of my body, as if "I" and my body shared the same location. But this is an appropriation: the self that experiences pains and thoughts, is not anywhere in particular, nor nowhere. Being embodied, it is inseparable from the body but not identical with it. To that last point we shall return.

6.6 FATTENING THE SELF: THE UNITIES REVISITED

We have tried to identify what is valid in the idea of the ownership of experience, what there is more than a merely grammatical connection in the idea "I *have* a pain", or "I *have* a view of a landscape", or indeed "I *have* a thought", or "I *have* a memory of such-and-such". There appears to be a relationship between the relevant experiences (one relatum) and an owner (the other relatum). What, however, constitutes the second relatum if it is not a body, a soul, a Cartesian spirit, or a Kantian transcendental "I", or "an '*I think*' that must be able to accompany all my representations"? What is it that makes something substantial enough to be an apparent "owner" of experiences, something at any rate more substantial than a grammatical subject?

The answer is not straightforward, but we have already assembled some of the elements. The substance of the self is woven out of a synchronic unity (a connectedness between explicitly distinct, or even disparate, elements at a given present moment), a diachronic unity (connecting the individual's life at explicitly different moments, days, and years). In both cases, the self is the common factor in the connected elements. In addition, to links between occurrent elements, there are certain standing elements – such as beliefs, attitudes, prejudices, skills, familiarities with aspects of and parishes in the world, bodies of knowledge, and acknowledged duties. There are other components of an external scaffolding that we shall discuss in Section 6.7. And then there is the living, and lived, body which we shall discuss in 6.8.

In this section, I want to return to the synchronic and diachronic unities. In doing so, I shall enter territory that I have examined in more detail elsewhere.[22] I revisit this, and other territory in the present chapter, less because it is new in my writing than because it is here illuminated by the central preoccupation with explicitness.

Synchronic unity is that in virtue of which the self at time t_1 is in some important respect "one", not in the sense of being mathematically unitary than of holding together with distinct parts interpenetrating each other. *Diachronic unity* is that in virtue of which the subject or self at time t_1 and at time t_2 are numerically the same, despite constant – usually, but not always, gradual – change over time. Both unities are modes of connectedness wherein the elements that are linked are both together and apart. They are explicitly *co-present* in a way that is unique to personal identity and fundamentally different from the co-existence of parts that characterizes the objective or apersonal identity at and over time of material objects such as sticks and stones, of collections of material objects such as patterns or landscapes, or of institutional entities such as clubs and nations.

6.6.1 Synchronic unity

A stone at time t_1 is "one" in virtue of cohering, within a boundary such that it is registered as a single item in the eye of a beholder, whose visual acuity is not sufficient to see the spaces between the atoms, and who is committed as we are in daily life to the folk physics of manifest reality. At the level of the macroscopic gaze and of our interactions with it, it is continuous within its limits, its outer surface. By contrast, a person such as RT at t_1 has not only an external, visible unity but also an internal invisible unity. His external unity is evident to a naïve observer and is underwritten by the continuity of flesh. "RT came into the room" implies that "RT's body came into the room" and that body is assumed to be associated with a single subject. His internal unity – his experiencing himself as "one" with a world as his cognate object – is, however, more complex and more interesting.

It is more complex because his consciousness at any given time comprises, or at least is divisible into, a multitude of profoundly diverse mental items – perceptions, thoughts, memories, emotions, intentions and so on. They are distinct – they are available, for example, to be attended to or recalled separately – and yet they belong to the same, irreducibly rich conscious moment of the subject RT. An itch on his face, the sound of music in his ears, and a thought about New York can belong to a conscious moment. Understanding this unity-combined-with-multiplicity, which has been described as "the binding problem",[23] has proved an impossible challenge for those who would wish to identify consciousness with neural activity that is necessarily scattered over the brain and parceled off in different pathways and modules.

Neurones are intrinsically neither together nor apart. They are separated in space and may be side by side in bundles. While this separation may be cancelled by convergence at a synapse (where nerves contact each other) it will result in the loss of the individual character of the activity of the converging neurones. Large numbers of discharges add up to patterns only in the eye of some beholder – typically a neuroscientist who extracts and combines the elements in question, though leaving the basis for this neuroscientific activity unexplained by his own brain activity.

The problem of togetherness-and-separateness, or merging without mushing, is encountered at quite a basic level; for example, bringing together the size, shape, colour, spatial proximity or distance, etc., of material objects in a visual field while they remain available as objects of separate attention. Even if modularity of the brain were able to account for different modes of perception and different cognitive functions (which, as we have seen in Chapter 3, it doesn't) this would only exacerbate the problem of unification. The greater the number of discrete or functionally specialized brain parts that are postulated, the greater the problem of the unity of the conscious moment. If neurones responsible for signaling the edges of objects, their colour, their location, and the category to which they belong, are distinct, we are presented with a challenge of understanding how we can experience, in a single state of consciousness, an entire scene populated with a vast number of intelligible objects set out in a meaningful relationship to each other, or at the very least cohering in a scene, in which they can nevertheless be attended to separately.

In short, the binding problem of the unity of the conscious moment is also an *un*binding problem. We experience things together and yet keep them apart or distinct; or more precisely we experience together things that we also experience as distinct and hence apart. We see that the cup is green *and* of a certain size *and* filled with coffee *and* on a saucer *and* over there *and* ... *and* ... *and*. The itch on my face, the sound of music, and the thought of New York, do not lose their identity in their co-presence in a moment of consciousness. The idea that specialized neural activity converges on a single set of neurones somehow "integrating" their activity leaves unexplained how the elements of the visual field – ranging from a cup to an entire landscape – are not merged into a mush of givenness, a splatter of thatter. When the landscape is gathered up as one, it still retains its status as many. This merging-without-mushing, what is more, has to take place in the context of a rather noisy brain, most of whose activity is not associated with consciousness.

When we go beyond the synchronic unity of perception, and the integration of looking and thinking, hearing and remembering, all of which may take place at the same time, and incorporate and be incorporated in the higher-level unities of our bodily self-presence – such as body image and the body schema – the plot thickens further. The coherence of my sense of being here, thinking this, remembering that, being related to things around my body and recalling things from the past or mediated to me by discourse, while I am weightily, bodily here, is unlike anything we see in the natural world.

Irrespective of whether at any given moment I can be said to have two separate experiences, or even two distinct conscious states – and some philosophers challenge this[24] – within that moment are *separable* experiences. I can see a scene or choose to focus on an element in the scene or on an element within that element. When I observe the landscape (as a whole) I remain aware of the possibility of attending to the field, the cow, the birdsong. When I see three cows, I see three cows not an agglutinated threesome without individual members. The whole does not supersede or blot out the parts. The aggregation in which we still retain distinct consciousness of the elements – which is difficult enough to understand – is not merely a heap but a genuine synthesis into something that has a coherent meaning. The composite nature of the moment of experience is not lost in, superseded by, the composition: the components are still there, up front, available for me to point out to myself or to others, or to recall separately.

In short, the sensory field reveals the entities that populate it as not merely individually present but also collectively co-present. For the most part, they are not present to each other. A pebble and a bird's song and a cloud may exist in the same patch of spacetime but it is their common presence to me that makes their separate existence a co-existence. Our ability to bring together elements of experience into a scene and yet focus on them individually as required rules out physical processes, even exotic ones such as quantum entanglement, as the basis for the unity of the moment of consciousness.[25] The ability to quantum disentangle at will does not, I think, lie within the capacity of macroscopic subjects such as you and me.

But the explanatory challenge goes beyond this. In daily life, I move not only from scene to scene, and from physical setting to physical setting. I pass from situation to situation. Their evolving intelligibility draws on my past experience, declarative (factual) memory, procedural memory (retained skills – including observational skills), autobiographical memory, and a variety of less tangible inputs to deliver the necessary sense of familiarity, which licenses my sense of this being "my hand" in "my study" in "my house" in "this neighbourhood" in "my life" and in "my world". My confident and usually justified awareness of where (in the widest sense) I am at this moment, of what is happening, what might happen next, and what is expected of me, all have to be in some way united in a moment of consciousness, while being available to be separately registered or attended to, remains utterly baffling.

One final point on synchronic unity. The separateness and connectedness of the elements of the moment of consciousness – such that I am aware of this room, *and* its individual contents, *and* the way it is illuminated, *and* where my being in this room at this moment fits into my life – replicates at a higher level the separateness and connectedness that is evident in the intentionality of simple consciousness directed at a particular object. My being aware of the cup on my desk connects me with, and maintains my separateness from, it. The intentionality of the consciousness that makes both of us explicit opens a space in which it is over there and I am here; I am I and it is it.

Thus, the fatness of each moment of the self: not so much a Cartesian/Kantian *I think* but an *I am*. The moments are sufficiently substantial to be able to take on the role of ownership of individual experiences. The seeing "I" is substantial enough to gather up, and makes sense of a landscape, and to possess the elements of which it is composed. We are already far from Hume's trickle of successive perceptions.

6.6.2 Diachronic unity

What I have described as the synchronic unity of the self in fact draws on past experience. Indeed, some diachronic reach is necessary for any experience that extends beyond a notional moment. Without unity over time, the experience of a symphony would be a mere succession of experiences of notes, and a meal would be broken down to separate mouthfuls and the mouthfuls experienced as a succession of isolated gustatory or masticatory sensations. In other words, the self is not divisible into moments or time-slices as, at any given time, it explicitly or implicitly overflows the boundaries of that time. What makes sense at a particular time, makes that sense because it reaches across moments of time. In the case of a changing material object, its state at time t_2 obliterates its state at time t_1. This is not so in the case of a person or self.

There are, however, particular challenges posed by the explicit identity-over-time of the self.

The most explicit inner markers and sustainers of the diachronic unity of the self are memories and, among these, the most striking are autonoetic memories – memories of personally experienced events recalled as having been experienced by one's self. They are memories of our own past which incorporate the knowledge that the past *is* our own past, a past in which we (re-)locate ourselves: I recall that "That happened to me", "I was there when that event happened", "I did that action" without recourse to records of objective facts.

In autonoetic memories the self, and its temporal depth, are made explicit. They have a triple intentionality:

1. There is the initial intentionality of the experience that subsequently forms the content of the memory. The experience is packaged with an awareness that it is *about* something other than itself – for example, an object in the weighty sense that is incompletely revealed by experience.
2. The secondary intentionality is that of the memory-now being explicitly about the experience-then. That intentionality is manifested in the explicit sense that the remembered event is at a temporal distance and belongs to the past. In the hinterland, there is the sense of the (past) self-world in which the remembered events happened.
3. This is highlighted in autonoetic memory whose third-order intentionality has as its intentional object the self whose memory it is. (And when we tell others about our autonoetic memories, there is perhaps a further, fourth-order intentionality opened up by the utterance in which we report it – to ourselves or to others.)

Autonoetic memories come tagged with "pastness", indeed my-pastness, which is how they reinforce diachronic unity, and the sense of temporal extendedness or depth of the self, of the person who remembers them. The that-it-is of experience becomes that-it-was of memory and that-I-was who remembers it. This makes them interestingly awkward.

Suppose the event I autonoetically remember took place on Wednesday. I recall it (perhaps with some considerable emotion) on Thursday. If we go with the standard naturalistic story (criticized in Chapter 3) the reason I experienced the event on Wednesday was that the event had a causal impact on my nervous system. The resultant effect in turn set in train more events in my nervous system, causing the event to be "registered, encoded, and retained" – to use the standard terms. When I recall Wednesday's event on Thursday, my direction of recall – the secondary intentionality of memory – reaches *backwards* along the causal trail to 24 hours previously. Mysteriously, I leapfrog the many other intermediary events in the 24-hour causal trail connecting the present with the experience which I locate in the past world in which it happened. When, on Thursday morning, I remember an experience I had on Wednesday afternoon, I look through Wednesday night.

I can of course leapfrog any temporal distance, in a session of reminiscence, of intimate togethering-for-its-own-sake. It may be driven by factual memory. You say to me: Do you remember the day we visited such-and-such last summer? I do remember it; and I may then unpack the whole day, like a pop-up shop, with a series of mental images shot from the angles from which I experienced the view. The countless experiences intervening between now and last summer obligingly efface themselves so that the interval does not get in the way of my recall.[26]

Yet more impressive is the way memories precipitate out of blankness when I do something that is usually called "racking my brains", though it should be called "racking myself". Even those who embrace neural accounts of consciousness would probably find it difficult to think of one part of the brain racking another part of the brain at the behest of yet another part of the brain. Such voluntary, effortful recall is a curious activity: we adopt a holding position, feeling ourselves to be in the vicinity of the things we are trying, at our own or another's behest, to resurrect, readying ourselves to pounce when the cognitive prey stirs in the thicket of forgetfulness.

Of course, there are false memories, but they do not detract from the true memories that can be requisitioned to order; or from the miracle of shared mental time travel. You say, "Do you remember when we ...?" and we converge and convene on an event that is no longer. It is an even more dramatic reminder of how memory is a kind of time reversal, looking temporally upstream from the present state of things to a prior one.[27]

The enigma of the explicit sense of pastness underlines the challenge of trying to understand diachronic unity – or at least connectedness. This is exacerbated by the further challenge of explaining how I have *ownership* of the bit of the past, or of the past me that experienced that which is remembered. I reach upstream to a self that I was, in order to affirm that the self that I am now is the same self. RT at time t_2 reaches back to RT at time t_1, who had the experience that is remembered. That RT *had* – enjoyed, suffered, owned, sought, made possible – the experience is perhaps a background assumption, but it cannot be too far in the background because there is an element of self-affirmation in the sense that that experience, and the world in which it took place, was *mine* and the person to whom it happened was *me*. The explicit past tense, and a connection forged between a phase-moment-world of someone who is now and a phase-moment-world of someone who is no longer and hence not in any sense available to physics, adds to the deep mystery of our self-affirmed unity over time. Not the least aspect of this mysteriousness is that the memory, which refers to an experience no longer directly attached to an embodied subject moored in space, located in time, is nonetheless individuated, retaining its status as a singular experience – experienced by a singular individual. It retains its place in a world-slice anchored to a past self living and breathing at a particular time.

Relying on actual memories to look after the diachronic dimension of the self may seem to render it vulnerable: we remember so little of the past. We can look back at yesterday and exhume a few of its experiences, though we may have an exaggerated idea of what we remember because its trillions of sensations are gathered up in a factual (general, discursive, communicable) record of what I did or what happened. This preserves little of the moments of consciousness. The thought that "I went for a lovely walk in the woods", possibly illustrated by a few scattered images, captures a minute portion of the experiences that constituted going for that walk. The movement of a particular leaf on a tree, the sound of a bird singing at time t, the similar sensations of my feet in my shoes as I pass 5,000 times through the cycle of left and right footsteps – most of these will be lost.

The further I look back, the less I remember, though I may alight on distant experiences corresponding to specific events – or factual accounts that masquerade as memories of such experiences. Even this patchy past may be only intermittently present. We have limited time and inclination to revive the contents of past wakefulness, since the present and its future have first call on our consciousness. If this were not true, our past would place an insufferable burden on the present: no present moment could keep alive a million previous moments. What is more, if, at any given time, much of the past were not consigned to temporary or permanent oblivion, it would not be possible to have, or to revive, distant memories – to look through the intervening thousand days and trillions of sense experiences to remember something that happened three years ago. The path to the more distant past would be blocked by the many-layered, thickly populated, nearer past.

All of this justifies the concern that explicit memory is inevitably a bit too patchy to carry much of the burden of holding ourselves together over time. The recall of memories is somewhat capricious, subject to the accidents of the present that may trigger recall. RT *memorius* as the curator of RT's temporal depth seems somewhat capricious and this may have alarming consequences. John Locke, for example, who famously argued that the continuity of the self was to be found in continuity of memory, concluded that things we cannot remember are not part of us. As the past as an object of recall evaporates, so the person whose past it was is no longer the same person as the present self.[28]

Thomas Reid challenged this counter-intuitive conclusion with the example of a soldier who was flogged as a boy at school, exhibited conspicuous bravery in his early adult life, and was promoted to being a general in old age.[29] The young soldier remembers the boy who was flogged and so is the same person as him. The general remembers the young soldier and so is the same person as him. The general cannot remember the flogged boy and so is not the same person as him. From this it seems to follow that the boy and the elderly general are and are not the same person. There is clearly a problem with Locke's making memory the metaphysical ground of the self, of personal identity. It does, however, provide first-person evidence of personal identity. My memory of my first (terrifying) day as a junior doctor is empirical evidence that I was that junior doctor over half a century ago. H. P. Grice responded to this by suggesting that each total temporary state of a self contains an element – a remembered experience – shared by the preceding total temporary state. There is, that is to say, an overlap between successive moments of the self. This will not be true of the total temporary state of different selves – such as me and Queen Victoria.[30]

This is not entirely reassuring. The idea that the continuity and continuation of the self is assured by overlap of successive moments reduces the self-over-time to something like the overlapping threads of a rope. I am not sure that being like a load of old (or ageing) rope does much justice to the rich connectedness of first-person being.

This concern can be addressed by reminding ourselves of the extent to which the past is sedimented and present in standing elements, rather than mere occurrent experiences and episodic memories. These will include our body of factual knowledge (knowing that), acquired skills (knowing how), assumed duties, circles of familiar friends and colleagues, and the implicit journey that has taken us to our present situation. These will supplement recollections of specific events that took place in our own life and associating them with a personal past. My performance in middle age as an experienced doctor running a specialist clinic stood on the edifice of the successive pasts in which I studied to qualify to enter medical school, worked to qualify as a doctor, built on my experience as a practitioner, engaged in refresher course to bring me up to date and so on. And my acquired skills, such as they are, will be made more explicit, and indeed reinforced, by my parallel role as a teacher of medicine, sharing what passes as my expertise in lectures and practical demonstrations. While the present expression of my standing characteristics incorporates the past, it does not swallow it up. I can still explicitly recall some experiences and events that marked the journey. These episodic memories mark the distances I have travelled, keeping alive the headlines and small print of the curriculum vitae that led up to the present moment.

This sense of oneself as an enduring entity can be further validated by the sharing of experiences we have referred to. "Do you remember such and such ...?" enables us to point at a distant moment which another can also look at, affirming the reality of a moment of the past and hence temporal depth of both our enduring selves. The past that is my past (overlapping with "our past" where the co-owners of the past in question will vary from occasion to occasion) can also be curated indirectly by the permanence of the objects that form our surroundings, most importantly our possessions. So long as we have a memory, we are located in a world of mementos, of souvenirs, and their endurance over time is a mirror of the endurance of their possessor. Their current presence stands for the presence of the past. And the signs of wear and tear, their connection with a particular era of our lives, add to their capacity to signify an outer past of material objects that is also our individual and collective past. The clothes I used to wear, preserved in photographs or a wardrobe, the street I used to live in, are external correlatives of my own past.

As I remember them, I am remembered, and my temporal depth and continuity is affirmed. There is perhaps more *je me souviens* in the everyday objects – the desk on which we write, the armchair in which we sit, the carpet on which we tread – than in the explicit souvenirs we acquire to commemorate high days and holidays.

It is worth developing this a little, if only because it would be a corrective to our earlier focus on memory as specific memories, as experiences explicitly remembered – either spontaneously or effortfully – as guardians of the past. We may think of specific memories as too focal, episodic, too much like thin strands connecting us with our past. There is something that goes beyond discrete memories, though it is connected with it. It is the broad, deep feeling of familiarity from which a boundless nexus of connectedness between present and the past may from time to time precipitate out as specific memories. That familiarity is most deeply and widely rooted in our companions (particularly our nearest and dearest), our workplaces, our routines, and the rooms, houses, and streets we dwell in. Their relative stability binds together its, and our, days.

Each time I enter my study I am enclosed in a space that stands for my continuity, reinforced by a familiarity so ingrained that I take it for granted. In taking it for granted, I take myself – my continuing, time-deep self – for granted. Even where I am required to "rack my brains" to remember where I have put something (usually a mobile phone or a notebook hidden under a sheet of paper), specific memories are sought against a background of items that deliver their payload of my past without my having to work to extract it. My familiar environment (persons, places, and things) amounts to a sea of aides-memoire that underwrite my continuity over time and do so without protesting the fact too much. And shared spaces stand surety for shared lives.

The ultimate memento is our lived body whose enduring public and private features – ranging from our face in the mirror, to the sensations in our legs – reassure us that we are what we have been. There are (usually gradual) changes but change is typically trackable over time. Even in the case of catastrophic injuries which may render us unrecognizable in the mirror, there is a continuous story that connects our bodies with their earlier states. The continuity, that is, is the continuity of a lived body, with no moments left out, such that even when we say we no longer recognize ourself, we recognize the self that cannot recognize itself.

To recap, the past with which we are connected, and which is the prior phase of a continuing self whose substance goes far beyond the successive time-slices of a flow of experience, and way beyond autonoetic memories, important though these are, is curated by the world of familiar things, a world which takes us for granted. (Perhaps, they are the gold that underpins the paper currency.) While the past is filled out by factual memories preserved individually and collectively, the sense of the who that I am – and what I might expect of myself and what might be expected of me by others – is enormously widened and deepened by the familiarity of people, places, and things, which makes of them a trove of largely unspoken memories and holds open the connection between the past and the future. This temporally deep and existentially wide sense of connectedness between my no-longer and my not-yet adds substance to the self, something which will be extended, deepened, through the role of the self as an agent in a sea of duties, rights, and responsibilities– to which I shall return in the next chapter.

When highlighting the continuity of the self over time, the focus quite properly tends to be on links with the past. Mental time travel, however, also goes forward into the future. Indeed, "we remember the past to imagine the future".[31] The past we draw upon gives specificity to the future for which we prepare or which we aim to shape. It is from the past that we derive the instructions, hopes, and obligations that have to be fulfilled in the future. This can work at many scales. Remembering that I have just turned left will place the next (right) turn in my immediate future as I follow a route. Knowledge that I have completed the first year of my studies will designate the year to come as the second year of my studies with all that follows from that. Each moment of our lives makes sense in the light of the afterglow transmitted from the previous moments of our lives.

There are some particular striking instances of the presence of the past that go beyond memory and touch on our nature as conscious, responsible agents. We have *ownership* of our past and indeed may own up to it. Mistakes and sins remain ours however remote they are. Seventeen years after retirement, I still feel responsible for the things I got wrong as a junior doctor. And we can still be brought to a halt in the street as we are struck by a memory of something stupid, crass, or embarrassing that we said or did half a century before. And, less dismally, we own our achievements.

6.6.3 Synchro-diachronic unity

It will by now be obvious that that there is no sharp distinction between the synchronic and diachronic unity of the self. It is of the very nature of the self that they interpenetrate:

1. Scanning what is present (for example a landscape) will generate a train of successive experiences that have to be tied together. That which is explicitly co-existent is necessarily experienced successively and those successive experiences are temporally united.
2. Something as basic as a sense of the fundamental stability of our surroundings, manifested in the inertia and constancy of the objects in it, also grows out of successive experiences united into a synchronic reality. Stasis builds up over time.
3. Likewise, the elements of the awareness of an object mediated by sight, sound, and touch are both simultaneous and successive.
4. The meaning of a present scene or of a present object draws on an understanding nourished by past experience and points to future possibilities. Immediate and extended familiarity converge in the seemingly unmediated sense we make of what surrounds us – as when I see an object as "my cup" or an eyeful as "the view of my garden".

5. The sense of the past – of a past event or of objects that have a past (mossed stones) – has to be experienced in the present: it is the presence of the non-present.

Perhaps the most striking localized fusion of the synchronic and diachronic self is in the realm of emotion. At any given time, an emotion may draw on a complex nexus of experiences, memories, knowledge, and reflections. This is clearly true in the case of rich, profound emotions such as love; but it is also true of spikes of feeling such as anger. We get angry for reasons that go beyond a provocation understood as a mere stimulus. Our anger may draw on our sense of who we are, our entitlements, our memory of things that have happened in the past, and general principles as to how things should be that justify our feeling – all aspects of the way in which we articulate a world and our place in it and the place in that world of things and above all people that are the objects of our emotion.

The depth and breadth of emotions are, of course, more obvious in the case of sustained feelings – as when we chew over things, feed our enduring resentments as to what is our due, or anxiously or joyfully anticipate future events. A refusal to forgive another is sometimes felt to be a marker, an affirmation, of integrity. Because what happened mattered then, it must matter now: thus, the past and present are co-present in a unified self. This may, however, be equally true of those cases where I choose to forgive and forget, to "move on" – where the self who moves on is more profoundly true to himself than the one who endlessly licks old wounds.

The fundamental and universal connectedness of the self ultimately resides in the ubiquitous fact that experiences are made sense of and that sense is made in the light of other experiences. We may describe this as an internal transillumination of any given moment of our lives by other moments in our lives.

6.6.4 Failure of unity

There has been considerable interest in individuals who, for a variety of reasons, seem to have *dis*unified consciousnesses – split brain subjects and the like. These are philosophically less significant than many seem to think. After all, for most such patients, there are still impressive, explicit connectivities across space and time. Indeed, it takes rather clever experiments to uncover divisions in consciousness. What is more, recent studies have suggested that, while dividing the corpus callosum connecting the hemispheres splits visual perception, it does not create two independent conscious perceivers within the brain.[32]

Others have emphasized the extent to which unified consciousness is undermined by unconscious mental forces, whose influence is hidden from us. And yes, it is true that we are not entirely transparent to ourselves, that we cannot always account for our feelings and behaviour, that we may be unaware of much that impinges on us from within. However, there still remains an overwhelming majority of connectivities of the kind we have been discussing. Indeed, they form the multi-layered background of every ordinary moment.

Our identity – evident at many levels from experiencing the unity of the scene before us, to grasping a coherently unfolding situation, to the internal stitching of the subject through memory – remains the robust miracle we have been thinking about. And this is true even of those split-brain subjects who took part in the studies that brought Roger Sperry his Nobel prize. They had sufficient residual self to agree to participate in his experiments, to make their way to the laboratory, to understand the purpose of the study, and to report their experiences.

6.7 THE EXTERNAL SCAFFOLDING OF THE SELF

So much for the internal connectedness of the self. We have travelled a long way from the thin trickle of perceptions that Hume encountered when he looked in vain for a self corresponding to David Hume: "For my part, when I enter most intimately into what I call *myself,* I always stumble on some particular perception or other, of heat, or cold, light or shade, love or hatred, pain or pleasure".

His philosophical gaze picked apart the contents of his consciousness and reduced it to a series of atomic perceptions that were, what is more, homogenized so that "heat or cold" were lined up with "love or hatred" as if the latter, like the former, were mere tingles of consciousness, disconnected from anything else – from the history and dispositions of a person who loves or hates.

What is most strikingly absent from Hume's account of the self are what we have called "standing" elements of the self-conscious subject – "the internal scaffolding" that holds it together. They include a body of factual knowledge, capabilities, embraced, affirmed, and enacted beliefs, habits, personality traits, long-standing projects and ambitions, commitments, and relationships with others. This "internal scaffolding" cannot, however, be separated from the "external scaffolding" that is created by our relationships with others

– individually, collectively, and institutionally. It is this that gives the self stability, solidity, a shared reality. Internal and external scaffolding in many cases interact, as do synchronic and diachronic unities. Obvious standing elements include beliefs – most developed in political, ideological, secular and religious convictions – hopes, desires, fears, dreams, plans that translate into enduring projects, long-standing relationships with others, ways of conducting ourselves that reflect us back to ourselves

Something as innocent and seemingly trivial as building up a stamp collection, developing expertise at a particular computer game, or bagging the Munros (Scottish mountains over 3,000 feet for the non-cognoscenti) will bind days "each to each" and thereby locate the successive moments of the self in a larger framework. Less innocently, though perhaps more compellingly, an obsessive love affair will connect the lover over time through the many different expressions of his love. The painstaking acquisition of a skill will likewise connect the hours of the tyro with those of the accomplished performer. The past is present in muscle and other aspects of procedural memory that makes us the beneficiary of the hours of practice that account for our present fluency.

Beyond this, the sustained duties we have to others – our partners, our children, our parents – will bring non-illusory consolidation of our sense of our self. Where this contractual dimension of the enduring self is formalized in an office we hold, a role we are signed up to discharge, our days are even more clearly stitched together, even if at times we feel stitched up as a result of placing our signature to a promise. The diligent schoolboy, the hard-working medical student, the harassed junior doctor, and the conscientious consultant, make sense of each other, as possible futures becoming remembered pasts. We are forever becoming inner or outer emeriti.

The external aspects of these modes of connectedness may provide an independent scaffolding for the self. Our self-recognition (or re-cognition) – "this is me", "this is still me" – is reinforced by the acknowledgement and recognition we receive and sometimes solicit from others, as we are mirrored to ourselves in shared worlds.

What others expect, in some cases require, of us is an important part of this. This is most explicit, formalized, indeed concrete, in the duties and various offices already mentioned. It encompasses a variety of settings ranging from the domestic to the occupational. To return to an earlier example, I enter my house and am greeted by my wife and children into whose continuing story I enter. I am the opposite of a stranger, the police are not called, conversations, and cooperation on a thousand threads of activity, are resumed. Familiarity in the eyes of familiar others reinforces my familiarity with myself. It is hard to overestimate the self-affirming importance of being taken for granted.

At work, I am likewise taken for granted and I make connected sense to myself in the sense others make of me, in the instructions I receive, in the duties that are assigned to me, in the instructions I give to others, and the duties I assign to others. All of this is reinforced by convergence in family groups, circles of friends, and – most explicitly – teams working together on projects or, over a longer period, delivering services and in the process acquiring, as a bonus – or sometimes as a burden – shared histories.

The role of acknowledgement by others in maintaining one's sense of one's self, of sameness over time, is brilliantly explored in Emmanuel Carrère's *The Moustache*.[33] A man shaves off his moustache for the first time in 10 years for a joke. He braces himself for others to comment on the transformative impact this has on his facial appearance. When nobody seems to notice what he has done, and his wife denies he ever had a moustache, he is deeply disturbed. This disturbance becomes more profound when she – and their mutual acquaintances – fail to affirm important facts about their lives. The scaffolding of his sense of his identity is gradually dismantled. His mind starts to disintegrate.

We acknowledge those who acknowledge us. There is a loop of reinforcement that begins with the return of the gaze. "You" and "we" reinforce "I"; and "I" reinforces "you" and "we". It takes a village to raise a child; and it takes a world, or a series of overlapping worlds, to underpin a self. This can be transient as when another's gaze acknowledges my selfhood or, when mocked, I feel skewered to a self of which I may feel ashamed. Even being ignored, when I think I should be attended to, can reinforce my sense of self, through outrage at entitlements that have not been respected.

One of the most profound modes of acknowledgement of, and by, others is the making and keeping of commitments. It was not for nothing that Nietzsche characterized man as "the promising animal" – a creature that is answerable for its own future.[34] Behind the making and keeping of promises is the strongest reciprocal sense of the reality and continuity of oneself and of another's self. Honouring the most trivial of commitments involves both parties in the time of the clock, the calendar, and the diary. And there is a shared sense of structured space; of a communal "out there" formatted in a map both parties share. The topic and purpose of the meeting, the preparation for it, and the journey to it, converge in affirmation of each other's existence. Our plan to see each other in a few months' time is built on and builds up our assumption that we are enduring selves who have a continuing existence in the interim, validated when we meet and recognize and are recognized by each other.

While "we" is central to the I, the I does not entirely dissolve into the "we". There are obvious reasons for saying this. Raymond Tallis is continuously in the company of Raymond Tallis, while he is only intermittently in the company of any other member of the "we". And there is an ever-changing membership of the "we" to which he belongs. He is the only common element, the only person who belongs to all the clubs and circles to which he belongs. Finally, there are still the private, sealed off elements of his consciousness that have to be communicated in order to be known by any other member of any we. What is more, communication of some elements may prove beyond the capability of the "I". While any communication about the nature of his experiences, even when he articulates it to himself, requires him to use the language of the collective, this does not mean that there is not a private dimension to the language he uses. In short, "I" depends on "we", but it does not dissolve into a sea of "we".

I have skated very superficially over a territory that has bottomless depths and boundless width. One has only to think of the interactions freighted with meaning that are delivered by gestures and greetings and – of course – by facial expressions. We literally and metaphorically face each other in so many ways and the resulting "inter-faces" – woven by smiles and scowls, looks of puzzlement and agreement – add to the "who" that each of us thinks we are. Doing justice to these modes of mutual affirmation – when "my face in thine eye, thine in mine appears"[35] would be sufficient to occupy the pages of a book even fatter than the present one. I will simply acknowledge them without attempting to do them the beginning of any kind of justice.

External scaffolding of the self allows a distance between the synchronic and diachronic unity of our perceptions, thoughts, memories and other things that make up the inner layers of the self and the (explicit) perception of that unity. My inward glance at any given moment does not have to penetrate to the limits of the unified self; it is – at least in part – curated on our behalf by the scaffolding. Even if all that was available to Hume were the slim pickings he reported as being delivered by introspection, there would be a vast hinterland behind the flow of experiences, most obviously in the "love" and "hatred" to which he refers in the passage that prompted this defence of the self.

I have talked about "projects", "duties", "offices" and so on, as if they were primarily elements that are merely *experienced*. I have implicitly marginalized the agency that goes into their realization. I will correct this in the next chapter when I discuss the self as agent. For the present, it will be sufficient to offer a brief correction of the idea that the self is a kind of spectator not only of what happens to it but also of what it does. The self is not merely a given that I receive. It is the scene of endeavour, effort, and struggle towards visible and invisible goals. To that extent it is self-affirming. While it is obvious that genuine agency requires a self – to make sense of what is done, why it is done, and, indeed, to provide the "becausation" that accounts for actions – it is equally true that agency *reinforces* the sense of self. This is true both at a particular time and over time.

At any given time, the sense of effort associated with action, nails me to myself, to this person that I am, not the least through highlighting the body from which the self is inseparable. Agency, however, operates in many different timescales, ranging from that of ducking to avoid a missile, to hurrying to fulfil an appointment, enjoying a holiday planned a year before, training as a doctor and pursuing a career in medicine, living a fifty-year marriage. These are external and visible expressions of the different timescales within which the self-as-agent operates, over which it is sustained.

Although it is inappropriate to regard them as separated, it is not unreasonable to imagine a dialectical relationship between passivity and agency in selfhood, in which each reinforces the other. As I assert myself through action, so my actions – particularly those that are extended in time, to the point where they qualify to figure in my CV or life story – add to my sense of myself. The extent to which actions are rooted in a temporally deep self can be reflected in the degree to which they are remote from mere reactions, mere responses to present external stimuli.

Explicitness lies at the heart of genuinely voluntary actions and in the temporal depth of what shapes them wherein lies their claim to be free. The agent who acts in the light of experience made explicit as memory is acting on his own behalf – even if under compulsion – and hence truly acting.

The present discussion of the modes of unity of the self is intended as a corrective to underestimates of the connectedness of the self and hence of its substantiality at any given moment. A conscious experience is permeated by a foreground and hinterland of self- and world-sensemaking: it is located in a landscape of meaning and significance. This may or may not be made explicit by the subject who connects, unites, stands on, reflects on, interprets, explains, talks about, her experiences. In either case, there is more to the self than what, at any moment, it makes explicit to itself or others, or would be evident when, by Humean introspection. Unpacking the self enables us to see how it can be substantive without being a substance, a thing, a ghost, or a spirit.

6.8 THE EMBODIED SELF

It may seem that my defence of the self from the autocides has paid insufficient attention to the most obvious source of its substantiality, continuity, fidelity to itself, and consequent re-identifiability: the body. After all, the body and the self are inseparable. When I say, "I am in my study", it follows that my body is in my study. Likewise, when you say, "Raymond Tallis is in his study", you imply that "Raymond Tallis' body is in his study". When Raymond Tallis is walking or falling down the stairs, his body is realizing that trajectory. When he boards a bus, so does his body. When he moves to live in Chester, his body will be found much of the time in Chester. This correspondence extends even to those states when Raymond Tallis is not awake. "Raymond Tallis is asleep in his study" implies that "Raymond Tallis' body is in his study". We can even seem to extend this identity to encompass certain manifestly physical aspects of the body. That Raymond Tallis' body is 5′ 10″ tall is seemingly inseparable from Raymond Tallis being 5′ 10″ tall. And Raymond Tallis and his body share the same weight (roughly 10 stone for the curious).

As we saw in the previous chapter, the relationship between Raymond Tallis as self and Raymond Tallis as body is, however, far from a straightforward instance of token identity, of "A is A". "Ambodiment" is not a simple material identity, as in the case of a statue and the marble of which it is made. This is signalled in the examples just given, where the am/is difference is not a mere linguistic idiosyncrasy. The claim made by Raymond Tallis that "I am 10 stone" and the assertion by a third party that "Raymond Tallis' body is 10 stone" have the same truth conditions – if one is true so is the other – but do not quite say the same thing. "I am 10 stone" corresponds to a complex nexus of experiences and plays out in many consequences relevant to my lived existence. "Amming" ten stone and "it-being" ten stone are not the same. In the space between the meaning and implications of the two sentences, there is an ambiguous territory of "am-tinged" is and "is-tinged" am, captured in, for example, my awareness of the physical interaction between my body and the physical objects around it. Walking on tiptoe to prevent my heavy footsteps making the floor creak and so betraying my presence, lowering myself gingerly into a fragile chair, deciding that I should not walk on a stretch of ice, reflects this ambivalence. I live my being 10 stone in different ways and to different degrees: more when I am puffing and panting up a hill, less when I am (as I am now) sitting and thinking.

One major truth highlighted in the investigation of ambodiment in the previous chapter is that the light of selfhood does not transilluminate all of the organism that makes it possible. As discussed in that chapter, most of what goes on in my body, consists of apersonal processes of which I am unaware. They usually have as little to do with my lived life – other than the small detail of making it possible – as chemical and biochemical events elsewhere in the material world. It is possible to go through life without even a brushing acquaintance with many of one's organs and tissues. As a consequence, the processes that add up to a living body and those that count as the experience of being one's self, culminating in something corresponding to a CV, are widely divergent. What goes on in the body over the years between birth and death does not add up to the story of an enduring self. And the body may continue relatively unchanged after the embodied subject has descended into coma, so long as it remains alive.

Of course, my body may be pressed further into service as the warrant for the continuation of my first-person being by being so in the eyes of others. It is the most fundamental identity certificate. I propel my body into a room, and someone says "Hi, Ray. Good to see you". And such retinal ID is supplemented by numerous forms of photo ID. It is worth noting that the latter is usually satisfied with a part of my body – namely my face. Much of my body beyond my face is "Anyone". While "the back of my hand" is the paradigm of familiarity, it would not surprise me if I could not pick it out in a line-up. As for my knees, shoulders, neck ... My point, I think, is made.

The attraction of the body as the guarantor of the reality and continuity of the self is obvious: it is a substantial, solid, enduring thing. But for that very reason it does not quite deliver what is needed. Yes, embodiment is a necessary condition of selfhood, but not all of its material objecthood can be placed in the service of the self because our idea of a self is not that of an object. It will be obvious from preceding sections that the modes of connectedness of the self over place and time are not those of a material body.

At the most basic level, Raymond Tallis' body at t_1, unlike Raymond Tallis at t_1, is confined to t_1. And the unity and multiplicity of the self – such that perceptions, memories, thoughts, emotions can be connected as a whole of mutually sense-making parts while still retaining their distinct identity – is unlike the unity and the multiplicity of the body, that has distinct and connected parts. The side by sidedness of my head and neck, or of my left foot and my right foot, is nothing like the side by sidedness of an itch in my foot, a memory of the last time I felt that itch, my irritation with it, and my reporting it to a friend who had reported a similar itch.

Nevertheless, there is no self without a body to give it the means of subsistence and the primordial viewpoint which ultimately makes for the possibility of a distinct life story. And, yes, the continuous trajectory of the body provides the fundamental glue of a coherent existence. This, however, is only a starting point and my connection with Raymond Tallis yesterday, a week ago, decades ago, is not captured by the connected path traced out by my body, leading up to this spot at which I am now located. The self-glue is, as we have discussed, fashioned out of acquired habits, skills, protracted concerns, preoccupations, duties, ongoing tasks, the familiarity of possessions, of locations, recognition by those I recognize, loved or hated by those I love or hate, private, shared, and collective memories, and so on. While none of these would be possible without the body and its connected states and locations the latter are not sufficient. That I am courtesy of my body (and my body being in a certain range of states) does not imply that I am my body or a succession of those states.

Of course, I cannot be mistaken as to which body is mine or which part of the world is my body. It is worth mentioning this because philosophers have made much – far too much – of the fact that under certain unusual circumstances we may be mistaken over some of the parts of the world that we identify with as our body. Individuals who have amputations may have phantom limbs, continuing to "am" limbs that they no longer have.

Neither phantom limb pain nor other illusions such as the rubber hand illusion[36] tells us that our relationship to our body is entirely illusory, that I am mistaken in believing in a complex identity relationship with Raymond Tallis' body. Mistakes are highly unusual and local. They take place against, occur in the context of, my correctly identifying the rest of the body as mine and indeed being unmistaken in relation to my hand for the remainder of my life. The rubber hand illusion, for example, highlights our lifelong fleshly hand non-illusion.

While there is, as we have noted, a divergence between the curriculum of my self and the *vitae* of my body, I and my body share a continuity. My intimate, inseparable relationship to Raymond Tallis' body is italicized whenever it is in pain or pleasure, whenever it is engaged in action, especially effortful action, and consequently suffers tiredness, shortness of breath, and so on. Even at such times, however, it does not seem as if the "it" of the body threatens to close in on the "I"; or the "I" that "ams" the body and the body that "I am" move towards fusion. And this distance between my body and I is even more apparent when I am engaged in less physically taxing actions. My flesh fronted by my typing hand may seem a little distanced from me by being the agent of my agency. Even so, my action in typing is inseparable from me and, as such, thickens my sense of myself. My self, as it were, borrows substance from my body, even though I distance myself from it by calling it "my" body. Thus my agency – which we shall discuss in the next chapter – adds a thick layer of ambodiment to the self. The exercise of that agency is, of course, suffused with the self whose aims and intentions, rooted in and transilluminated by many layers of its past, present, and future, it aims to fulfil. Effort – whether physical or mental – places in italics how "amming" is poised between being and doing.

This position is far from identifying the solid reality of the self with the human organism. We cannot outsource the enduring, rich, complex identity of the self to the endurance of the (relatively) stable body – not the least because the body as a material object does not exist in tensed time while tensed time is central to the very nature of the self.

We are therefore a long way from the notion, advanced by Eric Olson, whose "animalism" claims that our human being is our organic being.[37] That said, he acknowledges that selfhood and organic being are separable: while "each of us is numerically identical with an animal, [O]ur having mental features of any sort [is] a temporary and contingent feature of us" and that "any of us could exist without having any mental properties at that time, or even the capacity to acquire them".[38]

Before we leave the body, it is perhaps worth reminding ourselves of its special status as a scaffolding for the self on the borderline between the internal and external realms. It is more faithful than anything else in my life. Unlike other bodies, including those of our fellows, it is continuously present so long as anything is present. I am where it is and it is where I am. It underwrites the connections and continuity between literal viewpoints. And it is the irreplaceable primordial mediator of my agency. In all these respects, no-one can stand in for me. As such, it prevents the "I" dissolving in a sea of "we".

6.9 THE INCONSTANT SELF

We have identified a rich deposit account of selfhood, available at every moment of our lives and necessary for those moments and their worlds to make sense. We are a long way from a Humean current account of perceptions and feelings that are somehow bundled together. And that deposit-account self does not owe its reality entirely to the material object that is the human body, even though there is no self without such a body.

Some readers, however, may remain unimpressed, arguing that the idea of an enduring, substantial self is still an illusion, rooted, perhaps, in the grammar of the language that attaches everything that happens to and within us, and everything that we experience and do, to "I". The fact that "I" can be used by anyone of themselves – when I say "I" it means Raymond Tallis and when you say "I" it means you – suggests, it is argued, that the status of the "I" as a point of convergence of all the contents of a conscious subject is unearned. It is reification of an indexical and no more exists than "this" or "here" does. Just because I always use the word "I" when I refer to myself, it does not follow that the I that I refer to is always the same entity.

"Grammaticalization" of the self – a reduction of the self or an exposé of its emptiness – is a spectacular example of overlooking explicitness. When I say "I", this is a *performance* that affirms, asserts, highlights, myself. When

Raymond Tallis refers to himself his referent is a complex individual, with a singular history and a unique collection of qualities. That referent cannot be meta-grammatically deconstructed any more than the contents of any particular referent of "here" can be erased by the fact that the referent of "here" depends on where the word is uttered. Likewise, "I" cannot be dismissed as an airy nothing, an artefact of grammar. There is a world of difference between "I" as a subject of a sentence – as in "I am cold" - and "I" as a subject conscious of itself and its world. Besides, the cashing out of other indexicals such as "here" as definite referents depends on the self that makes itself explicit as "I". "Here" in its fundamental sense is where "I" am. There is no "here" without an I whose here it is. Far from deconstructing the self, discovering its relationship to (far less rich) categorizing "I" as an indexical only highlights its rich reality.

Given that the self is woven out of explicitness – that makes itself and other things explicit – not out of material objects, we can accept that it is real without having to reify it. The "amming" of the embodied subject as a self is not a ghost added to the bodily machinery.[39] Just because "I" is not an it does not mean that it does not exist: not to be a thing does not mean to be nothing. Indeed, if it were a thing – a what-is of the kind things are – it could not be that in virtue of which what-is becomes that-it-is, what-is-the-case, and so on.

Even so, some still feel dubious about the idea that the I, the conscious subject, amounts to an enduring self. The most obvious basis for suspicion is that the self is inconstant. Its surface is constantly in flux; indeed, its very being is dynamic. This is of course true; but nevertheless, while we are bombarded with ever-changing experience from without and streams of thought from within, we are not swept away by, never mind dissolved in, it. We do not lose ourselves.

During a train journey, sights and sounds and other sensations succeed one another in rapid succession. Even so, I am still on a particular train journey, it is my journey, and that journey remains connected with its (my) purpose – for example, to attend a meeting. That meeting will itself have a purpose that will connect with other, longer-term purposes. And even when other purposes intervene in the form of events that have nothing to do with the meeting, as when on the journey I take a phone call about another matter, or bump into a friend, both purposes will maintain their integrity. We may indeed "get lost in the moment" but we are found – found by the next moment or recover ourselves as we remind ourselves of the larger concerns that define us. I may be absorbed in a philosophy paper as I am travelling down to my meeting, but I am available to be reminded of, and recover my preoccupation with, the day's work, the week's purpose, the overall aim, as the train announces that it is reaching my destination.

My privileged relationship to the flow of my experiences, so that I am not swept away by them, relates to the other aspect of selfhood: my sense of being situated, of having an agenda. That which makes my present experiences mine also makes my past experiences mine and shapes the desire to regulate a future which is also mine. Against the dynamism of the present moment, sweeping me forward to the future, marginalizing or erasing the present as it becomes the past, therefore, there is a counter-current of reaching backwards to the past and, based on that, a reaching forward to the future which establishes a connectedness that is outside of the frantic unfolding of events in physical time. In short, the dynamic nature of the self does not justify denying its identity. The Niagara Falls is in a state of constant change and the water of which it is made is always being replaced. This does not, however, disqualify it from having identity. What is more, the self, unlike Niagara Falls, connects its present state with a many-layered past. My ability to make sense of the present moment is, as we have discussed, dependent on an ocean of sense that connects me with the layered, multi-dimensional worlds of my remembered past and my envisaged future.

But, for some thinkers, there is a deeper concern than that the self is hostage to the dynamism of its experiences. It is our infidelity to the purposes that motivate our lives and gather up our experiences into unfolding sense. Things that once mattered hugely cease to matter at all. Dreams fade, fears cease to haunt, desires wither, loves, interests, and preoccupations are replaced by others or by indifference. We put away childish things – and adolescent, adult, middle-aged, even elderly – things. We "move on". In some cases, this is because projects are completed and so they no longer require our attention. There is, for example, a constancy over change linking the medical student who worried over a forthcoming undergraduate examination and the consultant approaching retirement worrying over a promised development in a new patient service, even though the content of the worry has changed radically. To take another example: next to me is a row of 30 or more books authored by Raymond Tallis. Each of them was The Book so long as it was a work in progress. None shares the importance (to the author) of the present Work-in-Progress, nor would it be expected that it should. Arrival at a destination does not prove that the journey, which is now over and is no longer a matter of concern, was meaningless and the self that undertook it is now dead. The fact that, have I had a tooth removed, I am no longer preoccupied by the toothache that had filled my days, doesn't demonstrate my infidelity to myself.

Even so, it may seem that "getting over something" that has been all-consuming, coming to terms with disappointment – and this happens to all of us all the time – betrays a profound discontinuity or splintering of the self; an intimate mode of self-betrayal. While "moving on" seems appropriate when a project is completed or a dream is realized, forgetting a once-obsessional love affair, or "completing the grieving process", seem nothing of the kind. The abruptness of change may be shocking, as in the passage delineated with peerless clarity by Shakespeare, where he talks about sexual desire:

> Enjoy'd no sooner but despised straight;
> Past reason hunted; and no sooner had,
> Past reason hated, as a swallow'd bait,
> On purpose laid to make the taker mad[40]

seems to mock the very idea of a coherent self.

There are reasons for rejecting this conclusion. The passage from passion to indifference is often a long, reflective, and self-reflective process (setting aside the scenario described in Sonnet 129) in which the elements are connected, and any conflict is conducted through the sense-making capacity of the same self operating on its unique materials. There is, that is to say, a personal story of the transition which, however inaccurate, salvages a connectedness of meaning and hence of the self. What is more, the self can remember not only the process of change but how things looked before the transition. The recovered lover can recall the nights of anguish, anger, hope, disappointment, and delight. The claim that "I just can't see what I saw in him" is not always entirely truthful. It is more often an active rejection of a past – that is still *my* past – than its total annihilation, or mere evaporation. The fact that so many things, large and small, that once mattered so much ceased to matter is not proof therefore of the disintegration of the self into separate fragments. Or of its consisting after all of a succession of disconnected self-slices.

It would be unreasonable to require that the self should be unwaveringly faithful to the projects, hopes, dreams, desires that are important at different times of life. We should be allowed to grow up, grow away from, get over – as well as to fulfil and complete – those things that may have seemed to us to be the most important things in our lives. If our circumstances, including our inner experiences, change, it makes sense to change our preoccupations. This is no more the end of, even the betrayal, of a self than is ceasing to strive for a goal once it is achieved. The one who strove and the one who strives no longer are still of a piece. While fulfilment renders as "last year's story", the struggle for a particular mode of fulfilment, our self is tattooed with "the map of days outworn".

Towards the end of his life, Thomas Aquinas had a revelation, and he gave up writing. When his secretary asked him why he ceased to write, he pointed to his books and said "All I have written seems to me like so much straw compared to what I have seen and has been revealed to me".[41] This was not however, to cancel the 20 years of ceaseless thinking and frantic writing that led up to his moment of revelation. It seems unlikely that he would have approved of his secretary's burning his collected writings. Arrival does not empty the journey of meaning; even less does it disconnect the arriver from the journeyer. If that were so, all journeys towards fulfilment would be self-cancellation.

So much for the inconstant self. Beneath the changing narrative surface, there are more connections and constancies than disruptions.

6.10 THE INTERRUPTED SELF

> I cannot seriously suppose that I am at this moment dreaming. Someone who, dreaming, says "I am dreaming", even if he speaks audibly in doing so is no more right than if he said in his dream "it is raining", while it was in fact raining. Even if his dream were actually connected with the noise of the rain. Wittgenstein[42]

There may be a deeper challenge to the continuity of the self than our necessary infidelity to our own preoccupations and projects and our evolution through time. I am referring to the interruptions of consciousness to which we are all subject, most obtrusively in the mandatory intermissions of sleep and in the dreams that possess us while we are asleep. Let us begin with dreams.

"I had a dream" we say; but strictly we should say "A dream had me". In the dream, we imagine ourselves in a world that does not exist except in our imagination. More worryingly, the self that we imagine in that world is frequently quite unlike the self that inhabits our waking hours. We get up to all sorts of things that the waking self would not countenance. Some of those things are distant cousins of the things we do in daily life, but they take place on a stage that is a jumble of fragments populated by a random cast of others whose co-presence is insufficiently explained. Our behaviour often seems merely to happen rather than to be enacted. Admittedly, the dreams may not be entirely disconnected with waking life. They may be fueled by anxieties that trouble us, or used to trouble us, in daily life. An academic such as myself might have recurring dreams where, perversely, I keep failing to catch the bus that would take me to the place where I must sit the exam on whose outcome the fulfilment of my ambitions will depend. But the scene of the dream is disconnected from the rest of my life, unsupported by the modes of coherence that pervade waking life.

So long as we are dreaming, we don't usually doubt the reality of the crazy "I" in a crazy ersatz world. For some, most famously Descartes, this opens the question of whether our waking self is also an illusion; that the mini-madness of a nap exposes the groundlessness of the world of everyday life. I say this as someone who, just a few hours ago, was "deceived by dreams" populated by people whom I have not seen for years – one of them is dead – and events that (thank heavens) have not happened. Those dreams have inserted themselves between the writing of successive paragraphs in this text.

One response to this concern is that the individuals who populate our dreams and the events that happen in them are plagiarized from waking life. My panic-stricken dream that I was being expected by X and Y to give an hour-long talk on a topic for which I had neither notes nor (God help me!) slides was clearly woven out of real-life material. In short, the self of my dreams is parasitic on that of my wakefulness. The traffic is one way: dreams steal from wakefulness, but wakefulness does not (usually) steal from dreams though the latter may leave a hangover. And wakefulness looks back on dreams – judging them to be false – while dreams do not look back on wakefulness judging it to be false. While "we are close to waking when we dream that we dream",[43] it does not seem that we have been dreaming when we wake up to, or even out of, our wakefulness, as is the ultimate ambition of much philosophical thought.

There is another possible source of reassurance: the dream is "my" dream because I can recall it. I cannot recall anyone else's dream without having had it related to me by them or some third party. Even if at the time of the dream the dream had me, when I recall it – and worry about the reality and continuity of the self on account of what I recall – it is true to say "I had a dream". What's more, my waking recall of the dream, when I recount it to a waking (if reluctant) auditor is a testament to my capacity to reach across the interruption of sleep and assert a connectedness that sleep seems to break. In saying "I had a dream", I turn any tables that need to be turned on the dream if, indeed, it had had me. Any suggestion that I don't know which is me – the self who had the dream or the self who reported that he had had a dream – is existentially insincere.

So much for dreams. Do these reassurances also deal with the profound discontinuity that is the nightly experience of sleep, irrespective of whether it is dream-haunted or dreamless? Perhaps, dreamless sleep is more troubling given that there is nothing in it with which we can connect. Yes, we say "I had a sleep" just as we say "I had a dream"; but there seems to be an even stronger case for saying "a sleep had me" than "a dream had me". Sleep overcomes us. Our nightly sleep, rather than any dreams we may have had in it, seems to be an interruption to ourselves and, given its frequency, it is a challenge to the idea that we are unified over significant periods of time.

A challenge, however, that can be met. When I wake up in the morning, after a sound night's sleep, there is both external and internal evidence that I am the same individual as the person who fell asleep. The bed I wake up in is the same as the bed in which I dozed off. The body that awakes is hardly different from the one that fell asleep – something I can check in case of doubt by looking at photographs of myself that lie to hand. Ditto the room and its contents, (including the clothes I took off the night before), the house, the surroundings, and so on. The alarm that wakes me up is the same as the one I set to ambush me and the me that is woken up by it is the same as the me that set it to wake me up. The reasons for which I set the alarm – I mustn't miss that appointment, I have to finish something to meet a deadline, I am due at work in an hour's time – are rooted in the same situation (in the broadest sense) as that in which I went asleep. It is the same sense of who I am and what I ought to do, supported by pretty well the same body of knowledge and skills, and an awareness of similar duties reinforced by the expectations and assumptions of other people. There is the to-do list awaiting me, the sheet of paper on which I can check the telephone number I have to ring. In short, the self I doffed when I fell asleep is there to be re-donned when I awaken.

My sense that I am the same person is reinforced by the many-layered nexus of recognition, expectation, acknowledgement, of the individuals with whom I share the room, the residence, and the neighbourhood. I resume my life stories as they resume theirs, most notably those which overlap.

Thus, the many-layered internal and external support of the self that has remained in place as a standing state of what I am and does so irrespective of the interruptions in the flow of experience and the gatecrashing of my consciousness by dreams. Most of what lies above the ground level self of the flow of phenomenal experience attuned to material and social surroundings is intact: it has survived the interruption. The miraculous process of rapidly reassembling ourselves is supported, indeed driven, by the familiar world we wake to, with its familiar people, and their familiar expectations. That world is a dense realm of cues speaking to a self in touch with its past. Every night a different dream and every morning the same surroundings or – if I am in a hotel – a story linking present temporary surroundings with the more permanent ones.

Of the two interpretations of awakening – that the world of the sleeper, and the self that holds it and its many meanings together, is recovered intact; or that what he wakes up to is a simulation created by a conspiracy of others – the former seems (to put it mildly) more likely. The final nail in the coffin of the second view is that over breakfast we may share the fact that we were dreaming, and you may confirm that I was calling out in my sleep about the talk I had to give for which I was unprepared. And I, of course, know that my day does not require me to give this talk, and you can confirm this. And we may confirm to others reporting their dreams that they were asleep last night. Most tellingly, perhaps, we can refer to the time, and measure the period in which, our being ourselves was interrupted. None of this would be possible if we identified the self with a Humean trickle of experiences.

That is why, when you come upon me asleep, you can say "Raymond Tallis is asleep" not that "No-one is asleep" or "Someone other than Raymond Tallis is asleep". Which is why we do not fear the "little death" that is sleep – and not only because we are often tired and long to be embraced by sleep. Settling down to sleep – especially when we do so in order to make sure we are on top form for something we have to do the following day – is consistent with the self that the sleep will interrupt.

There are, of course, interruptions that are not terminated. There is the gradual erasure of dementia that might result in my not remembering the man who wrote this sentence – or not remember him in an organized way. I may descend into a coma from which I do not wake. Then I will have ceased to be a person, and entered into the state of an organism maintained on life support. Those who care for me may continue to treat me as a person in the hope that I might return one day. And for all of us there is the permanent interruption called death.

None of this, however, retrospectively cancels the substantive, continuing self that resumed itself after the interruptions of sleep.

6.11 MAKING OURSELVES

While it is implicit in my account of the self as having many layers beyond the flow of experience, I have not sufficiently addressed our own (active) role in constructing our identity. The nineteenth-century German philosopher J. G. Fichte famously asserted that the "I" was self-positing, and that the original unity of self-consciousness was an *act* before it was a *fact*.[44]

This might be something of an over-corrective in the light of the extent of the sometimes helpless dependency of the "I am" on the "It is" of the body, most notably when the latter is being assembled *in utero*. That dependency is not, however, a simple causal dependency, though it has material conditions, and is, as we have seen, bafflingly complex. Most importantly, it is not one way. Much of active life is addressed directly and indirectly to curating our bodies, maintaining them in the healthy state – fed, watered, sheltered from mineral, vegetable, animal, and human threats – necessary for them to underpin our continuing existence as conscious subjects. Importantly, Fichte's over-corrective is a reminder of something that autocides from Hume onwards seem to overlook: our part in creating, or at least shaping, ourselves. This increases over time, as we develop from self-poor infancy to the complex, robust self of typical adulthood. Active self-development is central to the very idea of the self. Inevitably, this implies change that, as we acknowledged in the previous sections, may be profound. After all, our aspirations for our future point to a future self, different in at least some important respects from that of the present.

Hume's conclusion that, if we cannot find the self in the flux of impressions, it must be a fiction, rests on the false assumption that the self is the kind of entity we might expect passively to encounter, a mere "impression" on all fours with other impressions. Such a vision of the self renders it susceptible to being understood as something whose origin and curation could be entirely outsourced to the connectedness of the physical states of the body or the brain. To the contrary, our identity as a self is something we affirm, embrace, or cultivate, out of ingredients in part given, in part chosen, in part received, in part generated.

The part we play in shaping ourselves is particularly evident in the way we prepare, teach, and train ourselves and, more broadly, *position* ourselves to exercise our agency.[45] Behind this, there is something deeper. To a significant degree, we are self-identifying, and our *identity is self-stipulating*. This is best understood through contrasting personal identity with the examples used by Derek Parfit to defend a reductionist – that is to say an emptying – account of the self.[46]

Parfit argues that there is nothing corresponding to a "self" underlying the experiences, beliefs, skills, dispositions, etc., that may be attributed to a given individual at a time or over time. He arrives at this conclusion because he believes that there is no "further fact" corresponding to "personal identity" additional to the body, brain, and succession of psychological contents, of the individual. Persons or selves, he argues, are like nations or clubs.[47] Clubs consist of members and premises and rules, whose criteria are loosely defined. There are no further facts that amount to the club itself. That is why we cannot say precisely if and when the club ceases to exist or, if it had ceased to exist, what circumstances would determine whether it had been restored. After how many changes of the premises, the membership, and the rules could we say that the club was no longer? The same, he argues, applies to personal identity: there is no particular fact corresponding to Raymond Tallis and so it cannot be a matter of fact when he is no longer. There is no sharp boundary between RT-existing and RT-no-longer-existing.

The analogy misses a fundamental difference: namely that RT is self-identifying, self-stipulating, or at the very least self-affirming, as if from within. People are not like social clubs. A club does not gather its own components, achieving clubhood by something like unaided apperception. It is the members – not the club – that stipulate when a club has ceased to exist. What is more, its present existence is primarily in the consciousness of its membership and secondarily of those who know of it. It is not the club itself but those who have gathered to reinstate it who will decide whether the Athenaeum in 1900 and the Athenaeum in 2000 are the same club. There is nothing, in the case of a club, like the synchronic and diachronic unities that justify my belief that you at time t_2 are the same person as you were at time t_1.

My identity is primarily embraced, affirmed, from within rather than being entirely externally imposed according to some verbal or other convention. A club does not own itself, know itself, judge itself, or act as an agent by acting on itself. Behind this difference between selves and clubs is something fundamental: "That I am" or "That *I* am" or "That I *am*" is not an assertion that can be questioned: it is what we might call "an existential intuition"[48] that lies at the heart of the self-consciousness and the synchronic and diachronic unity of the self. It isn't granted by others on the basis of my meeting certain criteria.

If I were relying on others to grant me a personal identity, for me to have selfhood, I would not have a personal identity nor would I be a self – notwithstanding the role that interaction with other selves play in shaping and maintaining my sense of who and what I am and the importance of being acknowledged by others. As such, selfhood has a performative dimension, as I embrace, or enact what I am in pursuit of what I might be. While I do so with the support of a scaffolding of a familiar world, I have had a hand in building up that scaffolding, not the least through my part in familiarizing myself with it and learning the knowledge and skills necessary to negotiate it. Unlike clubhood, selfhood is inseparable from consciousness of itself. The club is not gathered together by awareness of itself as a unified totality.

To look for, and fail to find, and consequently to reject, a non-reductionist account of the self on the grounds of not being able to identify it as a "further fact", additional to psychological and bodily continuity[49] is to make an error similar to that made by Hume. Like Hume, Parfit, looks for the self in the wrong place. The absence of a further fact, like Hume's failure to find a perception of the self in the stream of perceptions, does not demonstrate that selves can be reduced to a Humean stream of psychological phenomena plus a part of the body – according to Parfit part of the forebrain. Parfit's requiring "a further fact" distinct from physical and psychological continuity – in order to demonstrate that personal identity is irreducible to something impersonal is entirely misconceived – indeed self-contradictory. For facts are objective, or impersonal. What is more, personal identity is upstream of, prior to, any facts, including, or especially, any facts about me: it is that in virtue of which the truths of my case are *mine*, are *me* and they require the Janus-faced experiences of that-it-is and that-I-am. I would not expect my identity to be non-reductive courtesy of a fact – "further" or otherwise – about me. Further facts are not going to ensure that there is something about me additional to streams of consciousness and the continuity of the body. Nor something distinctive or individual. If it were beyond streams of consciousness and the physical world, it would be featureless.

The existential intuition – "I am this" – withstands the objective changes in our selfhood over time as we have discussed. Identity is not, however, a mere tautology: "I am Raymond Tallis" is fundamentally different from either "A is A" or "I am I". My being (or "amming") who I am is among other things a sustained embrace of the contingent truths about me, beginning with the fact of my being born. Of course, the truths that I embrace are constrained: there are limits to my self-definition; I cannot qualify as Napoleon, or an angel, on the basis of a personal *fiat*. While I cannot be mistaken that I am, I can, however, be mistaken in thinking that I am Napoleon or an angel.

There are, that is to say, limits to the extent to which we can appropriate ourselves or define what counts as "I", what we can "am". The unchosen agenda, preoccupations imposed by events, circumstances, situations, will direct what we pay attention to, the specific contents of the world, our bodies, and our past, which we embrace as ourselves. Our capacity to choose ourselves is, to put it politely, limited. Our "coming to" out of deep, dreamless sleep or coma is to encounter ourselves as a given which we receive, as something that seems to have preserved itself in the absence of our awareness of ownership, without any help from ourselves. And, as we have discussed in the previous section, thank heavens for that.

The mark of identity, of selfhood, then is not a "further fact" about me – nor can it, or need it, be. Its elusive reality is flagged up in the intimate agency of self-embrace. This is revealed in the "amming" that has, however, a complex relationship to the being of the body. As we have emphasized, the "I" of Raymond Tallis and the "it" of Raymond Tallis' body are far from identical, as becomes clear when we think of the kinds of facts that are true of Raymond Tallis' CV and those that are true of his body. The life of the body – animation – and the life of the person – "amimation" – are widely divergent.

Selfhood, then, remains real, but not as either a Cartesian ego, or as an elusive "further fact" given to us as our personal identity. Nor is the self a hybrid made out of the animal organism that makes it possible added to the successive moments of its stream of consciousness. Selfhood is forged in the realm of a mode of being – "amming" – that fuses an "I am" (the subject) with an "it is" (the needing, interacting body). The many subtle modes of that connection are transformed by the active self that is at the same time shaped by it. We embrace that which embraces us.

It is here we find what truth there is in Fichte's assertion that the "I" is an act, not a fact. It is most certainly not a fact – nor does it need to be – but it is not *entirely* an act.

6.12 CONCLUDING THOUGHTS

We may seem to have drifted a long way from our central theme of explicitness. Examining the self, however, brings us as close as we may to its mystery. While the rather slimmer notion of the conscious subject unfolding over time reminds us of the vehicle of the transition from what-is to that-it-is, teasing out the elements of the self takes this further. It points in two directions: the self makes the world explicit as *its* world – its situation, its surroundings; and it makes itself explicit as something more than a succession of experiences of entities other than itself.

Making oneself explicit involves the kind of internal connectedness that is described in this chapter, as a result of which we encounter ourself in a past that is our past on behalf of a future that is our future and highlights present that is our present and what it requires of us. The self of the embodied subject *faces* the world and is faced by it; has a full-blown relationship with what is out there in which there are two relata – albeit of a different kind – the "I" and the "it" or the "in here" and the "out there". The self has enough presence as self-presence to be able to experience itself as anchored to its body and, as such, distanced from the material world out there. The being of the self is not reducible to that of the body, though it is inseparable from it.

The claim that there is no such entity as a self but only the illusion of one created by (for example) neural activity is not only pragmatically self-refuting, as we saw in the case of Hume and Metzinger, but also, importantly, misunderstands what is necessary for something to qualify as a self. The actual features of ourselves and the overriding faculty of making a world and itself explicit – through internal transparency, reflection, connected sense – are self enough. Such explicitness, generated by the self, of the universe as its world and of itself in that world, could not be a property of what-is, even when the what-is in question is a nervous system. Hence the emptiness of Metzinger's claim that "a self-model is an entity in your brain".[50]

When revealed in all its complexity, a lifelong "illusion" that I am a self is not distinguishable from the real thing. As we have observed, it requires a lot of self to think about the self sufficiently to claim that it is a fake and set out why nothing real corresponds to it. Likewise, for an autocide to publish his autobiography – as Daniel Dennett has done recently, looking back at a long life – suggests that he doesn't take his own views on this matter very seriously. The title – *I've Been Thinking* rather than *There Have Been Thoughts* – says everything.[51] He may have killed his self in theory while he embraces it in practice, perhaps having recognized that a lifelong (or nearly lifelong) illusion of a self – a self, what is more that has the power to narrate its life – is not too different from what is ordinarily thought to be a self. At any rate, if it is dead, it won't lie down.

The existential intuition – that I am, that I exist – runs almost as deep as the lifelong sense that-it-is. We have seen how the two realms of the "I" and the "it" are connected through our being (and explicitly not being) and consciously being (and consciously not being) our body. The body is the deep connection between "that-it-is" and "that-I-am". Nevertheless, while the I and the it are intimately connected through ambodiment, there is a complex, variable, multidimensional gap between the "I" and any particular part of the body, even if the part in question is the brain.

To be a human being is to inhabit a location in space and time and yet to engage with realms that lie beyond literal surroundings. We are often, as the phrase goes, "miles away" – and, of course, hours, weeks, and years away. And those miles are defined not by tape measures but by proximal and distal realms of meaning. Nor are they merely outer distances internalized. The worlds that we face as an unfolding array of material objects, often in the guise of affordances inviting, demanding, motivating, or justifying action, sites of opportunities and obligations, classified according to the behaviour they might justly evoke, acquire much of their meaning from what, for the want of a better phrase we call inner depths. The miles and weeks of our multi-layered world are defined not by tape measures and clocks but by proximal and distal realms of sense.

The senses we make of what is before us and of what we make of ourselves are interwoven. Perhaps this did not need to be said. I do so as a pre-emptive strike on the idea that the self is in a different realm from the world it confronts. I say this, in other words, to address a residual suspicion of dualism remaining after the self has been unpacked from the embodied subject. For reasons that will already be clear, this does not mean that the self is identified with a material body wired, through law-governed causal connections of the kind we ascribe to physical reality, to the remainder of the material world. Rather it *faces* that world and engages with it from a virtual outside, created out of explicitness.

Most importantly for the present inquiry, that which makes things explicit makes itself explicit as the variously summed, synthesized, and reflected upon accumulation of its experiences, of its making things explicit. In making things explicit, we explicitness-makers make ourselves explicit – as other than that which we make explicit. As such we are a long way from the self as an absence in a Humean succession of experiences that – to use Hume's laden term – we "stumble" upon as if it were a hidden inner stream. The bundle theory of the self is ruled out by its complex modes of coherence, its unities at and over time. In referring to plans, duties, and obligations we remind ourselves how the exercise of our skills and habits reinforces our sense of the very self that is their substrate. The self is not something revealed to introspection but someone who is lived.

Which brings us to the self as agent, the theme of the next chapter.

CHAPTER 7
Agency: explicitness in action

7.1 THE REALITY OF AGENCY

Selves – embodied subjects – are agents. They *do* things and the things they do, actions – notwithstanding that they are typically physical events – are fundamentally different from the mere happenings occurring throughout the physical world, including their own bodies. "Amming" opens up a territory within being, occupied by doing.

The relevance of actions to the present inquiry is that they are made possible by explicitness – by an explicit self, making the world around itself explicit. Before I develop this point, I shall briefly summarize the case for the distinctive nature of actions in a world of happenings, for their being genuinely free.[1]

Free will seems impossible in theory. Nevertheless, it appears real in practice: there really is a fundamental difference between things that someone *does* and things that merely happen to or around that person. If there were no such difference, our lives would lose much of their meaning – though this concern should not be taken as evidence in favour of free will. Our lives, after all, may well be meaningless when they are considered *sub specie aeternitatis*. More to the point, seeing how free will is possible requires a radical rethink of our relationship to the natural world, a critical look at the concepts of causation and of the so-called laws of nature, and reflection on what prompts and guides actions and how those actions are put together.

What are we talking about when we talk about agency and free will? The definition by Robert Kane captures what matters: "the power of agents to be the ultimate creators (or originators) and sustainers of their own ends and purposes".[2] "Ultimate" is perhaps a little too strong. The point, however, stands:

as genuine agents, we are responsible for certain events that are deemed to be actions. We are capable of being *initiators* and in this respect we are fundamentally unlike the overwhelming majority of entities in nature which go with the flow, are conduits merely passing on what they inherit, without fundamentally changing the course of events.

The case against the reality of genuine agency is easily stated. Actions, so we are told, are material events in a material world governed by the so-called laws of nature. Moreover, all events seem to have a causal ancestry that reaches back beyond anything over which agents can have control. This case is lucidly summarized by Peter van Inwagen: "If determinism is true, then our acts are the consequences of the laws of nature and events in the remote past. But it is not up to us what went on before we were born, and neither is it up to us what the laws of nature are. Therefore, the consequences of those things (including our present acts) are not up to us".[3]

This traditional case for determinism has recently been supplemented by "neurodeterminism", a consequence of the view discussed in Chapter 3, according to which persons are identified with their brains which are material objects causally wired into the rest of the material world of which they are a part and subject to the laws of (physical) nature. If we are our brains, and our brains are the immediate causal ancestors of our actions, then we can no more control those actions than we can stand outside of our brains and manipulate them from without. The claim that we do so would seem to be analogous to Baron von Munchausen's claim that, when he and his horse ran into a marsh, he was able to rescue them both from sinking in the mud by pulling himself up by his hair. Neurodeterminism, what is more, seems to have been given empirical support by studies of brain function conducted over the past several decades. It has been claimed that, when we make voluntary decisions, the brain seems to signal the decisions we are about to make before we are aware of them.[4]

What's more, agents seem to *rely* on the laws of nature and causal connections for their actions to be possible and, more importantly, for them to have desired, or at least predictable consequences. If action and reaction were not equal and opposite, I would not be able to push myself out of bed in the morning or walk to the shops to buy what I need for breakfast. At every moment of our lives we depend on a dependable physical world. Without the universal enabling constraint of what we call the laws of nature, we would be helpless.

Game, set, and match, it would seem, to determinism.

A defence of the reality of free will and of the difference between (physical) actions and mere happenings is rooted in an appreciation of the unique nature of human consciousness – its "aboutness" or *intentionality* – and of human beings as embodied subjects – the subject of Chapters 3–5. Intentionality, as I argued in Chapter 3, cannot be understood as the product of the material world interacting with the human body or, more specifically, the human brain. My vision of an object "out there", for example, is not identical with neural activity triggered by light energy, though the latter is its necessary condition.

The intentionality of consciousness, in virtue of which the conscious subject is in contact with what is around her but at a distance from her, opens a space between the natural world and conscious agents. That space of that-it-is is vastly extended by the sharing or joining of intentionality to create a public, human world, a community of minds, of selves, facing nature, as discussed in the previous chapter. Courtesy of this human world, the thatosphere, agents can act upon the natural world from a virtual outside.

The force of "virtual" here is to flag up the difference between this outside and spatial outsides and to highlight its difference from the relations of inside and outside seen in the material, natural world. When I reach for a cup, the cup is outside of my body and my reaching closes the distance between my hand and the cup. But there is another distance between me as a conscious subject and the cup as a material object. That distance of intentionality remains even as I lift the cup to my lips. It is evident in the asymmetry of my being aware of the cup that is not aware of me; of my making explicit a material world that does not reciprocate by making me explicit.

For many writers, the most powerful case against free will has come from science – in particular, as we have seen, neuroscience. It is a particular irony, therefore, that the most striking expression of the ability of conscious subjects to stand outside of overwhelmingly inanimate nature and to act upon it from a virtual outside is to be found in the practice of the experimental natural science whose findings are supposed to have reinforced the belief in our subordination to the forces of nature. There could, however, be no more compelling expression of the human distance from the "law-governed" universe revealed by science than science itself, in particular our ability to conduct experiments that reveal seemingly unbreakable regularities in nature that are expressed in the laws of science. This, as we discussed in Chapter 4, is a highly unnatural enterprise, beginning with the unnatural nature of even the simplest measurement. Even further from nature are the procedures that lead to our being able to identify general patterns or causes of events and to apply the acquired knowledge in driving spectacular technological advances. The practice of science – in which we choose which questions to address, which experiments to perform, which outcomes to focus on, and which events to designate as signals and which as noise, is a supreme expression of free will.

This distance from which we uncover "how things work" – both inside and outside the laboratory – reflects the privileged position occupied by human agents in the order of things. Even more compelling is our extraordinary capacity to exploit the discovered laws, directly or via technology, to deliver desired outcomes, that has transformed our world into a landscape of artefacts used to support and vastly increase our agency. Experimental science helps us to put the habits of nature in their place, categorizing them as laws, and to exploit them in a manner characterized by John Stuart Mill as using "one to law to counteract another".[5]

I employ the phrase "the habits (in the sense of consistent patterns of unfolding, uniformities) of nature" because the idea of *laws* of nature is haunted by the ghost of the idea of a legislator acting from without. The definite article at the beginning of the phrase is not meant to indicate a definite number of discrete habits. It is only when we extract them as the laws of science does the uniformity of nature become divided into separate tendencies. How many laws are extracted depends on the division of labour between the sciences and their individual scope and ambition. The very idea of The Theory of Everything suggests that the completion of science will be a unification and at that point, the habits will become one habit – an overall direction of unfolding. Such a theory, as suggested in Chapter 2, may be indistinguishable from a Theory of Nothing-in-Particular – of, indeed, Nothing.

That we need to stand back from the habits of nature to make them visible to us – sufficiently visible for us to use them to create the technology that transforms the world in which we pass our lives – helps us to see what the so-called laws are. That we *can* stand back in this way shows that we are not imprisoned in unbreakable physical mandates that exert unchallengeable power over what happens, including what happens in our lives.

What is revealed by scientific inquiry are not laws of nature but laws of *science*, teased out of the habits of nature, by different stances on its tendency to unfold in a uniform way. That we can tease out, and then exploit these laws, in the laboratory demonstrates our special relationship to them. This is of course true not only within the laboratory. The latter is not a metaphysically privileged space, otherwise our spectacular exploitation of the laws of science to direct the course of physical events - evident in the wall-to-wall science-based technology that we deploy to manipulate the world in which we live – would not be possible. Both inside the laboratory where they are revealed and outside the laboratory where they are applied, we bring the habits of nature on our side.

The artefactual nature of the laws of science – the fact that they are not simply a passive image of the habits of nature – is evident from what takes place in the laboratories where they are discovered. Extracting the laws involves examining the effects of certain highly purified, controlled interventions on an aspect of a piece of nature – a parameter – while holding steady other aspects of what is being examined. The *ceteris paribus* conditions that hold in relation to laws of science are unnatural – just as is the distinction between that which is relevant and that which is irrelevant, between signal and noise, foreground and background – or indeed between the physical actions that count as an act of measurement and the events within that which is measured (also an artefact of regulated attention) that counts as the outcome of measurement or the datum.

We can acknowledge this without embracing an anti-realistic, purely instrumentalist, view of the laws. The evolution of laws that have increasingly impressive predictive power and extraordinary, empowering practical applications indicates that they are not disconnected from reality. What they have to say about the physical world out there captures important truths about that world without rendering what-is entirely transparent or articulating its intrinsic nature without remainder. From the standpoint of our capacity to act freely, it is sufficient that the laws of science should be cognitive instruments which enable us to act more effectively on the world in pursuit of our goals. And the processes by which we acquire those instruments – described in Chapter 4 – are a striking reminder of the distance between physics and the physical world; between the latter and the exercise of the skills that generates its scientific portrait.

We may complement our undermining the supposed determinism of the laws of nature by a deconstruction of the very idea of causes. Hume claimed that causation understood as a necessary connection between events – a *de re* necessity – exists in the mind, not in those events.[6] Our regularly fulfilled expectation that thunder will follow lightning leads us to feel that the lightning is a kind of oomph that necessitates the thunder. The status of lightning as a cause and of thunder as an effect is, however, dependent on where one defines the beginning and the end of a sequence of events. After all, the lightning is also an effect of prior causes and the thunder a cause of subsequent effects. The definition of a cause and its effect and the cause–effect relation are, that is to say, dependent on the direction and grain of attention. Lightning and thunder do not draw their own boundaries. There is no beginning at the beginning of the lightning or end at the end of the thunder. Indeed, it seems that the physical world unfolds as a continuum rather than as successions of discrete and separate events

If the physical world were made of discrete events that constitute beginnings and ends, the invocation of causal connections would be a response to the need for a glue to close the separation between events. While it may be problematic to claim that any breach in the continuity of the unfolding of events and processes is due to the irruption of a conscious subject[7] transforming what-is into the scene of her world, it is nevertheless clear that what are identified as objective or natural causal links are in fact interest-dependent and cannot be understood outside of the context of human agency, whereby events are requisitioned as means of shaping the flow of events to realize chosen goals.

At any rate, it seems justified to think of actions as *interventions* in a continuum, elevating the state of what-is at a particular time in a particular place to "a beginning" pointing to another state that is envisaged as "an end". It is this that accounts for the transformation of the monotonous, universal habits of nature into the rich mosaic of everyday life.

All of this points to the conclusion that, far from being an obstacle to freedom, a "law-governed", causally connected, natural world is its sustaining condition. The laws are the habits of nature teased out first by prescientific consciousness and subsequently by scientific investigation. And causation, far from being an additional source of legislation, or inevitability, can be seen as the local operation of those laws and a place where they could be exploited. In short, we can replace "laws of nature *and* causal necessity" with the habits of nature as revealed to quality-controlled inquiry as the laws of science operating at a locus of interest.

None of this is to deny that we operate within constraints. There are things I can do and many more things I cannot do. Courtesy of our science, many of things I can do – talking to a friend in Australia without even raising my voice, using a signal bounced off a satellite circling the Earth – lie beyond constraints that would have seemed insuperable a century ago. But many constraints are firmly in place: I cannot jump unaided to the moon, lift a mountain, be in two places at once. And I have many more subtle limitations, not the least cognitive ones – as will be evident to anyone who is reading this page.

7.2 ORDINARY AGENCY

To appreciate the operation of freedom in practice it is necessary to scrutinize everyday actions. The most cursory inspection shows how they are utterly unlike the sequences of events that are seen in the natural world. They are put together differently: the succession of events that take place in the fulfilment of something as straightforward as a hair appointment booked for a week hence would have a near zero probability of occurring by chance as an expression of the habits of nature. As for the events that make up my pursuing a career in medicine, they would be even more improbable, given that they are spread over several years and the sequence of occurrences is fractured by numerous interruptions and resumptions that may range from seconds to weeks.

One way of capturing the distinctive feature of actions as opposed to the trajectories of events in parts of the physical world is to see that the elements of which they are composed occur because they have been *requisitioned* and shaped, most typically by so-called "propositional attitudes" such as intentions, reasons, hopes, beliefs, and bits of knowledge. Propositional attitudes have a complex intentionality that, like the intentionality of all conscious states, cannot be understood as the mere law-governed effects of material causes nor, on the other hand, as being themselves causes comparable to those supposedly operating in the natural world. They occur because they are meaningful, and their occurrence is an articulation of a meaning.

There are many grounds for this claim. What we might call *becausation* of action draws on envisaged, imagined, anticipated, future *possibilities*. Possibilities grow out of the individual and sometimes shared intentionality of human consciousness. As we shall discuss in Chapter 9, possibilities are pure explicitness: they exist only insofar as they are envisaged; what they envisage is general; and, unlike material objects, states, or events, they transcend the present, drawing on the past, and pointing to an imagined future – a future moreover that may not come to pass. In other words, the pull of the future is exerted by a potentially non-existent state of affairs: it has only to be envisaged as, for example, desirable and to be enacted or undesirable and to be headed off.

Even in those cases where the envisaged state of affairs comes about, the material events of which it is composed, are not prescribed precisely by the intention of the agent. Let us suppose that I plan to deliver an out-patient clinic tomorrow. The physical nature of the events that take place in the clinic will be prescribed only in the most general terms. The exact positions I adopt as I sit in my chair, the directions where my shadow is cast, the events that take place in the clinic – which patients arrive, what they say and do and need, what I say to them, what other staff do – are not foreseen. And much of what happens will belong to inescapable background states – in some cases noise surrounding the signal of the intended action. The connections between the present where I plan to carry out a certain task and the material events that fall with the envelope of "doing the task" are not connected in a remotely Laplacean way.

The other side of this truth is that the future towards which agency is directed is a pull embodied subjects exert on themselves. Inasmuch as they are made explicit, possibilities are not *inserted* into the mind as effects of material surroundings that then act as causes. The envisaged possibilities that are judged desirable or undesirable, prompt events that are directly or indirectly realized or (in the case of a future that is feared) prevented from being realized. I emphasize this to highlight how belief in agency does not reduce consciousness, or the conscious will, to a mode of causation added to, or competing with, the causation that is traditionally ascribed to the natural/physical world. We are dealing not with "causation" as triggering actions but "becausation".[8]

It is worth emphasizing how the future is not something that is given in, or belongs to, the physical world. This should not need spelling out but this passage from a serious philosopher indicates that it does: "there is", he says "at any instant exactly one physical possible future".[9] No, there isn't. There are *no physical* possible futures because there are no futures in the physical world. The physical world at time t_1 is the physical world at time t_1. Time t_2 does not exist at time t_1. Nor do its contents. They would, after all, be in conflict with those of t_1 – and those of any number of other futures filling the space between the present and the end of things. The future time t_2 at t_1 is a construction, an invention, of the conscious subject, as are t_3 and all its successors.

This is most obviously true of the structured future of the diary and calendar. Next Wednesday does not exist except as a cluster of possibilities, in the mind of someone envisaging them. Whether the events that the subject anticipates as populating the future come to pass is neither guaranteed nor ruled out. Agents bet on the future, having certain expectations of events they would like to head off or to bring about. And, to the extent that they have agentive powers, they rig the future they are betting on.

To summarize, actions differ fundamentally from other happenings in the physical world in respect of the way they are put together, the holistic interconnectedness of the intentions, beliefs, thoughts, etc., that prompt, inform, and shape them, and their relationship to time which is made explicit as tensed time. This is especially obvious in the case of complex actions that take place over a protracted period – as when for example I fulfil a plan to bring up my children in a way that will make them good citizens, campaign to improve public services, or train to be a doctor. The countless events that correspond to the realization of these aims, or the path to their attempted realization, are not specifiable in terms of sets of physical movements or energy exchanges described by natural science.

While these are particularly complex examples, the principle also applies to relatively short-lived, specific actions, such as those that realize a plan to meet someone sometime next week, tidy a room at the weekend, or finish reading a scientific paper. We are still remote from (say) spontaneous responses to stimuli, given that even something as simple as honouring an appointment is built on a complex story of our life (what we prioritize, what we owe people) and various bodies of information (what I can fit in, what else I have to do, where the venue is, what the transport arrangements are, what date or time it is). While the fulfilment of a commitment is not defined by a specific set of physical movements, there is a broadly defined range of physical movement that would count as fulfilling that commitment. My journey to our meeting would have to take place at a particular time and finish at a particular place – the agreed venue; but the bagginess of these constraints is utterly unlike the precise unfolding of a physical system. The physical environment in which I enact my meeting will be realized materially but only as a token of a very ill-defined type and we will stipulate whether it fits the bill. Or rather the type-token relationship is utterly different from that seen in the case of natural kinds.

It would be difficult to exaggerate the importance of tensed time in agency. We may think of tensed time as time made explicit and of a dimension of the self made explicit – in a past that is my past, a future that is my future, together with placing time in italics as "now". The invention of the future populated by possibilities that are drawn from the past – either as explicit possible states of affairs or as the implicit sense of the present – creates a space towards which the agent reaches. That future gets a purchase on the present, enhancing the capacity of the agent to act upon herself to deliver on her intentions, to fulfil freely chosen goals. The viewpoints we adopt on or towards successive time points in the physical world are that in virtue of which we stand outside of ourselves, our current state and physical situation, enable us to act on them. Tensed time, sedimented into clock or calendrical time, lies at the heart of distinctively human agency.

Agents, then, exploit the actual, physical world to realize envisaged possibilities generated by their individual or shared intentional consciousness. It is from the standpoint of these possibilities that the habits of nature can be exploited, individual law-abiding events – traditionally seen as caused and as causes – transformed into handles to bring about desired states of affairs, and material circumstances become platforms from which we can spring to a goal. The discovery of handles and platforms will not just be a matter of passively registering the intrinsic properties of the physical surroundings. Rather, it will be the product of an actively cultivated mode of seeing-as. The virtual context constructed by the mind will present what-is as the preface to what might be, to a range of possibilities.

This is evident in permissive causation where we choose which events we allow to happen. It is even clearer in what we might call "the set up" where we position ourselves or arrange elements in the material world to make certain events more likely. I point a camera at a bird's nest. It does not take pictures until I see the bird feeding its young, when I will press the knob to open the shutter and thus permit an interaction between the nest and the photosensitive surface of the camera. Or I arrange that things will remind me of what I have to do – setting an alarm, getting someone else to nudge me, and so on. Even settling down to sleep, and thus to embrace passivity, may be an act: I make myself comfortable, close my eyes, turn off sources of sound, to increase the chances of it happening.

When we appreciate the significance of the space opened by intentionality, structured by propositional attitudes such as intentions, knowledge, desires, hopes, etc. – which, to repeat an earlier point, should not be thought of as an effect of certain causes or as a cause of certain effects – we can understand how free will proves, after all, to be real in theory as well as in practice, to be real rather than illusory. The downstream consequences of our actions are not merely an expression of an undeflected transmission of upstream influence. And actions are not obliged to be the passive expression of laws evident throughout in nature, though they exploit the habits of nature sometimes revealed through inquiry as the laws of science.

Actions are physical events that would not have occurred had they not been intended and their features are realizations of the possibility or possibilities it is envisaged that they would fulfil. They happen because they are justified – at least in the eyes of an agent for whom they fulfil an intention – rooted in a range of mental attitudes such as hope, fear, desire, a sense of duty. Nothing is justified (or indeed unjustified) in the physical world: mechanisms happen, period.

Of course, not everything agents do is preceded by explicit deliberation. Repeated, much practiced, actions as part of well-worn routines may be an almost effortless, near thoughtless expression of a habit acquired in the past, when the action *was* deliberate. And not all the components of a deliberate action are deliberate. At a certain level, what we do involves elements that are so close to reflexes that they seem to be mere happenings – as when I withdraw a limb from a painful stimulus or avoid a physical threat before I have explicitly registered it.

Even so, these happenings are often in the service of a larger goal that is deliberately aimed at. Prompted to run for the train that is to take me to London for that important meeting by the knowledge gleaned from consulting the time table, I put one leg in front of the other, deploy fancy footwork to avoid dog dirt and other obstacles, and steady myself as I slither on an icy pavement – all without thinking or obvious deliberation. These seemingly automatic events are taking place because I want them to. When they unfold fluently, with so little effort that doing seems to ride on a river of happening, this may conceal how they were once painfully executed. Walking, driving a car, playing the piano, may seem to the well-trained agent to take place of their own accord; but this seeming automaticity is the exercise of painfully acquired expertise. By contrast, no happenings in the realm of what-is are *rehearsals*, even less ones governed by the idea of perfection or at least improvement.

The point remains, however, that there are no actions that are entirely uncontaminated with happenings. Actions cannot be entirely transparent to, or put together out of, volition. When I walk, there are many layers of the action that I do not *do*. I do not do the muscle contractions, the orchestration of those contractions to steps, the spinal regulation of the steps to allow me to concentrate on my goal when I am heading towards it. I am even less responsible for the blood flow to my muscles, the oxygen delivery to the contractile elements, or even larger-scale regulations such as the maintenance of my posture as I move from place to place. This notwithstanding, it is perfectly clear that there is a fundamental difference between the movements of my legs when I am having a seizure and their movement when I am hurrying towards a destination. While doing is embedded in happening, there is a genuine contrast between what regulates them.

Skilled action is never entirely dissolved in aconceptual, narrative-free routines. There is still a guidance from explicit thought when skills are exercised in particular circumstances – as they always are. There is the pre-task of getting to the place or circumstances where the well-rehearsed routine can be performed. There is a necessary background awareness of the reason for, and the goal, of the act. There is the continuing need to take account of the particulars of the material circumstances in which the act is performed. Even when I do something that seems automatic such as running for a bus, it is at most *semi*-automatic. I run for a particular bus for a particular reason. The conceptualizing "I" is not entirely given over to a doing that is so automatic as to be a mere happening. A mist of explicitness envelopes and permeates the doing. The practice of routines can never be entirely routine, if only because it is always at the risk of being tripped up by particulars.

Our powers are vastly amplified by the technologies we have collectively created, and other manifestations of our capacity to enhance each other's agency through ways of working together within teams and institutions that exploit our cumulative knowledge and skills. Nevertheless, as already mentioned, there are limitations to our freedom. Those limits begin with the fact that we do not choose our own existence and our ability to live and develop past early infancy when we had little control over our capabilities. Much of what I am – a privileged individual at a time in history when, in some parts of the world, it is normal to have a long, healthy, reasonably prosperous existence, free of persecution, has been contingent on accidents, on mere happenings or others' doing rather than my own agency.

That we do not bring ourselves into being or choose many aspects of our lives is, however, the other side of the fact that, as agents, we have to have something to be free from, free about, and free for. Without unchosen givenness, without a context into which we are (to use Heidegger's term) "thrown", our agency would lack specific content and meaningful goals. There would be no starting points and no ends which, scaled to our being, are necessarily local and, for this reason, seem contingent. Only if our freedom is limited can we be free from or free towards. This said, much of our lives consists of acquiring the skills, creating the situations, that will provide platforms, and the many-layered supportive material environment, for assisting and enhancing the expression of our freedom.

It will, I hope, be evident from what has been said that agency is possible – indeed real – and that it owes its possibility to explicitness. Just as explicitness is not a material effect of, or part of, what-is, so it is not, of itself, a material cause acting on what-is in competition with patterns of the unfolding of what-is most accurately and comprehensively captured in physical science. It is not a mysteriously muscular ghost.

7.3 THE AMBODIED SUBJECT AS AGENT

At one level the agent is the embodied or ambodied subject, and the body is the primary agent of our agency. All the modes of movement in space, the countless ways in which we physically shape the world, and our communications with our fellow creatures (facial expressions, speech, writing) begin with our bodies. Our legs, hands, and our mouthy faces are the primordial tools. Our legs take us to our destination, or to the transport which will take us to our destination, to the place where we can perform the actions we wish to perform. Our gripping, grasping, grubbing, groping, holding, cupping, catching, pressing, pulling, pushing, squeezing, pinching, tweaking, twiddling, plucking, prizing, picking at and picking up, carefully or idly fingering, leafing, tousling, dabbing, caressing, scratching, slapping, punching, embracing, stroking, patting, smoothing, tapping, shaking, drumming, clapping, poking, prodding, button-pressing, pointing (at, out, to), enumerating, gesticulating, threading, scraping, insulting, bending, twisting and stretching, glad-handing, signaling, hand-shaking, hands help us (literally) to manipulate the world. How just was

Aristotle's description of our hands as "the tool of tools"![10] And our face – with its thousand expressions, including a hundred varieties of smile, and its capacity to emit verbally mediated meanings whose referents range from a speck of dust on a table to the universe – enable us to respond to, engage with, what is out there at many different levels of concreteness and abstraction.

The cooperation between parts of our body is a prototype of the first-order cooperation between individuals that will fill, shape, and define our lives. I hold on to the surface with my hand while I pull myself up into the standing position. I lift the food to my mouth with my hand, chew what has become the contents of my oral cavity, swallow – and then my body takes over. Hand-with-hand cooperation has countless manifestations. It might be direct – as in clapping (rapid to show admiration, slow to show contempt), scooping water, or handwashing – or indirect as when one hand holds a nail and the other wields the hammer. It is in this use of parts of our body as implements we are most aware of our bodies as possessions; as entities other than us - though damage to them and effort exercised by them are lived by us as modes of being ourselves. And when our carnal instruments fail us, as when we sicken, the sense that they are other than us dominates over the feeling of our being ambodied in them.

The relationships implicit in the use of parts of our body as the primary agents of our agency are, to say the least, complex. Behind my running for a train may be the training I performed on a treadmill as I brought my legs up to the speed I anticipate I may need on some future unspecified occasion. The purposes to which I put my hand range from the simple – as when I grip a banister to steady myself – to the slightly more sophisticated as when I insert a thread into the eye of a needle – to the most sophisticated of all as when my fingers, dancing to the silent music of abstract sense, deliver this sentence on to the computer screen. And as for my face, reference to spontaneous grimaces, carefully timed and toned smiles (the product of the work of who knows how many muscles acting in cooperation), parodic copies of doubt expressed in wrinkled noses, and its utterances ranging from grunts to sonnets recited with attention to timing, tonal envelope, and rhythm, hardly begins to capture its boundless versatility and the complexities of the relationship between conscious subjects as ambodied agents and their bodies.

It is, however, a reminder of what is slowly unpacked from the first inklings of ambodiment awakening in the newborn infant who has not yet discovered her toes, seen her face in a mirror, or learned that her hands are hers and that they will be the most faithful and versatile servant of the myriad expressions of her will. The interactions between the conscious subject, her body, and the innumerable types and scales of surroundings that she learns to inhabit as manifestations of her world – a world that she comes to understand that she shares not only with visible others that are present but also with invisible others. Those others are implicitly present in the traces they have left – ranging from footmarks on the floor to the artefactscape that surrounds and assists her in her days and the voices transmitted from afar courtesy of electronic devices or mutely deputized in the endless acreage of written pages or screens.

It is in this trialectic between the conscious subject, the body, and the world in which we find ourself, that agency develops. While agency may be expressed as the realization of discrete but interwoven intentions, and the propositional attitudes connected with them providing "becausation", it is not to be understood as puffs of mentation, of will, mysteriously pulling above their weight, driving the heavy objects of the world. From the beginning, the agent is an ambodied subject, able to stand up for itself, and take hold of its state, transformed into "now" nourished by the past, in order to bring about an envisaged future.

Agency is not confined to actions that move extracorporeal objects. There is the manipulation and enhancement of our sense organs to enable us to ascend from passive experience to active perception, and thence to scrutiny, searching, and formal inquiry. Experience as an activity may be exemplified when the head scans round or the ear is cocked, observers take themselves to a vantage point or a position from which the invisible becomes visible, or when the sense organs are assisted by instrumentation. Or, more subtly, when a text provides a metaphorical overview on a cognitive terrain. The active, quality-controlled and ultimately dephenomenalized perception that is measurement, discussed in Chapter 4, has taken our collective agency to unimaginable heights.

The concept of initiation, of a start, a beginning pointing towards an end, central to that of agency, is inseparable from that of the "am" of ambodiment which lays down a here and now, a personalized beginning, in a universe that has only one remote, theoretical beginning long preceding the life of any agent and which, being prior to conscious subjects, has no here or now. The agent puts her foot down and propels herself in a direction she defines as "forward", rather than being helplessly propelled to an invisible, because non-existent, future by an invisible, non-existent past, that being without goals, has no forward or backward direction. The very idea of a beginning is connected with agency – with *initiation* by a being who intervenes, appropriating the flow of events, turning their energies into the realization of meanings.

As human subjects mature, so they are unpacked into the self whose complexities we described in the previous chapter. Consequently, we can act with increasing expertise on behalf of that self, transforming our body from a fate, a given, to something we direct along a course we increasingly lay down for ourselves. There is an iterative cycle: the temporally deep, complex self underpins sustained agency and sustained agency affirms and builds the self who knows what it is doing and why; and does what it is doing *because* it knows what it is doing and why. There is an increasing range of possible freedoms though our fellow humans may crush those possibilities by a variety of physical, social, and psychological means.

As we suggested in the previous chapter, we may think of "amming" – embracing this body as me, asserting my identity – as proto-agency, as lying at the root of freedom. To "am" a particular body is to care for its fate. This is often complex and indirect, as when I go to the doctor, receive and cash a prescription, and take the medicine, or when I sign up to a gym to keep myself fit. "Amming" straddles being and doing, between a portion of what-is, registering what-is, and trying to shape what-is. To "am" this body is not simply to acquiesce to a received, imposed, objective identity. It is an assertion, an embrace, of identity on the borderline between passivity (receipt of a given) and activity. It is to this extent that *ego* is *ago*; that first-person being is enacted being.

Nevertheless, as we discussed in Chapter 5, the transformation of the it of the body into the am of the subject is incomplete and variable. Most of my flesh is not ammed most of the time; but, importantly, I am some of it all of the time. And the "am" relationship is italicized when I use my body as the primary agent of my agency in the way described earlier. Even in those circumstances, however, the relationship can deteriorate into "I-it", as when I find my hands or legs clumsy, not obeying my instructions. This, however, further highlights the extraordinary relationship I have to my body when it serves my intentions. It is after all remarkable that my arm goes up just because I want it to do so in the service of, say, the intention of voting at a meeting. By contrast, any attempt to move a table telekinetically, by pure willing unmediated by physical interaction between it and the body, would be in vain.

Ultimately, the duration and scope of our agency lies beyond our control. As already noted, I did not put myself together. There is a sense, after all, in which – notwithstanding our subordinating the fleshly givens of our being to our individual purposes – we do not entirely escape the unchosen fact of our existence or the usually unchosen fact of our exit. I can take only limited credit for this sentence I am now writing not being the last sentence I shall ever write or think.

7.4 THE POINT OF AGENCY

Although human behaviour is deliberate and to a significant degree elective, there is a background assumption that body-mediated actions will be programmed by evolution to promote the survival of the body and hence of the self. The question then arises why explicit agency is necessary; why more is required than that which can be built into the organism to maximize the chances of survival and successful reproduction. Why should the organism – as it does in human beings but also, though to a much lesser extent, in other living creatures – embrace itself and have behaviour expressing care for itself? Are not witless mechanisms enough? They were, after all, sufficient to drive the evolutionary process. Why should (organic) being be supplemented by amming? Why should the maintenance of the organism be more secure if aspects of it are *performed*? Why should the human creature be driven through its life by an agenda that it has explicitly, a personal agenda rooted in the specific details of its individual life, in which there is such congruity between the meaning and significance of its actions and their meaning and significance to itself? After all, mechanism has served living tissue in many forms very well and it is only recently that *H. sapiens*, or indeed organisms with the slightest hint of conscious agency, have come to dominate the earth? Natural selection not only fails to explain how sentience awoke in insentient living tissue, how unconscious mechanisms delivered the transition from lifeless chemicals to conscious organisms, but also what consciousness brings to the party.

It is easy to exaggerate the benefits of being conscious if we overlook what is achieved by organisms in the absence of consciousness. Nothing in our portfolio of voluntary actions can begin to compare with the complexity of the processes that synthesized us out of the zygote, and the intra-uterine journey from this starting point to the howling, nappy-filling apple of our parents' eyes. And if we had to *do* all of the things that made our continuing life possible we would not survive beyond the moment of our birth. In short, what is delivered by unconscious happening – including (if you accept the evolutionary account of the generation of consciousness) consciousness itself – makes the achievements of conscious agency seem very small beer.

The most important consequence of being a conscious self is that you are explicitly located in a world that is yours and it is this that makes your actions bespoke. When I meet a friend at 10 a.m. tomorrow, it would not be sufficient to mobilize a friend-meeting-reflex, acting on instinct shaped, if necessary, by prior operant conditioning. A general friend-meeting-reflex would not address the many singularities of actual meetings which make sense only in the light of the temporally deep biographies of the friends, the recalled intersection between their lives which has a temporal depth of its own. Much of this, of course, is buried in implicitness, in the taken-for-granted, in the going-without-saying. The many layers of explicitness buried beneath what has come to be regarded as perfectly commonplace create a misleading sense of banality. The everydayness of the repetitions of days makes the miracle of our human days invisible. Familiarity breeds if not something as positively negative as contempt, it does extinguish wonder.

There is, in a planned meeting, a unique cross-roads between intention, purpose, the caring for the other and/or one's self that is, prospectively at least, to be realized. While this unique place may have some physical specification – if we agree to meet in a particular pub – that specification does not define the meeting. Indeed, there is no reason why, at the last minute, I shouldn't suggest a different venue because you have told me that you want to share something that is worrying you and you have heard that the music in the chosen pub is rather too loud. Even the most basic elements of the meeting cannot be captured in their physical specification.

Conscious agency, in short, enables actions to be tailored to the unique, and uniquely valued, actual past and possible future of the agent and those she knows. That is one of the reasons why the exercise of free will is less a matter of choosing between A or B and more a matter of getting into a place where A vs B is a decision point. We play a major part in framing our lives such that, from time to time, it presents itself as being the node of a fork in the road.

Reminding ourselves of the bespoke nature of conscious behaviour, however, brings us no nearer to seeing benefits of consciousness compared with the unconscious mechanisms that drove the evolutionary process and maintain evolved organisms. Indeed, a dependence on consciousness and all that is required to maintain it, would seem to add an extra dimension of vulnerability to the organism which has not been compensated for until quite recently in the evolutionary tree. Even less does it explain how consciousness arose, even if it were of benefit. All sorts of things might be useful for an organism – such as the ability to disembody when faced with a predator and re-emerge a mile away – but this does not make them more likely.

7.5 WHAT-IS AND WHAT-OUGHT

If there is a passage in the writings of David Hume that has an even higher citation index than the one we quoted at the beginning of the previous chapter on the self, it is the one from the same work – *A Treatise of Human Nature* – in which he asserts (or is interpreted as asserting) that you cannot derive an "ought" from an "is": how things ought to be from how they happen to be. A reason should be given, Hume says, "for what seems altogether inconceivable, how this new relation [of ought] can be a deduction from others [of is], which are entirely different from it".[11] No such reason is forthcoming if we look for it in the natural realm of happenings rather than the extra-natural human world of "doings".

There is, admittedly, one sense of "ought" seemingly applicable to the realm of happenings. It becomes apparent when things unfold in a way different from that which we have come to expect. We shake our head in surprise, in disbelief. We think of things that "shouldn't" or "should" happen. Even this sense of ought is unnatural because expectations are not part of the physical world. There are no standpoints from which the future to house expectations can be projected.

Nevertheless, this is not the kind of "ought" we are talking about: the moral ought that guides our judgement of our own and others' behaviour; of doings that would not have happened if nature had been left to get on with itself. Unlike happenings, doings can be classified (however we may contest this) as "right" or "wrong". While happenings may be desirable (rain in a dry climate) or undesirable (a plague of locusts) they are not judged as moral or immoral – or not at any rate in the absence of a God deemed to be omnipotent and therefore responsible for all happenings. As agents we have responsibility for the decisions we take, the choices we make where we have the power to choose, the actions we perform and some of their consequences – whether we recognize it or not.

Defining the limits of that responsibility – determining when we genuinely have a choice, when our choices are fully informed, and which of the consequences of our actions for which we should be fairly held responsible – is a matter of endless dispute. Our duty of care to others may extend to foreseeing and preventing certain unintended consequences of our actions, however well-intended, or failures to act however this may seem to be justified.[12] There is no clear division between those ripples we have set in motion for which we should be blamed or praised and those which are beyond the point at which doings give way to happenings. What matters is that we subject some of the events that take place in the world which are the result of human agency to moral judgement, in accordance with general principles which we apply (or should apply) to ourselves as to others; or to ourselves as instances of humanity. It is as doers who initiate or deflect or prevent happenings, we can be held to the standard of doing unto others as we would wish others would do unto us. The realm of "ought" is that of doings in a universe of mere "is" happenings.

Hume's problem is sometimes presented as that of closing an unbridgeable gap between matters of value and matters of fact. To some extent that is true: if happenings or states of affairs were not valued differently – as desirable or undesirable – there would be no "ought", pointing agency in one direction rather than another. The story, however, goes deeper than that. Desirability or undesirability is not intrinsic to the states of affairs that are so classified. Rain in a drought or a plague of insects simply "are" unless they are experienced by a conscious subject whose happiness or unhappiness they promote – and indeed who gathers up states or happenings into "a drought" or "a plague". In short, value requires beings who value themselves and others and the gap between is and ought is opened up by the emergence of such beings in a universe that overwhelmingly does not value itself, not the least because it is entirely unaware that it is, never mind that it has a future populated with rival possibilities. And moral judgement as to what ought to have been done requires beings who not only value themselves but are capable of agency that can influence the course of events.

The is-ought gap, therefore, is not simply between matters of fact or the propositions in which they are stated and matters of value or the propositions in which they are formulated. So-called matters of fact and matters of value judgement are both late descendants of the awakening of beings to themselves, to that-I-am, such that they have the ability to generate doings rather than merely being localized sites of happenings. The *universalized* or *universalizable* moral principles into which "ought" is unpicked are later descendants still. With agency comes responsibility and with responsibility comes moral judgement. We are capable even of standing outside of ourselves in order to pass moral judgement on our own actions and feel shame ("ambarrassment") or pride. This is yet another mode in which we make ourselves explicit. Both matters of fact and matters of value belong on the far side of the transition between what-is and that-it-is.

7.6 CONCLUDING THOUGHTS

In this chapter we have built on the inquiry in the previous two chapters into the nature of the self with the aim of making the unique nature of human agency more visible, and of what it is that makes it possible. We have exposed the point-missing stories that gather up agency into the flow of physical events passing through the body and erase the fundamental difference between happenings and doings. If we were just a conduit for a stream of causes and effects, understood as expressions of a *de re* necessity whose ancestry reaches back to the beginning of things, it is difficult to see how we could even identify individual causes, let alone exploit them as handles by which we could manipulate the course of events. If we were entirely dissolved in a sea of material causation, it would be even more difficult to understand how we could entertain the *illusion* that we are agents, genuine originators of our actions, exploiting causes to bring about desired effects.

Agents do not do what they do by breaking the habits of nature (upon which, after all, they rely for their actions to be possible and for them to have predictable, desirable consequences) but by exploiting the habits of nature made explicit as knowledge of "how things work" – and hence how they can be worked on. The most sophisticated expression of this is the science that teases laws out of the habits or uniformities of nature. This does not require acting on the physical world using disembodied mental content, ghostly entities mysteriously fiddling with the machinery of the world. The ambodied self is beefy enough to engage with the material objects on equal terms. Nevertheless, it engages with that realm asymmetrically from a virtual outside, opened up by individual, shared, and communal intentionality of conscious subjects. The physicality my body shares with the table explains why it moves when I lean against it; and my complex self generates the explicit reason that prompts me to move the table – something of which neither it nor the material world of which it is a part has any inkling.

The reasons why mere happenings happen are not entertained by those happenings or the objects that are involved in them, while doings would not happen without the reasons for them being entertained by an agent. Actions are a special category of happenings illuminated from within by purposes and intentions; by whys and wherefores that express the self-concern of a "I" (as an individual or a member of a collective) explicitly present to itself as a creature extended at a time and over time. Doings occur *because* of their purpose that is made explicit as they are bathed in the light of possible futures salient to the agent; they are not so much caused as *becaused.*

The self as agent and the agent as the self reinforce one another. We begin our lives as almost selfless and almost helpless neonates and subsequently mature to the point where we are conscious agents acting in and on the world to fulfil the quotidian needs, and the dreams, hopes, and fantasies of the self. Our mature selves and our developed agency drive each other forward – my strivings, successes and failures put me in italics – though both can be erased if the body is sufficiently damaged, when they cease to be what they are and I cease to be a fount of, the possibility of, events that happen or might happen as the realization of envisaged possibility.

In our journey from neonatal helplessness to the end of all our striving, it is easy to overlook how much of our agency is devoted to developing and preserving our agency. The most obvious manifestations of this are the multitude of ways we look after our bodies – from toothbrushing, to regulating our diet, to keeping fit, to subscribing to a healthcare scheme. And then there are the many skills we endeavour to acquire, to maintain, and to improve. Our self-curation also involves the myriad of ways in which we regulate the many layers of our environment – with the help of a wardrobe of clothes, systems of heating and lighting, housing, and voting for regulations that may make our world safer. And behind all this is a limitless menu of artefacts available for our use.

Self-curation is a particularly striking expression of the double relationship we have to our body: as an object of which I am aware and as something that I am. In this ontological littoral zone, our bodies are not only the primary agent of our agency but also that on behalf of which agency acts, and consequently the primary shaper of the agenda of our agency. The am opens up the distance from which we act upon the world on our own behalf and the is of our body allows us to engage with the entities in that world as an object among objects, exploiting the ontological democracy I discussed in Chapter 5.

The fruits of our individual and collective agency are all around us. While, when we discussed the supposed adaptive value of consciousness, we questioned whether the deliberate action of an isolated individual would favour survival more than would the spontaneous unfolding of the natural world, of mechanisms of unwilled happenings, there can be no doubt that the explicit collaboration of many individuals in small groups and vast collectives acting with (shared) deliberation, has in recent millennia – an eye-blink in the history of what-is – delivered for the human race. The evidence of the power of agency is all around us in the planet on which we live and which we are in danger of working to its, and our, death. Thus, the consequences of the explicitness that transforms happening into the possibility of doing.

Envisaged possibility lies at the heart of agency. The next two chapters will focus more directly on possibility: on thought as the entertainment of possibility (Chapter 8); and on possibility *per se* (Chapter 9).

PART III
Thought and possibility

CHAPTER 8

Free-floating explicitness: thinking about thinking

8.1 INTRODUCING THE THINKING ANIMAL

> Our creatures are our thoughts, creatures that are born Giants: that reach from East to West, from Earth to Heaven, that do not only bestride all the sea and land, but span the sun and firmament at once: my thoughts reach all, comprehend all.
>
> John Donne[1]

> We might say that thought is to human beings what flight is to eagles and swimming is to dolphins. Tim Bayne[2]

That-it-is encompasses many modes of consciousness from the most passive sensory reception – things we can't help seeing or hearing – to facts that are discovered as a result of active scrutiny and investigation and, beyond this, to the products of the most intense, brow-furrowing thought. Thought may propose not only things that exist but also others that don't. It may entertain the fact that something didn't happen, that such and such is not the case or does not exist, or reflect on the fact that it did happen or did exist. It is this that makes thought – pure that-it-is – seem a disembodied manifestation of the embodied subject. It is a mode of that-it-is that seems only loosely tied to what-is.

While we might be reluctant to locate thoughts in a mind understood as a portion of an immaterial substance, it is difficult to ignore their ghostly nature. Nevertheless, for many philosophers, thought is far from ghostly: to have a thought is to be the site of pulses of neural activity. As we saw in Chapter 3, however, the attempt to identify mental entities with neural discharges in the brain runs into insuperable difficulties. These difficulties are particularly obvious in the case of thoughts. It is no wonder therefore that thoughts play such a crucial part in Descartes' arguments for dualism. The nature of the Cartesian "I" which thinks, what it is in addition to the present flow of thought, is however uncertain. Even so, thoughts seem to transcend the material world, including the body of the thinker. They seem to have a special mystery.

Before we examine that mystery, it is important to flag up the distinction between a token or occurrent thought and a thought type. What follows slides past many difficulties in defining the nature of thought types and token thoughts and the relationship between them.[3] It will, however, be sufficient for our present purposes simply to point out that token thoughts happen at a particular time while the types they instantiate are timeless; and the former are had by only one person, while the latter can be had – be instantiated in a token thought – by any number of people. The thought that "Paris is the capital of France" may occur to me, or be expressed by me, at a quarter past four on 27 October 2022, when I am sitting in my study; while the fact that Paris is the capital of France is not something that belongs to anyone. To put this another way, that Paris is the capital of France may occur to me, but it is the thought, not the fact, that occurs. Indeed, its token instances will occupy a place in a temporally sequential train of thought.

Most strikingly, the token thought has only a tenuous connection with any phenomenal experience had by me now. A particular experience associated with it is irrelevant. A mental image of the Eiffel Tower has a difficult relationship with the relationship captured in the fact that Paris is the capital of France. After all, there is probably no mental image corresponding to "France" when I have this thought and certainly not to "capital" – not at any rate that is prescribed by a necessary connection with the thought. And any feeling of certainty I might have as to the capital of France, or of triumph over those who incorrectly thought it was the capital of Germany, is equally irrelevant. The contrast between knowing that Paris is the capital of France and having a sensation such as toothache reminds us of how we are not obliged to *live* our thoughts in the way that we live our experiences.[4] It is a reminder, too, of how entities are crash-dieted, stripped of so much of themselves, in becoming the subject of sentences.

Some individuals with synaesthesia associate the sounds of language with colours, even where the sounds are the imaginary sounds of thought.[5] However, the colours are incidental and have nothing to do with the reference of thought, which remains colourless. Even the thought that daffodils are yellow should not be coloured, any more than the incorrect thought that they are blue or the correct thought that they are not blue. Daffodils are yellow but *that* they are yellow is not.

The reader may have noted that I have slipped from talking about thoughts to talking about facts without clarifying the relationship between them. I shall return to this. Our job for the present is to reflect on what we might call the *stand-alone explicitness* of the realm of thoughts which is typically that of verbally articulated experience. This is the realm of possibilities being entertained and given an existence of their own in discourse; of the space of assertion and denial, and, with this, modality (that such-and-such might or might not be the case, might be necessary or contingent); and – a topic for Chapter 10 – that of truth and falsehood.

When I have an experience, that which I experience – its intentional object – must be directly or indirectly present, if the experience is to be veridical. When I recall an experience, that which I remember, if the memory is not false, must have been present to me at some time in the past. The mystery, the ghostliness, of thoughts is compounded by the fact that what they are about does not need to be, or to have been, present to me. Their objects are invisible, inaudible, intangible, etc. This follows from their objects not being actualities but, primarily, *possibilities* and those possibilities can just as well be realized as not realized. That Teotihuacan is the capital of a Mesoamerican realm is for me as much a fully-fledged thought as that Paris is the capital of France, though I have not been to the former while I have had several unforgettable trips to the latter.

8.2 UNWIRING THOUGHT

While the relationship between token thoughts and their referents shares some features with the intentional relationship between perceptions and perceived objects there are important differences, the most obvious of which is that thought is unwired to the surroundings of the thinker. Thought is also boundless. Our thoughts see so much because they overlook so much. My capacity to think about Paris is not limited by the distance between my body and the City of Light or the quantity of material between the city and where I am now. And to think of something is to simplify it; to make vast stretches of it mind-portable. No thought could match a landscape in all its details or even an individual tree, never mind the city of Paris.

It is this that makes a causal theory of the reference of thoughts, where the referents are states of affairs that are entertained in thoughts, even wider of the mark than causal theories of perception or of memory. Perception, as we discussed in Chapter 3, cannot be understood as the mere physical effect of something impinging on the body, or the brain, because perceptions reach back to the events taking place in a certain object to transform them into a revelation of the object. The perception is "about" that which justifies it, which may or may not make it true. Embracing the causal theory would require that we accept that, when we perceive a table, the latter is both the cause and the object of our perception. Thus, the fundamental problem of causal explanation of the intentionality of perception. And as we discussed in Chapter 6, memory has double or triple intentionality: the memory reaches out to an experience that itself reached out to the spatio-temporal slice of world in turn connected with the self whose memory it is. Causal theories of memory therefore do not hold water. It seems even less likely that the referents or intentional objects of thoughts are their causes.

Before we proceed, it is worth noting that the use of the concept of intentionality in relation to both perception and thought, while justified, may conceal a fundamental difference beneath what they superficially have in common. In the case of thought, we have a token thought that may have been prompted by something present, but it is very often about something not actually present. I may be more likely to think about entities in Paris when I am in Paris, but the things I encounter when I am in Paris do not add up to Paris, however Paris is defined. More typically, I think about things that could not be characterized as "present", when – as is often the case – my thoughts take a general form, applying general predicates to individual entities or general predicates to general kinds, some of which may be abstract. The absence of their referents does not undermine the validity of the thoughts, especially if we are questioning whether, or denying that, those referents exist or are intending, hoping, or dreaming that they might "one day" exist. In contrast, a perception to which there corresponds no object out there is an illusion or an hallucination. It is perfectly okay to wonder whether there is an oak tree in 6, Garth Drive, irrespective of whether there is one, but not to perceive such a tree if there isn't one. Or to think that "All people are mortal" even though there is no particular experience corresponding to "all people" or the generality captured by "mortal".

Thoughts, of course, may be triggered but not by the presence of their referents but by an association of ideas. Catching the scent of a flower may prompt me to think of spring and then to remember a particular event in spring and subsequently to think that I must write to the people who shared a cottage with us last spring. The scent of the flower cannot be held accountable for this thought. So even where thoughts are triggered by something that is present, the connection between trigger and thought is typically tenuous. That is why a photograph of me, or my image in the mirror, may give no hint as to what I am thinking about at the time. Guessing my thoughts would not be helped by including my surroundings in the photograph. Indeed, if I were to look at someone's photographic portrait, I would have no idea *that* he was thinking, did I not know this from the immediate experience of my own thoughts which tells me that people tend to be thinking pretty well most of their waking hours. When it comes to thinking by appointment, as when someone clears a space to focus on a particular topic, the disconnection with the surroundings becomes closer to completeness. The contents of this paragraph have little connection with the desk on which my computer is located, the view outside of my study window, or the pressure of my socked feet on the carpeted floor.

Thus the evidence against a causal theory of thought. There is, what is more, a further problem with the very idea that a thought could be caused by its referent. It is clear that the referents of most thoughts are not the kinds of entities that exert causal powers as they are usually understood. This applies even to the simplest thoughts: the referent of the thought that Paris is the capital of France – the city itself – does not seem to be able to gather itself up into something that acts upon the subject to make her have the relevant token thought. What would be the causal agent? Clearly not Paris or France: it would be difficult to think of either of these working as one to have a collective effect. And the *relationship* between Paris and France such that the former is the capital of the latter seems an even less plausible candidate as the cause of the thought. Indeed, a more likely candidate would be another thought corresponding to a rival claim – that (for example) Paris is the capital of England or Avignon is the capital of France. In short, as is most usually the case, thoughts are prompted by other thoughts – often in dissent, or as elaboration, or by mere association.

What is fundamentally wrong with causal or quasi-causal theories of thought, according to which thoughts are the effects of their referents, is that those referents do not, in the absence of thoughts, exist as entities that might have causal powers. It is only as the subject of a sentence, as the referent of a name, that Paris is gathered up as an entity. Even then, that entity is not an object understood as a quantity of material occupying a bounded stretch of space and time. "Paris" as the subject of a sentence is not set out in space and time in the way the city, defined by certain stipulated boundaries, is. More generally, the objects of thought are not objects analogous to those that spatially surround us.

This remains true even where the thought refers to a discrete material object. The table on which I am writing this sentence is not identical with the table to which I have just referred in the last paragraph but one. As a subject of a thought, it is not present as a time-slice in a spatial location. When I refer to the table, I do not refer to it at a particular place and a particular time in the way that I perceive it at a particular place and time. That is why its availability to be referred to has nothing to do with my location and why, also, it is available to be united with predicates that, even more obviously, have no location. And while I can walk through Paris, I cannot walk through the Paris that is the subject of a sentence in which it is asserted that it is the capital of France – or more to the point that it is not the capital of England.

One way of highlighting the reason for this is that when we visit Paris, and walk through it, we do not visit the city but part of the city; indeed, we follow a narrow path through some streets in the city and our journey occupies a small time-slice of its history.[6] It is only as the referent of a word that Paris becomes a mind-portable unity. After all, "Paris" could continue to be a referent of a sentence even after it had been entirely obliterated. And non-existent objects such as Pegasus can be perfectly acceptable subjects of an intelligible thought even if they do not exist. Indeed, the thought "Pegasus does not exist" makes perfectly good sense. We know what is being talked about. Or think we do.

While reference, therefore, points intentionally from a token thought to that which it is about, the intentional relation is not the causal effect of the referent of the thought. Paris being the capital of France does not cause the thought "Paris is the capital of France", any more than the non-existence of Pegasus causes the thought that Pegasus does not exist.

All of which may seem obvious and not require to be spelled out. After all, causation is a wobbly concept even as applied to the relationship between material events in the material world. Since Hume's critique (referred to earlier), *de re* necessity has been regarded with suspicion. It is a manifestation of the mind's tendency to project its own sense of inevitability – such that events that usually follow one another are perceived as inescapably following one another.[7] Nevertheless, such spelling out seems to be required in view of the popularity among philosophers of Saul Kripke's so-called causal theory of reference, despite that theory not really being a causal theory in the way that makes it attractive to those who embrace a physicalist account of the explicitness that is captured in discourse.[8]

It is for precisely this reason that I shall devote some time to criticizing the theory. After all, the fundamental mission of the present book is served by gnawing through the causal threads – however problematic they have come to be seen for other reasons – that are assumed by those of a physicalist persuasion to connect conscious subjects with their worlds. "Causation" is often a proxy for physical interaction.

Kripke's theory was introduced to explain how the elements of thoughts such as proper names and natural-kind terms come to have the referents they have and how they retain their connection with their referents. Kripke envisaged an initial or original "baptism" directly witnessed by at least one person at which an entity is named – as when we name the dog "Rover" and subsequently call it by that name.

In the case of natural-kind terms such as "water", it might be thought that there was no original baptism. Nevertheless, they function more like proper names than is usually supposed: "What a natural-kind term refers to is determined not by its associated properties, descriptions, or concepts but by the actual nature of a particular thing (e.g. water *around here*) by means of which the term's reference was initially fixed".[9] This original baptism of water (somewhat a reversal of roles) liberates a term's reference from any description or cluster of descriptions. It is the point at which discourse escapes from being confined to a bubble of language.[10] The relevance of this to our present concern is that the journey from the original fixation of the reference to its subsequent appropriate use is complex but, as with proper names, it is described as *a causal chain*. The causal chain connects the observers of the original baptism with any- and every-one who then uses the name.

The nature of the causal element in the so-called causal theory of reference is rather complex: it refers to the process by which after the initial act fixing the reference of the word, the name is passed on from speaker to speaker:

> Speakers succeed in referring to something by means of its name ... because underlying their uses of the name are links in a causal chain stretching back to the initial dubbing of the object with that name. Subsequent speakers thus effectively "borrow" their reference from speakers earlier in the chain, though borrowers needn't be able to identify any of the lenders they are in fact relying on. All that is required is that borrowers are appropriately linked to their lenders through chains of communication, chains of passing on the same name.[11]

Such complexity does not dissuade many philosophers from thinking of the referential capacity of thoughts as being based on a causal process in the traditional sense, that mental contents (including thoughts) "pick out their *referents* by virtue of causal relations between the representation and the referent".[12] This is, however, precisely in the wrong direction for any putative causal theory of reference. The intentional relation between mental entities such as thoughts and their referents is from the former to the latter rather than the other way round. If the direction of causation were from referent to the token thought, we would not be able to entertain the thought that such-and-such is *not* the case or infer from the fact that something is the case that other things cannot be the case; that if I was in London at 12 noon on a particular day, I could not be in Paris, Prague, on the moon, all day in bed at home in Chester, and so on, at that time.

There is thus no causal relationship between what is out there – Paris – and that in virtue of which the word "Paris", in a thought about Paris, refers to the city in France. And this is consistent with the fact that we can encounter the name "Paris" as a puff of air or a squiggle of ink, before we encounter Paris by, for example, walking its streets – as is true of most of us who use the word. In fact, Kripke's "causal theory of reference" does not deliver what it seems to say on the tin: it does not refer to any putative causal connection between a name encountered as a token and an experience of Paris as a referent. The theory is much less ambitious: it refers to the process by which a convention – which gave Paris the name "Paris" – is propagated through a linguistic community.

The communication of a name – of, that is, a convention of reference – seems even more remote from anything that might be gathered up in the notion of the "causal". When you teach me the name of your dog, so that I am equipped to call it back from a fight with another dog, your action involves many modes of intentionality. There is a managed convergence of our consciousness so that we focus (a) on the same object (your dog); (b) on the name of the dog; and (c) on the connection between them. You might in addition also focus on the sound of the name as when you correct my pronunciation. All of which is remote from any connection we might think of as a causal connection – especially between a thought about the dog and its referent.

What the so-called "causal theory of reference" highlights is how calling things by names in an original baptism so that we can subsequently refer to them is far from anything that could be assimilated into a naturalistic picture of a world understood as a network of causally connected material events. As we discussed in the Overture, when something is named, the entity is effectively detached from itself and then reattached via words that enable it to assume the role of the subject of a sentence which talks about it. It is relevant that this process of affixing referents to terms is remote from a causal process. The baptism *initiates*, rather than being part of, the causal chain. Causal chains, as usually understood, do not have a beginning, as we discussed in the previous chapter. Those who claim that lightning causes thunder accept that the causal chain does not begin with lightning, since the latter is itself an effect of prior events or states of the atmosphere.

Most importantly, when I refer to Rover, the direction of reference is from word to thing and, as already noted, it is clearly in the opposite direction to any supposed causal chains that delivered the word to me to use, so that I could refer to Rover, to the natural kind "dog", and consequently be able to think to myself that Rover is a dog.

There are many other reasons for denying that the reference of a thought is its cause. As we have noted, it is possible to have a thought about something that does not exist. Or for one to think about something that will exist only in the future – for example the Christmas tree I have not yet bought. In short, the arrow of intentionality goes from the thinker to the object of thought rather than from the object of thought to the thinker. And thoughts about future possibilities, including those that are not realized, are in as good standing as are the thoughts we may have about things that are in front of us when we think about them.

The idea that thoughts, and their constituent signs, gain their referents from the causal power of those referents is manifestly indefensible. It is clear that the material stuff on the planet within the boundaries of Paris could not be the cause of "Paris" having a referent, especially since those boundaries are not defined by material discontinuities but by legislation.

It might still be argued that there is a causal element in discourse. Consider, for example, when my use of the words "Rover" or "dog" prompts you to have mental images of Rover or a dog intended as an exemplar. Your thoughts thus directed are effects of what I have said: your token thoughts have been prompted by my token thoughts shared with you. This does not, however, apply to me who is using the word to tell you – or to remind myself – that Rover is a dog, or that he is a naughty dog, or that he is lost. At any rate, when I think about Rover or communicate a thought about him, he is not the cause of my being able to refer to him. After all, any putative powers of referents to bring about changes in me will be limited if, like Pegasus, they do not exist or if they are ill-defined, as is the case with "Paris" which, even if what is referred to is precisely specified to coincide with its municipal boundaries, what is contained within those boundaries – streets, conversations, weather – cannot be gathered up into a single cause. Any effect that reference to Paris may have on my auditor will be entirely idiosyncratic: random images, personal memories, some recalled facts, accidental associations. This applies *a fortiori* to the (not infrequent) thoughts we humans have about the universe.

You may think that I have made something of a meal of distancing reference from any kind of causal connection, but it is relevant to our understanding of the nature of thought and its place in our reflections on explicitness. Our free-floating thoughts are especially striking manifestations of that in virtue of which we are *un*wired into the material world. Thinking marks the widest distance within individuals opened up by the transition between what-is and the that-it-is. It becomes particularly obvious when we recall that thinking is not a succession of isolated thoughts. Thoughts sail in a wide sea of cognition, much of it expressed in the form of other thoughts, even though they may be unthought at any particular time. Individual thoughts are part of a wider conversation I am having with present and absent interlocutors (as well as absent and present slices of myself), some who have entered my life directly, and others who are part of incessant exchanges between members of communities of minds in which I take up a small place, mediated by spoken or written words or other information-bearing modalities. To remember this is to see the absurdity of the idea of thoughts being causally wired into their referents.

The explicitly speculative dimension of many thoughts show how different thinking is from the (relative) passivity of sensory experience. Of course, we may, as noted at the beginning of this chapter, *seek out* sensations – scanning around us, snooping, positioning ourselves for future inquiry, setting up quality-controlled observations. And there is some top-down shaping of even the most basic of experiences as we make sense of what our senses deliver to us. But thoughts are top down all the way: they are products of our generative mind. The thinker may seem like a ghost in one's head, but it is an active ghost and not just when we are racking our brains and furrowing our brows.

Admittedly, we may sometimes seem to be merely a channel through which our thoughts flow and are surprised when we discover where that flow has taken us. But we are able to wake out of wool-gathering and take control, to concentrate on "the matter in hand", turning the tables on matters that have us in hand. This is sometimes assisted by others in their role as the external scaffolding of our selves.

The *objects* of thought are present as pure explicitness. I can think with equal facility about the cup in front of me ("Is that a crack in the handle?") or about Paris (which I cannot presently see, hear, etc., because I am in Prague). That we are not spatially related to the referents of our thoughts, *qua* referents, may seem less obvious when the referent is a particular, spatially defined, object. It remains true, however. Consider when I think about a household pet such as a cat that I know is in the vicinity. I am clearly not thinking about it as being located in a particular cat-sized place at a particular time, even less of the sum total of places and times it has occupied and the activities it has engaged in over the span of its existence. Even if I confine my thoughts to the-cat-now, it has no location so long as I am thinking about it rather than perceiving it. After all, I could be entirely wrong about where it is; and will certainly fail to be precisely right. While the cat's location at a particular time is spatial, *the fact that* he is in the room is not thus located – that is why I can assert this fact repeatedly, even when it is no longer true, or I do so in order to state that it is no longer true. *The fact that* the cat is in the room is no more spatialized than *the fact that* it is no longer in the room. More precisely, it is only at the level of fact that the cat can *be* "no longer in the room". Or, indeed, be "not" anything.

When I am thinking of Paris I may have a few mental images – the Eiffel Tower or the bedroom of the hotel where I stayed a few years ago – but none of this is required for the act of reminding myself or someone else that Paris is the capital of France or even declaring that it is my favourite city. The images can be mere clutter and there is no mandatory "what-it-is-like" to have this thought – a topic for the next section of this chapter. The words I hear in my head or say out loud may be necessary to specify the thought I am having – indeed to have one thought rather than another – but they are nothing like Paris. Indeed, "Paris", is not like anything anyone might experience in Paris.

It is a particularly striking aspect of the unwiring of explicitness-as-thought that I can think about things in general ("cups", "lightning flashes", "colours") as well as particulars (the cup in front of me, the lightning strike that just took place, the colour of that cup), concrete items (that cup) or abstract ones (courage, history). Perceptions, however, are typically of particulars and of concrete items. If this distinction is sometimes not as clear as it might be, it is because we *recognize* what we perceive: it makes sense to us as, for example, belonging to a general class. I see this entity *as* a cup, and thus may seem to gather it up into a general class which it instantiates. This is most obviously the case when we observe in order to report what we see or we look about us to find something picked out verbally. "I am looking for a cat that has gone missing" or "Someone has asked me to look for a cat that has gone missing". The object in question, when it has been perceived, is presented as satisfying a general description. This remains true even if the generality of the description is simply at the level of not specifying the object's state or location. Although perceptions and thoughts are so profoundly different in the way we have described, we encounter and engage with the world as thinking perceivers and perceiving thinkers. That may be why the free-floating, unwired nature of thought does not strike us as forcibly as it should.

Once translated into sentences, my experience of seeing a cup can become common property, a fungible item of information. The fact that I saw a cup on my desk at 1:30 p.m. 20 June 2022 is clearly of little interest to others, an unimportant part of my biography, nothing to broadcast, unless there is someone interested in the location of the cup, which may have gone missing. It will not count as general knowledge to be tested in a pub quiz. And that is true of the overwhelming majority of my experiences, which do not get elevated to the status of facts that I think and think about.

Are there not cases where the object of thought is located in space, such that thinking seems like a kind of pointing? What about thoughts employing indexicals whose referents depend on the circumstances – the physical location and the identity of the speaker – in which they are employed? Indexicals – "here", "now", "I", "this" – seem to anchor thoughts to particular places in space and time. They do so, however, through a route that is general: that is why indexical statements make sense and can be mobilized in a multitude of situations; why you understand what I mean by "here" even though I am not referring to any particular here; and you understand that when I use the word "I" I am referring to Raymond Tallis and not to the "I" you are referring to when you use "I". The indexicality of terms is a general means of attaching necessarily general discourse to particular situations. "Where is the cat?" you ask and I say, "It's here" and this delivers the cat's approximate location because you know that I am referring to my vicinity and you know what my vicinity is: it is where you locate my body through hearing my voice. If an uttered thought of this kind touches on an actual locality it does so only courtesy of the body of the thinker, from which speech is being emitted. That body may directly contribute to ensuring indexical statements land on the correct referents by pointing at them or nodding in their direction.

The general point is upheld: there are fundamentally different kinds of distances crossed by perceptions and thoughts. We cannot get (literally) nearer to or further away from an object of thought by thinking more or differently about it or repositioning our bodies. Moving my body south across the English Channel may bring me nearer to Paris but not to the "Paris" of thoughts such as "Paris is the capital of France". When I am in Paris, I am not in, or closer to, the abstract total that is denoted by the word "Paris" than when I am in Chester.

There is a more conventional way of thinking about the difference between entities as objects of perception and as objects of thought. It invokes what linguists call the arbitrary nature of language.[13] With a few unimportant exceptions – so-called onomatopoeic terms – words have nothing in common with their referents. There is nothing tree-like about the word "tree", "wet" is no wetter than "dry" (and the French "*mouillé*" and "*sec*" perform this referential function just as well), and "miniscule" is no smaller than "huge". How do they get away with being arbitrary? Simply this: language does not express what-is by mirroring it. Indeed, the very fact that how words look or sound – and they do not have a distinctive smell, taste, or touch – is not connected with their meaning, enables them to transcend the different sensory modalities, helping to piece the world together, to weave together different and novel currents in the stream of consciousness, in the transition from what-is to that-it-is. If the word "table" had to look like its referent, then it would have to encompass a limitless range of looks of tables seen from a multitude of distances and angles – clearly impossible – and also do justice to other features of tables, such as their weight, solidity, and surface feel. And there could clearly be no word that looked like Paris, politics, or the universe – or indeed the referent of the word "word". Nor, were phenomenal appearance important, could proper names successfully refer to fundamentally different entities – as in the case of Felix referring to Felix Mendelsohn, the operation by which Germany planned to invade Spain via Gibraltar in the Second World War, the fictional confidence trickster in Thomas Mann's *Confessions of Felix Krull*, or to the cat who will be invoked in this chapter.

Before we pursue this, it is probably prudent to head off a "Gotcha!" moment from those for whom explicitness is identical with activity in certain parts of the brain that "represents" what the thinker is thinking of. Part of our argument against this claim in Chapter 3 was that replication was not representation: it did not deliver the *presentation*, necessary to make replication the basis of *re*presentation. There is however a commoner response: that what is going on in the brain "encodes" the perceived entity – colour, shape, distance, etc. In this respect, we are assured, it is no different from what is true of language made of arbitrary signs. The word "yellow" is no yellower than is the word "blue" or "nothing" and yet it does the business. "Yellow" doesn't have to look like yellow to be about, or refer to, yellow. In sum, there is no natural connection between the form of words and their meaning; between what they look or sound like and what their referents look or sound like. And the same is true of the phenomenal appearance of neural activity and that which it is about.

Far from being a counter-argument, the example of the arbitrariness of linguistic signs helps us to see more clearly what is amiss with the mind-brain identity theory. As we noted in Chapter 3, that which is *en*coded has to be *de*coded and the replacement of neural discharges by other neural discharges would not translate neural activity into experience that "replicates" and hence "represents" the object such that it would be experienced by the subject as something present to it, something it encounters, makes sense of, and with which it engages. Language requires a conscious subject to decode it so that it should have meaning. It is as conscious subjects that we learn that "yellow" means a colour like that of a daffodil. Colourless neural discharges cannot be decoded *in and by the brain* into the experience of yellow in the way that the trained human subject – a conscious subject trained by other conscious subjects – learns what "yellow" means. This would presuppose a prior direct experience of yellow and a variety of other experiences, which is precisely what neural activity is invoked to explain.

Why, you may ask, have I spent so many pages and taken up so much of your time demolishing any causal theory relating to thoughts? It is because causal theories are part of an endeavour pursued by thinkers of a physicalist persuasion to make thoughts respectable, upstanding inhabitants of the natural world. Acknowledging that thoughts are not entities in nature in the sense of being part of what-is – on the contrary they are pure that-it-is, separated from the what-is that they make explicit – is central to the argument of this book. One way of capturing explicitness, most strikingly exemplified in the free-floating explicitness of thought, is to see that it is not an *effect* of what-is as a cause. Thought of what-is-the-case, that-it-is, is unwired from what-is.

The distance between perceptions and thoughts is replicated in that between perceptions and facts. It is as absurd to ask about the location or appearance of facts as it is to ask the same question of thoughts. The reason for this is obvious: what-is becomes fact only in the context of thought. Both facts and thoughts are the products of explicitness. In the Overture, we discussed the opening assertion in Wittgenstein's *Tractatus*:

1. The world is all that is the case.
   1.1 The world is the totality of facts not of things.

Insofar as the world, understood as the human world, is "all that is the case" it is the sum of thoughts or facts in which what-is has been transformed into what-is-the-case. Acknowledging the intimate relationship between facts and thoughts, and the nature of thoughts, justified our dissent, in the Overture, from Wittgenstein's further claim that "What is the case – a fact – is the existence of states of affairs" and that "A state of affairs (a state of things) is a combination of objects (things)". Insofar as the world "divides into facts", it is not a combination of objects. In short, "facts" are as remote from being portions of "what-is" as are the thoughts that gather them up and entertain and communicate them. They are not even replications of aspects of what-is, such that they are "about them" in virtue of mirroring them directly, indirectly, or in encoded form.

Most importantly, they are not wired into what-is understood as the material world, as effects of the latter understood as causes.

8.3 THE PHENOMENOLOGY OF THOUGHT

Thinking about thought is a strange business. It should be impossible – like water trying to navigate a stream in a boat made from water. We most often characterize thinking as "inner" (of which more presently) speech. And so, as we cock our inner ears, we hear a voice "in our head" – the head being an ill-defined location closer to the intracranial darkness behind our eyes rather than, say, to our feet or indeed the rest of the universe. The thinking head is the entity that most directly faces (strange verb!) us as we stand in front of the mirror.

It is tempting to think that this imaginary sound, this virtual or subvocal speech – is necessary so that we can inform ourselves as to what we are thinking. This hardly holds up: it would imply that we need our thoughts to be fully formed *before* we can tell ourselves what they are; that we have our thoughts before we know what thoughts we are having; and that if we tell ourselves what thoughts we are having by having them "in loud", we shall know what they are. At this point vertigo beckons. Let us temporarily retreat from philosophy to psychology.

Psychologists have pointed out that the silent soliloquy, footnoting our wakefulness from early childhood to our final days, serves many functions. As well as assisting us to plan, control, and direct our actions, thoughts also propose alternative realities. All in all, thinking to ourselves seems an instance of something that Wittgenstein said was impossible: the right hand giving the left hand money.[14] Talking to ourselves is a part of the process of registering and making sense of what is going on in, and around, us. Articulated, that sense becomes "we" sense, an internalization of the dialogue we have with our fellows. As Charles Fernyhough put it "A solitary mind is actually a chorus".[15] Our most private thoughts draw upon the language, upon ways of structuring the world, we have imbibed from our collective human consciousness, manifest in speech, writing, the artefactscape in which we pass our lives. Hence the range of cognitive productivity, encompassing private gossip, rehearsal of what we might say, micro-dramatization, in an endless monologue – a one winged buzz of the inner conversation that is an endless assertion, affirmation, and recollection of who, where, what I am.

That head of mine in the mirror is therefore to an important degree an echo-chamber and its echoes have drawn their originals from a thousand worlds, from our many modes of being with ourselves, with others, and with the rumours of others, with the wide and various cultures of the contemporary world, and with the chaos of history as represented in my mind. And it is no surprise that our thoughts come in a variety of styles, with tones of voice, as if they were meant to be spoken out loud. This said, we are not always so considerate of our imaginary audience: our muttering is fragmentary and frequently condensed or elliptic – something beautifully captured by James Joyce eavesdropping on the streams of consciousness of Leopold and Molly Bloom and Stephen Dedalus.[16] Many things are left unsaid; dots in a multiplicity of spaces are unjoined.

Of course, we do not usually have to explain everything to ourselves: our thoughts are primarily narrowcast to an audience of one – corresponding to Plato's definition of thought as "a discourse that mind carries on with itself".[17] *I* know what I am on about, though I may articulate more than is tailored to the needs of the single intracranial audience member because I imagine myself telling (others) what I am thinking. I slide into communicating or even explaining mode.

There are, however, many aspects of thought that do not amount to "sentences in the head" – not even ill-formed ones. Sometimes a thought is so far from a word-shaped, linear fragment of articulation with a beginning and an end. It can feel like a blush of realization: seeing that something is so without spelling out what that realization is. If we tend to exaggerate the extent to which thought is verbally expressed to ourselves in tidy propositional form, it is because thinking often takes the form of rehearsing something we are going to say, or recalling something we or someone else has said. Irrespective of how tidily sentence-like thought is or how untidy it is, the fundamental mystery remains intact. Thought is the most striking explicit expression of the consciousness that is present in all experience.

Thinking can afford to be almost featureless compared with the world that is thought about; indeed, it is important that the phenomenal aspects of thought do not get in the way of its transparent reference to its objects. It is necessary that the token should have only sufficient substance of its own to ensure that the thought should be experienced as actually happening and that what happens should count as the occurrence of this thought (about Paris) rather than that thought (about the whereabouts of the cat next door); that it should be individuated.

Some people report that their thinking is not only decorated with images but dominated by them. A flow of images would seem to be an appropriate alternative to one's native language, given the tangle of connections between cognition and sight: vision is the most epistemic, or knowledge-like, of our senses. Even so, the experience of thought is only patchily cinematic for much of it is abstract or has important abstract elements. The thought that tomorrow is Sunday would have little other than accidental connection with any visual images. And even the thought that Paris is the capital of France would not be delivered more completely if it were decorated with a mental image of the Eiffel Tower standing for Paris and another image (heaven knows what) to stand for France; for the key assertion – that Paris is the *capital* of France – does not correspond to any visual image. The designation of a city as a capital is a legislative, that is to say abstract, concept. If I am trying to think the thought that Paris is the capital of France, the less clutter (mental images of the Eiffel Tower, postcards from extra-Parisian France) the better. Even relevant images might get in the way, given that the Eiffel Tower, while an iconic image of Paris, has to be made to mean the city, which could equally well be signified by the Seine, the Place Pigalle, the Moulin Rouge, the Quay d'Orsay, or the hotel where we once happened to stay. If, in order to have this thought about Paris being the capital of France, I would have to do justice to its two protagonists, plus the full range of the connotations of the concept "capital", this relatively simple thought would be unthinkable. It is hardly surprising, therefore, that for most people, thought is overwhelmingly in words rather than images – not to speak of the already noted fact that thought is often a rehearsal of one's contribution to an anticipated conversation.

So, the phenomenal accompaniments of thought – the sound of words, images of referents – are only tenuously relevant to it. Contrary to what some philosophers think, the poverty of phenomenal surface is not an obstacle to ascribing intentionality to thoughts.[18] When it comes to cognitive phenomenology, it seems advisable to have as little phenomenology as possible – enough to make thoughts present and individuated – so that I know what it is that I am thinking. For the sake of transitivity from the tokens of a thought to its referent, of the transparency of the thought, the fewer phenomenological clothes the better. The referential function has to be quarantined from intrusive images and other clutter. Yes, token thoughts need to have phenomenal surface to be registered as, count as, occurring, as being *this* thought rather than that – or, indeed, to be present or to have happened at a certain time – but not in order to deliver their referential function.

And this applies with equal force to the homeliest examples of random thoughts as to Thought with a capital T. We might expect factual, abstract, and general thoughts to have fewer phenomenological clothes than thoughts that articulate desires, hopes, and fears. The latter might even be placed in italics by bodily sensations – for example, a fast heartbeat in the case of fear or other modes of excitement. And there will be a different cognitive phenomenology when thoughts are imported from others (via voices or written words) compared with what is the case when we deliver our own thoughts to ourselves. Tones of voice and styles of print will coat whatever meanings are intended by the words.

All of which underlines how the "what-it-is-likeness" of thoughts is less important than in the case of perception. Hence the (correct) emphasis on the contrastive rather than the positive features of spoken and written words. There is nothing red about the thought that plums are red, even though this may have been triggered by the experience of a red plum. It was this that lay at the heart of Saussure's insight into the nature of language. Words deliver what they deliver, not through representing what they express but through their opposition to other words.[19]

We have discussed how the explicit or implicit "that" of thoughts opens up a special kind of distance between them and what they are about. It might be expected that any kind of phenomenology borrowed from perception would get in the way of the thoughts accessing their referents. In the occurrent thought "that it will rain tomorrow", any picture of raindrops, or a downpour, is unnecessary. The word "rain" is sufficient to nail the thought to a referential object. If I think "It *might* rain tomorrow" I clearly cannot have an image of a particular rain shower, because this thought is not committed to there actually being a rainstorm. Besides, a necessarily particular image of a scattering of raindrops does not do the generic work that "rain" does. And the negative thought that it will *not* rain tomorrow – so I will, after all, suggest that you and I go for that long-postponed walk in the countryside – could not depend on some image corresponding to the absence of a shower. An image of a sunlit landscape would not specify the absence of a shower, or not at any rate without the guidance of words to indicate that it is meant to portray that absence rather than, say, a sunlit wood. And there is no image corresponding to the temporal container – "tomorrow". There is as little guidance as to what a sign of "tomorrow" should *look* like as to what it should smell or sound like.

The discussion in the preceding paragraphs overflies many hotly disputed issues in philosophy around the topic of "cognitive phenomenology". A recent authoritative account of the state of play classifies philosophers as being "conservative" or "liberal" on the key issues in this area.[20] Liberals argue that there is, after all, something it is like to think a conscious thought, and this is distinct not only from what it is like to be in any other conscious state such as perceiving, remembering, or being emotionally aroused, but what it is like to have another thought. Conservatives deny there is any such distinct cognitive phenomenology. It is easy to sympathize with both views. Distinctive cognitive phenomenology is necessary, or so it seems, for a thought to occur to one, for the thought to be *had*, for one to know or register one has had this thought rather than that, at this particular time, and for the thought to be about this rather than that.[21] On the other hand, it is important that the phenomenology should not be laid on too thick, to ensure the unobstructed transitivity of the thought with respect to its intentional object. Thought articulated as sentences seems to occupy this middle ground most comfortably.

Given that thought is a kind of ghost speech, a self-haunting, it is hardly surprising that people with schizophrenia sometimes suffer terribly from the experience of "thought insertion". They hear their thoughts as voices that they ascribe to an outside source, believing them to have been implanted by alien, often malevolent agents. It is a reminder of the mystery of our ability to assign our thoughts to ourselves, given that they are formulated in the no-one language of "the they", of anyone, of everyone.[22]

This tragic situation exposes the uncertainty surrounding the scope of the certainties supposedly delivered by the Cartesian *cogito* argument: "I think therefore I am". The conclusion "I am" seems to be *less* than the premise: "I am thinking" seems to say more than bare "I am" – which is, of course, why the argument seems so secure. My being is a necessary condition of my thinking. On the other hand, the I which I am seems to be more than a succession of moments of thought. The argument does not seem to deliver, for example, Raymond Tallis – with all his attributes, duties, lifeworld, etc., if only because the "I" that comes out of it is a general "I". There is, however, an essential indexicality in an "I" that is uttered in what is perhaps an essentially performative "proof" rather than a logical inference.[23]

Even my most original thoughts sew together, and unpack, concepts that I have had no part in creating. What is more, they may mean things or have reference to things with which I am not familiar – or not as familiar as I thought. Thus, the case for the so-called "externalist" view of thought, according to which my thoughts have a dark heart hidden from me, because they are articulated in terms whose meaning, or more particular referents, may not be fully grasped by me. At the very least, the intrinsic properties of the referents may be concealed from me.[24]

Ownership of thoughts, however, seems rather more solid than ownership of perceptions – at any rate transitive ones.[25] The thoughts I have about the landscape I am looking at seem more mine than the landscape as experienced by us both as we contemplate it. What is more, the ancestry, the environment, the particular salience of thoughts are internally determined. There is no equivalent even in perception to intellectual property: what I perceive over there seems to belong to the world; whereas what I think about it seems to belong to me. Nevertheless, while we are more likely to be responsible for our thoughts than our experiences, this is not always the case: we frequently actively seek out experiences (and even pay good money for them) while thoughts may just occur to us, as if we were their recipients.

There are, of course, different degrees of activity and passivity in thought. Thinking may be something we do or something that just happens. As we have noted, an unceasing flow of ideas, images, and bits of chatter spontaneously wells up in the cognitive spaces. We say "a thought just occurred to me". The stream of thought swerves between chaos, the low-level coherence of the association of ideas, and the more deliberately requisitioned and organized narrative, arguments, and wrestling with problems. There is idle recollection and fantasizing. And there is possession by thoughts that circle round the same miserable place, anchoring the thinker to obsessive guilt and fear: thought as the generation of "mind-forg'd manacles" rather than an expression of freedom.

The opposite of this passive possession is effortful thought, driven by a determination to examine, to solve, to explain; being pushed from within by explicit questioning, by cultivated doubt. The effort may not bear most fruit when our brows are most furrowed. We may have to be patient and our patience is often rewarded when we are least expecting it. Nevertheless, as has often been remarked, happy accidents occur only to the prepared mind and the mind's self-preparation may be a matter of months, even years. The "Eureka!" moment may be the result of earlier mental effort, honoured as we seize and build on it.

Active thinking of this kind may take the form of a search, of circling round a prey, though the predator and prey are not separated, and the circling is made of the same kind of stuff as that which is sought by the predator. Our effortful mentation draws on many aspects of ourselves. We fight our way against counter-currents of distraction to a thought that remains just out of reach. Something rather similar happens when we try to recall an item such as a name. We use various techniques to transport ourselves into the vicinity of the target by silently articulating something we associate with it, hoping this will cast light on its near neighbours.

One of the most astonishing of our cognitive capacities is our ability to *re*trace the journey that led up to a thought we are puzzled to find ourselves thinking. What got me thinking about that? Why did last year's holiday "pop" into my mind? The backward glance looking for the engine that set a train of thought in motion is an extraordinary act of inner attention and cognitive will; of our capacity to inspect our own connectedness. It is as if a drop in a current were able to swim upstream to a position it had earlier occupied in order then to track its downstream trajectory. At the very least, this undermines simplistic notions of the "triggering" of thoughts as if they were effects of causes. Thunder, so far as we know, does not trace its origin to lightning in order to explain itself to itself.

8.4 "INSIDE" MY HEAD

Since Gilbert Ryle's assault on Cartesian dualism, philosophers have been reluctant to think of thoughts as ghostly goings on in the head.[26] And yet it is difficult not to think of, or even sense, the thoughts I am having now as being "inside" my head. They are not of course inside my head in the way that my tongue is inside my mouth, or my brain is inside my cranium, or a neurone is located in the cerebral cortex. If they were thus located, it would not be possible to comprehend how they reached out of the head to connect with their referents; how a token thought supposedly a few inches from the tip of my nose would do its work of invoking a particular or a general referent through linguistic means. In short, localizing thoughts in space generates the same difficulties of accounting for their intentionality as does localizing perceptions in space. Given the nature of the referents of thought, which can include explicitly non-existent objects, the difficulties seem even greater.

Nevertheless, the intuition that my thoughts are in my head, and for that reason nearer to me than pretty well anything else – that they are a privileged, especially intimate, manifestation of self-presence – is difficult to shake off. If we are inclined to locate thoughts "inside" ourselves, it is not because they are in fact spatially located but because the folk psychologist in us believes that entities that are individuated must have spatio-temporal location and, given that thoughts do not belong to any sensible outside beyond the body, they must, by default, be "inside". As such, they can be audible to me though they are silent and inaudible to everyone else. This intuition of being "inside my head" makes the neural theory of thoughts easier to swallow and the latter lends its scientific respectability to the intuition. It is why neuroscientists might feel comfortable with the idea that thoughts originate in the cerebral cortex.[27]

It is, however, as inappropriate to think of thoughts being inaudible as being audible. Thoughts are not (even) invisible because they are not analogous to that which is potentially but not actually visible because hidden from sight or lost in darkness. It is no more appropriate to describe them as invisible than to describe them as odourless or inedible. It is not even true to say that they are visible when they are written down. What is written down are the *symbols* of thought, which of course are located outside the body, on a page or a screen. If visibility were of the essence of thought, they could not be thought until they had been written down.

Even then, the notion that they have spatial boundaries when they are inscribed is misleading. We are told that a sentence expresses a thought – with a beginning and an end. But the thought is completed for the originator before it is written down and for the recipient only after it has been read. The beginning of a sentence expressing a proposition is not the beginning of that proposition or the thought corresponding to it. "Paris" is not the beginning of the assertion that "Paris is the capital of France". It becomes the beginning only when the assertion has been completed. What we see on the page, set out in space and taking time to read, are embers, cold flames.[28]

8.5 TOKENS AND TYPES

We have been circling round one of the most mysterious aspects of our thoughts: their dual character as tokens and types. It is worth pausing on this duality because it helps us to define the distinctive roles of psychology that explores the empirical nature of token thoughts and of philosophy that endeavours to clarify the conceptual geography of thought.

While our token thoughts are embedded in me-here-now, and implicitly attached to the "I" who thinks, they instantiate type-thoughts whose objects are disconnected from any here and now. They are about timeless, general possibilities, whose realization would not be confined to, or even in part captured by, a sensory field, a discrete part of spacetime. While this is obviously true of "Paris is the capital of France", which is a legislative truth connecting two political entities, it is also true of "Felix (the cat) is out in the street", in which the connection between the cat and space does not specify the posture or state of the cat nor a particular cat-sized location in space.

It is therefore not without reason that we say of someone lost in thought that they are "miles away", though the miles are not those that separate physical locations. And while it is true that I might have the thought that "Paris is the capital of France" while I am waiting for train in London, the thought is not located in London, because it is the token of a type that has no location. Token thoughts occur in time, but they do not occur in space, except insofar as they occur to me who am and yet am not my body. Any location my token thoughts may have are borrowed from my body but I-as-thinker is less my body than I-as-walker or I-as-bench-sitter. My thought took place at the railway station only because the body of the thinker of that thought was in the station. It is a spatial parasite and for this reason it does not make sense to ask how much of the space of the station my thought took up, whereas the space occupancy of my legs can be a matter of practical interest to a fellow passenger. While it is precisely accurate to say that my body is on a particular bench (as well as having any number of other spatial relations to items in my vicinity), it is not true to say that the "I" is on the bench as an entity adjacent to surrounding objects; and the same applies to the thoughts I have. And it would be as untrue to say of you and me sitting side by side on a bench that our token thoughts are two feet apart as it would be to say this of the thought types the tokens instantiate. While you and I might be next to each other as we think, our thoughts are not next to each other. Our thoughts are closest of all the manifestations of our individual beings disconnection from the physical world. They are the least embodied of the contents of our lives.

The distinction between the token thought and the type thought may seem relevant to the assumption that token thoughts, like other mental events, occur in time. I can say I thought "Paris is the capital of France" yesterday afternoon when I was struggling with a general knowledge crossword. If token thoughts occur in time, they must occur at a particular time. It does not, however, follow from this that, as has been argued by Michael Lockwood, they must have a location in space, because space and time are inseparable.[29]

At a superficial level, it may seem that the relationship between a thought type and its tokens is not fully analogous to that between the class of cats and an individual cat. There can be classes without any examples to instantiate them – as in the case of unicorns – where the class exists only as a token forming the subject of a sentence. In the case of thoughts, however, the possibility of the existence of the token and of the type cannot be separated. If no-one had ever had the thought that "Paris is the capital of France", or "That cat next door is getting on my nerves", then there would be no type corresponding to these tokens. In short, there cannot be an empty class awaiting instances to take up their seats: there can be no "unicorn" thoughts – thoughts in the anteroom of existence – though there can be thoughts of unicorns. In the case of thoughts, the instance (the token) and the class (type) are inseparable because both are verbal: they are both on the same side of the class/instance division. Thus, another reason for believing that token thoughts can no more be spatially located than type thoughts.

The fundamental point is that thoughts are fragments of explicitness that cannot be located in the realm of what-is. "That x is the case" does not have a location among x's or among other entities such as y's and z's. This fatally undermines mind-brain identity theory, according to which mental events, including token thoughts, are identical with neural activity. Given that neural activity is located in space while mental events are not, the one could not be identical with the other.

Which brings us back to Lockwood's claim that a (token) thought has a definite location in space because (a) it has a location in time, and (b) space and time are inseparable in special relativity. According to special relativity, however, it is only with respect to a particular frame of reference that any event has a definite location in either space or time; and the determination of the spatio-temporal interval between one event (and that would include events such as token thoughts if one accepts a neural account of thought) and another to space and time is also frame-of-reference dependent. Frames of reference, however, are generated by conscious observers. We therefore must invoke an observer to locate mental events in time and space and the idea that the observer is constructed out of neural activity is, as we have discussed, to say the least problematic. What is more, there is nothing in neural activity to account for the two distinct faces of thoughts – as tokens and as types – or to underpin the difference between the general sense of a thought type and its particular expression in a token thought or, as in our minds, to allow them to co-exist side by side.

8.6 THE EXPANDING REALM OF PURE EXPLICITNESS: NON-EXISTENT OBJECTS

The development of thought expands the realm of that-it-is. And when thoughts are quoted, cited, communicated, connected with other thoughts, challenged, or argued over, the distance between what-is and that-it-is widens further. The gap between the that-it-is of thought and the proposed what-is to which the thought refers is highlighted when the thought is found to be false, when it proposes an entity, event, or state of affairs that does not exist. The gap remains even when the thought correctly denies the existence of the entity in question, as when we think or assert that "There are no such things as unicorns".

The relationship between thought and non-existent objects is not as puzzling as it is sometimes made out to be. How, it is asked, can a thought be referentially connected to something that does not exist? How can such a thought be about, targeted on, a something that is a nothing – or a nothing that is a definite nothing, so that (for example) I can have distinctive, meaningful, thoughts about unicorns rather than hamadryads and flag up the differences between them? The answer lies in the fact that the combination of properties in what-is is contingent. Let us return to unicorns. There is no reason why horses should not have horns. And if there is no such reason, it is acceptable to entertain the possibility – and even to assert the reality – that there are unicorns. And, given the extraordinary range and strangeness of things that do or did exist, (octopuses, dinosaurs) belief in unicorns seems entirely reasonable. We cannot rule out there once having been such creatures in the past.

Unicorns are not, however, the progeny of a single isolated mind. It takes a community of minds – bound by a similar heritage, language, collective memories – to curate such a non-existent entity. Mythical beings such as unicorns – not to speak of witches, devils, Brexit benefits – are the work of many conscious subjects that are, or have been, in communion with one another.

As a result, these non-existent objects may acquire an apparent substantiality and may, as a result, have considerable influence in human affairs. Indeed, the non-existent intentional object that goes under the name of God has perhaps had more influence on human affairs than any other object real or imaginary.

And what of thoughts that employ singular referring expressions whose referents are both highly specific and non-existent? It is equally a contingent fact that while there were inhabitants at 16 Baker Street in London, there was no-one living at 22b; that while there are detectives with different names, none was called Sherlock Holmes; that, while many people wear deer-stalker hats, none of those who wore such hats was a detective living in 22b Baker Street. If the referent of a thought had to exist for the thought to have a definite referent, then false thoughts about actual existing entities would also lack referents. The thought that Raymond Tallis (who is five feet ten inches tall) was six feet tall would have no target because there is no six feet tall Raymond Tallis. (I have deliberately chosen a plausible, indeed mundane, example – rather than say a 200-feet-tall Raymond Tallis – to make the point about honest, reasonable error rather than fantasy.)

That we can say meaningful things about fictional objects is the other side of the fact that we can say untrue but meaningful things about existent objects. Moreover, if we could not say meaningful things about non-existent objects that are highly specified, we could not meaningfully say that they do not exist. "There are no unicorns" would be as meaningless as "There are no gloogs" where "gloogs" is a word that belongs to no language.

It might be argued that talk about a six-feet-tall Raymond Tallis is less vulnerable to the charge of meaninglessness on the grounds of lacking a referent because at least Tallis has an existence manifested in his realization of many other properties – weight, etc. But this is equally true of unicorns, that have other properties than that of being horned. Indeed, if we think of an entity as an assemblage of properties then any claim to existence is relative. Unicorns would have 95 per cent existence – with the missing 5 per cent corresponding to the horn.

There appears, therefore, to be no sharp cutoff between the existence and non-existence of the referents of thought. To say that there exists a horse that has a horn is untrue; but there is a ground-floor truth in the assertion that there are horses and such things as horns. In other words, "There is a unicorn" can be unpacked as "There are horses, and there are horns, some horses have horns" in which the first two parts are true, and the third is false. In short, the assertion does not have an entirely empty reference, so long as there is a separation of the subject of the thought (the horse) from certain properties of it. The reference to unicorns is problematic only because there is a fusion of existent objects (horses and horns) resulting in the creation of a non-existent object. Non-existence encompasses the entire object only in the sense that "Six-feet-tall Raymond Tallis" is a non-existent object.

There is, it seems, no ontological difference between existing entities referred to as a subject via untrue descriptions – as when it is asserted that "Six-feet-tall Raymond Tallis is a man" and assertions about non-existent entities. A statement about six-feet-tall Raymond Tallis has a no more secure landing place than statements about unicorns or Sherlock Holmes.

This highlights the distinctive nature of the realm opened up when the transition from what-is to that-it-is goes beyond perception, and memories justified by perceptions, to that of propositional truths expressed in thoughts. In the realm of thought, explicitness is pure, free-floating; there is no limit to the that-it-is entertained by thinking subjects as possibilities. In this realm, what are entertained are possibilities, a category that includes things and states of affairs which happen not to be the case. Both possibilities and contingent impossibilities are equally unreal in the sense that they exist only insofar as they are envisaged. We shall return to possibilities in the next chapter.

8.7 FURTHER THOUGHTS ON THINKING ABOUT THOUGHTS

The daunting challenge – and the vertiginous pleasure – of thinking about thought is something we noted at the beginning of the chapter: the requirement to inspect thoughts by means of other thoughts. Those other thoughts have somehow to rise above their fellows to count as vantage points on thought-in-general. Particular thoughts are mobilized to represent Thought, in the endeavour to arrive at the essence of thought.

This brings us close to the place where we began: in the Overture, where we tried to ensnare explicitness – an endeavour like picking up water with a spoon made of water. Or, to change the metaphor, attempting to take hold of tassels of fog with tweezers made of mist. If we are insufficiently aware of the impossibility of this task it is because we are deceived by the facility with which we handle the relevant terms, just as we can glide past the fact that we use words to talk about language because we have the word "word". Here, as elsewhere, we take explicitness – and its many layers – for granted.

Given that we have to use thought to think about thought, it is hardly surprising that while nothing could be closer to us than our thoughts, it remains elusive. Relying on introspection is, as William James remarked, "like trying to turn the gas up quickly enough to see how the dark looks".[30] At any rate, thinking about thought is not a means by which one can land on oneself, as a place of arrival. The thoughts about thought are, as you are now discovering, just a passage in the flow of thought.

Perhaps things are not quite so challenging for the philosopher who deals with thoughts-in-general as they are for the psychologist who tries to take hold of thought as an ongoing actual process. While thinking about thought, philosophers have to utilize some instruments of thought – strategic recall, assembling reminders, and logic – they do not have to catch themselves catching themselves. Nevertheless, that this seems possible – and that thinking can transcend itself to think about "Thought" – is extraordinary. It is a feat akin to that of standing outside of our own body. This alone should enable us to resist the temptation to seek an account of thought that draws on one of the late, sophisticated, products of thought: objective science and, in particular, neuroscience. But I have said enough on this head. I simply record my astonishment that so much of the folded light of our days is caught up in thoughts which are even less substantial than the puffs of air and dances of ink in which we record them.

Our capacity to think about thought, thereby making explicit one of the most developed expressions of ourselves as explicit animals, belongs to a family of enigmatic faculties rooted in a fundamental aspect of our humanity: our ability to encompass ourselves – as when we talk about "matter" or "human beings" or "the self"; or try to get our head round (as the saying goes) the totality of things as when we talk about "the universe" or, indeed, about "the totality of things". It is the most developed expression of our capacity to make what-is explicit and to assert *that* it is. While we may argue over the logic of "I think therefore I am" and how much "I" it delivers, we must still concede that to think about one's own thoughts, to chase after this thought that I am thinking, is to place the astonishing joy-filled sense of our own being, the *that I am*, in italics.

Thought, whether in the form of an inner soliloquy, speech, writing, or some more technical communication, represents a major step in the transition from what-it-is to that-it-is – or, as with propositional thought, "that-it-is-the-case-that". The world revealed to me through, for example, vision is a coloured realm of particulars. The realm of items accessed in thought, using terms that have a general, shareable meaning, is featureless. It corresponds to pure "thatness" lifted off from any material item so that it can refer to it and assert that it is or that is a such-and-such or has certain properties. The thoughts we have and share with others weave the "Thatosphere" in which we have so much of our being.

While perception and thought share the property of awakening what-is to that-it-is, in virtue of making what-is an intentional object, they have differences that go deep. Most importantly, thought is at the further distance from what-is. It is explicitness liberated from what-is.

Whatever one might think of the *Cogito* argument it is easy to see how it seduced Descartes in the direction of a dualism in which thoughts are elements of an immediately accessible stuff parallel to the more distant stuff of a material world. But it is fundamentally mistaken. Thought is disembodied because it is the purest manifestation of "that-it-is". It should not be concluded from this that it is a manifestation of another kind of stuff, of another kind of what-is, additional to that of the material world.

It is easy to see why our capacity for thought is connected with the idea of freedom. My being able to distance myself from what is before, and indeed around, me – so that I can think about tomorrow or next year, articulate ideas about Paris when I am in my study in Chester, cast my mind back to the eleventh century, not to speak of talking myself towards solutions to anticipated problems – is an astonishing measure of my independence from the world as it is served up to me. "I think, therefore, I am (not quite) here".[31] Hence, "My mind to me a kingdom is".[32] Of course, we can be the prisoners of our thoughts where they express obsessions. As Hamlet says, "Oh God, I could be bounded in a nutshell and count myself a king of infinite space – were it not that I have bad dreams".[33] Even for those who feel unable to "shake off" thoughts, other thoughts will amplify the distance from the material world that is necessary for them to act upon it as agents.

Before we leave thought, I want to link back to the arguments in Chapter 3 against the neural theory of mind.[34] If thinking were (just) neural activity, what could "thinking about neural activity" consist of in neural terms? The intentional object, the referent, of thoughts about neural activity is not a particular item, not a particular neural discharge, but neural discharges in general and (to add to the difficulty) the general relationship of those neural discharges to the generality of mental contents, such as perceptions and their objects, memories, and – yes, thoughts. The fact that we can think about such things at the very least demonstrates that collectively we can take an outside look on our (generalized) brains to consider how it "works", how it relates to items which are entirely unlike what happens in the nervous system. More specifically, the assumption that thoughts (like other mental entities) are neural discharges presupposes that individual material events can not only have intentionality (the challenge for neural accounts of perception) but have as their intentional objects items that do not exist in a particular place in the material world – namely general classes of items (in the case of neural theories, the class of "neural discharges") – or items that are not only general but are also abstract (thoughts). It seems difficult to see how neural discharges could amount to thoughts about the relationship between thoughts and neural discharges. Thinking with my brain about my brain seems to require my brain to stand outside of, and register, and unpack, itself and its supposed grounding of the subject thinking about it.

If the mind-body identity theory were true, "I am thinking about my brain" could be translated without loss into "My brain is thinking about my brain"; or "Part of my brain is thinking about (all of) my brain"; or "Part of my brain is thinking about part of my brain thinking about itself – or thinking about thinking about itself"; or "Part of my brain is thinking about a generalized brain thinking about itself – or thinking about thinking about itself". Somehow these thoughts would not be in "neuralese" but in the English language. If any of these translations were true to what is going on we would have to explain the origin of "my" – the self-referring possessive pronoun. And to explain how the brain came to take a stance outside of itself in order to point to itself as "my brain", to connect it with the owner (which is presumably neural activity) and to place it under a general category – namely "brain".

Giddy yet? This is what happens when we try to make explicit the activity that is supposed to account for explicitness – identifying it with the neural activity in the brain, the putative brewery of explicitness; when we try to *think* the widely accepted idea that my thought about what is going on in my head is itself something quite unlike a thought going on inside my head. When we try to conceive of waves of it-happening becoming self-referring I-knowing of the it-happening as that which underlies the I-knowing.

This little riff acknowledges the special challenge faced by those who wish to understand thought in terms of events in the body, more particularly in the brain. My thought that "Paris is the capital of France" has a referential object that is neither in my body, nor in Paris or in France (for what I am describing is a non-spatial relationship between the concept of Paris and that of France), nor in some intermediate place between my body and Paris. This is why the idea that cognition is *embodied* – succinctly summarized as "the theory that many features of cognition ... are shaped by aspects of an organism's entire body. Sensory and motor systems are seen as fundamentally integrated with cognitive processing"[35] – doesn't take us any closer to understanding the nature of thought. It is difficult to imagine how even simple facts – such as that Paris is the capital of France – could be embodied. Facts are, after all, timeless as well as spaceless. And what could possibly embody the corrective assertion that "Paris, not London, is the capital of France" – never mind "$E = mc^2$" or "I am fed up with all these arguments over the nature of thought"?[36] Even thoughts related to immediate surroundings – such as "It is raining" or "The cat is in the garden" – would be difficult to translate into states of the body. Their subsequent availability to be reframed as timeless assertions such as "It rained at 10 a.m. on Saturday 12 June 2022" betrays their true nature and their non-translatability into bodily states.

When we encounter something via a description, we do not in any sense of the term engage with it bodily – "bump into" it or encounter it from a particular embodied angle. "I" and the referent of the thought – even if the thought refers to an object in the vicinity of the thinker – are not co-located. After all, a thought about something next to me – say an umbrella that I think might be useful to take with me on holiday next year – does not have as its referent ("next year's usefulness") an entity or a property that has a spatial location. And while material objects may be seen as affordances – as something that might be of use, or that could justify or demand action – as in the case of the umbrella, the thought that it might be of use on next year's holiday is hardly something that can be meaningfully embodied. Any "overlap between conceptual knowledge and motor intentionality"[37] is minor even in the seemingly straightforward case of an object lying to hand. The overlap seems non-existent in thoughts about quite elementary matters of fact such as "Felix the cat is in the garden", never mind abstract thoughts corresponding to the claim that "$2 + 2 = 4$" or "$E = mc^2$".

Even less amenable to such translation is the vast hinterland behind the simplest thoughts. My thought that Paris is the capital of France links the profound, complex idea of a particular city (gathered up into a proper name) with the equally profound and complex idea of a particular country (gathered up into another proper name) with whatever is buried in the idea of a city being the *capital* of a country. This idea will draw on a large body of knowledge and understanding. Taking the trouble to think of, and assert, the connection between the city Paris and the country France, will itself have a rationale that reaches directly or indirectly into various aspects of myself. The thought about Paris may be mobilized as the answer to a quiz question, the reason for moving house to live there, the regret that such a beloved location is so far away, the observation that Paris's elevation to the status of capital occurred at a particular date, or offered as a correction of the erroneous claim that it is the capital of Spain. At any rate, it cannot be translated into a particular pattern of bodily behaviour or a propensity to such a pattern. Any claim, therefore, that "the body thinks" must require clarification of what is meant in this context by the body.

Even if there really is such a thing as "embodied cognition", factual, propositional, thought is remote from it. Sensorimotor embodiment accounts of the mind emphasize the coupling of the organism to the environment at the expense of the undeniable extent to which the conscious, thinking subject is offset, uncoupled, from the environment with which she engages. Ultimately, for some thinkers of the enactivist persuasion, the relationship with the environment boils down to the thermodynamics of semi-permeable membranes. This overlooks the deliberations of the human agent and planning – often shared – that goes into action and points to an as yet non-existent but envisaged future.

Which brings us to a particularly explicit aspect of explicitness: truth and its inseparable partner falsehood, something we shall address in Chapter 10. We shall approach truth by indirection and first consider what truth (and falsehood) are about: namely, reality, something that becomes an issue when, courtesy of thought we not only cease to be immersed in what-is but approach it (and ourselves) from a distance. Before that, however, we need to deal with something that has crept up on us: possibility – and its numerate sibling probability.

And so we dismount from the endeavour to do something that may at the outset have seemed impossible: think about thinking; or make thinking itself the object of thought. Or, more fundamentally, make explicitness – in its purest form as thought – entirely explicit. We end with a rueful acknowledgement that the appearance of impossibility is justified.[38]

8.8 A POSTSCRIPT ON ENACTIVISM AND EMBODIED COGNITION

At the heart of the mainstream version of enactivism are the "4 Es" according to which mental processes are:

1. *Embodied*: they involve not just the brain but also bodily structures and processes;
2. *Embedded*: they function only in relation to an external environment;
3. *Enacted*: they involve not only events in the nervous system but also what the organism does; and
4. *Extended*: they extend into the organism's environment.

These attributes are open to a spectrum of interpretations ranging from the truistic to the revolutionary. As a corrective to the idea of the mind as a discrete, bounded, neural computer acting on the body from a privileged interior and, through the body, its environment, enactivism may seem attractive.

Nevertheless, the problem of how first-person being pitches its tents in the apersonal world of the organism does not seem to be made easier by redirecting attention from ionic currents in the brain to a smorgasbord of giblets bathed in various fluids maintaining its dynamic equilibrium, physiological support systems to ensure continuation of organic life, and, beyond that to bones and muscles, and ultimately to the environment of the whole organism. It is not at all clear, in short, what is meant by, what counts as, "the body" in embodiment and enactment, least of all when talk is of "embodied cognition".

As we discussed in Chapter 5, only a small proportion of my body seems at any one time to be *am*bodied to the point where it could underpin cognition. The body schema may be inseparable from the realization of my effortful journey up the hill but not from the plan that lies behind my going up the hill in the first place. What is more, there are certain interpretations of enactivism that collapse the distinction between the first-person "am" of the conscious subject and the impersonal "it" – much of the body, the environment, and the tools, with which the conscious subject interacts in pursuit of its conscious ends.[39]

For leading enactivists such as Ezequiel Di Paolo, there is a continuity between life and mind, such that mind is "the lived dimension of the living organism".[40] This might at first sight seem to be attractive to someone such as myself for whom our mode of being in the world as agents is that of an ambodied subject rather than as a neural computer operating on a material world inside the body and, through this, on the extracorporeal world. Nevertheless, enthusiasm for the continuity between mind and (organic) life, between mind and living processes, must be tempered by acknowledging that, as we have discussed in Chapter 5, only a small proportion of the life of the human organism is accessible to consciousness or plays any part in determining the agenda of an individual human life. Pretty well everything that goes on in our bodies is mindless and is remote from the thoughts and other propositional attitudes that shape the voluntary activities that fill our days.

This would not need to be spelled out were it not for the fact that, at least for some radical enactivists, the key concept linking mind and life is *autonomy*, or self-individuation, which establishes the identity of beings as living organisms and what counts as their environment, thereby transforming a physical milieu into a place of significance, salience and meaning. Granted, this may be the ground floor of what is necessary to be a conscious being; but it is a long way from what characterizes the distinctively human conscious subject.

These doubts are reinforced by the fact that the 4E's are, according to some mainstream enactivist thinkers, present even in single-cell organisms. Consequently, they would seem to offer nothing that would come close to the kind of ambodiment uniquely developed in, and enjoyed or suffered by, human agents. The claim by Di Paolo *et al.* that "the explanatory principles that help us to study the organization of life are continuous with those that help us to study the mind, without reducing the latter to the former"[41] does not seem to hold up, if only because most life seems to be mindless and the vast majority (perhaps all) of non-human sentient creatures lack anything comparable to the one mind of which we have certain knowledge – the human mind.

The reduction of *meaningful* engagements with the environment to that which is necessary for the organism's viability may seem to justify ascribing experienced meaning even to organisms that lack sentience. In most cases, however, any meaning discoverable in the interactions between organism and environment, exists only insofar as it is ascribed to the organism by an informed external observer such as a biologist. To suggest otherwise is to fall victim to the fallacy of misplaced explicitness – which is the other side of the tendency in philosophy to overlook explicitness.

For this reason, we should acknowledge that the object-consciousness (Chapter 5) and self-consciousness (Chapters 6 and 7) of the kind associated with human beings are latecomers. Intentionality as the mark of mentality, does not seem to be grounded in the basic organization of life.

We may grant that our negotiations with the material world are not purely intellectual exercises, discharged through a cerebral mind, without going to the extreme of seeing all agency as bodily guided bodily activity shaped by a material environment. Enactivism, with its talk of the "sensorimotor" system, closing any gap between (sensory) input and (motor) output and effacing any sense of *a world faced by a conscious subject*, threatens to collapse into a behaviourism which does not do justice to much of our (thought-driven) human behaviour. Hence the grounds for rejecting George Lakoff and Mark Johnson's assertion that "Thought requires a body not in the trivial sense that you need a physical brain with which to think, but in the profound sense that the very structure of our thoughts comes from the nature of the body".[42]

In a laudable endeavour to integrate life and mind, the organism and the conscious person, enactivism threatens to obliterate the mind and much that is characteristic of the conscious person. The claim that *any* system capable of sustaining its integrity in virtue of adaptive regulation, must be engaged in some kind of sense-making, is clearly false. What it overlooks or minimizes is explicitness and the present chapter, through an examination of cognition at its most *dis*embodied is a reminder of the fact that explicitness cannot be found in what-is.

CHAPTER 9
Pure explicitness: possibility (and probability)

9.1 THOUGHT AND POSSIBILITY

Thoughts express possibilities that may or may not correspond to anything actual. Articulated possibilities are pure explicitness: that-it-is is detached from what-is, though in the case of a true thought, its intentional object will be an aspect of what-is. The detachment of thoughts from, the independence of, what-is is most obvious in the case of the thought that such-and-such only "might be" the case or the thought that it is not the case. Possibilities, that is to say, carry the second-order possibility that they might not correspond to any past or present actuality or that they envisage a future which, after all, may not come about. This may be concealed in confident expectation until that expectation is confounded. Possibilities that are explicitly entertained as possibilities – "It is possible that …" – are haunted by uncertainty as to whether anything did, does, or will, correspond to them: they are asserted or denied against the background that they only *might* be the case.

Possibilities may be packaged as thoughts in many different ways.[1] These include: as singular factual assertions ("Paris is the capital of France"); as general facts ("Dogs are meat-eaters"); as speculations ("There may be life on Mars"); in the form of questions ("Is Paris the capital of France?", "Do dogs eat meat?", "Is there life on Mars?"); or as things to be denied ("Paris is not the capital of France", "Dogs do not eat meat", "There is no life on Mars").

Possibilities may be sensed, even acted upon – by higher animals as well as humans – without being articulated. They are present as unspoken presuppositions. Consider, for example, the implicit expectations that shape our journey from one moment to the next as when, for example, the apparent last step on a flight of steps proves to be the next to last and we fall over. They count as thoughts, however, only when they are formulated as propositions, typically either expressing or challenging claims to knowledge. Such formulated possibilities may be entertained without seriousness, sometimes being bracketed with imaginary inverted commas, as when we think of something we might say to others, particularly when it takes the form of a joke. The point is that the referential object of thought, as opposed to that of a perception, is not actuality but an articulated possibility that may not correspond to actuality. My articulated judgement as to what is going on now, what is "out there", may be incorrect.

Even when my thought refers to the past and hence is informed by a factual memory, it will have an aura of uncertainty. Some of this may be assimilated into a vagueness of the relevant intentional object. Memory is not as unreliable as anticipation, but it can often let us down. At their best, remembered objects and events have a blur of generality, falling short of the sharp-edged thisness of what is present. And this applies more completely to possibilities entertained in thoughts that refer to the future. All such thoughts are to a greater or lesser degree speculative. (I am writing this on a train which I confidently expected would arrive in London at a time when, it is now clear, it will not arrive in London.)

9.2 POSSIBILITY AND ACTUALITY

It is perhaps hardly necessary to spell out these obvious facts to underline the fundamental nature of thought: that it stands outside of the natural or physical world which consists solely of actualities. For this reason, it is appropriate to follow up our examination of thought with a discussion of possibilities – and of their (seemingly) more robust cousins probabilities.

When we envisage a possibility what we envisage, however specific it may seem – as in the case of "The cat Felix is in the room next to me" – will always fall short of the singularity of an actual object, process, event, or state of affairs. There are many states of affairs that would justify the claim that Felix is in the next room; and the number of ways this could be fulfilled would multiply exponentially as the grain of attention drills down to the position of individual hairs on Felix's fur or the disposition of atoms in his body in relation to those in the room.

There is, therefore, an incurable generality in possibilities, however precisely they are spelled out. The obverse of this is that they can be satisfactorily fulfilled in a range of ways. This is significant when the possibilities in question are ones that we wish, through deliberate action, to bring about. As we discussed in Chapter 7, there are many ways of realizing an intention to bring about something one hopes to happen or to head off something one fears. Just how many becomes evident even in the case of something as simple as actualizing the possibility of taking a walk in the woods. The footsteps one takes, the speed at which one walks, the precise trajectory, and all the sights, sounds, smells, etc., that one encounters or seeks out, are open to wide variation. There is much, that is to say, that is not prescribed by the intention. When it comes to something more complex, such as having a fortnight's holiday in Cornwall, or training to become a doctor, the number of different events corresponding to fulfilment of the intention is vast – and that number will depend on the scale of attention. If the possibility that "Raymond Tallis has a walk in the woods on Wednesday afternoon" is defined at an atomic level, there are trillions of ways in which it can be fulfilled. What is more, the fulfilment of the intention will always entrain events that have no relevance to it. It is not possible to become a doctor without leaving footprints on many surfaces, trailing one's shadow in innumerable rooms, being part of the background noise of other people's lives, that have nothing to do with being a doctor. Possibility-fulfilling actions, in short, will be signals in widening ripples of noise that are the unintended features or consequences of the physical events that comprise the action.

The same openness applies to anticipated happenings unrelated to agency, even simple ones. For example, I may expect that it will rain tomorrow. The possibility corresponding to my expectation can be actualized in a wide variety of ways, from a modest rain shower over a small space lasting 10 minutes beginning at 4 p.m. to a downpour lasting for hours encompassing many square miles. In short, there is a wide range of events that will prove an anticipation to be true.

The examples highlight something fundamental about possibility and its distance from the natural world. The latter (at least at a macroscopic level) has properties that may be shaded in limitless ways and is susceptible to almost limitless variation. Possibilities, which exist only insofar as they are entertained, are discrete and, as the obverse of the fact that they can be fulfilled in many different ways, artificially divide what happens into "Yes" and "No": I took a walk in the woods, or I did not take a walk in the woods.

Consider the famous example from Aristotle regarding a future sea battle and the argument for logical fatalism.[2] If I predict that there will be a sea battle tomorrow then, so the argument goes, my prediction today will already be either true or false. The die is already cast. Or so it may seem. Let us suppose that there was indeed a sea battle on the day after my prediction, then it will appear that this was inevitable because my prediction yesterday was already true. In which case there is nothing anyone could do to head it off and nothing I need to do to make it happen.

This alarming argument, according to which all propositions about future events already have a truth value, depends on our seeing the future in yes/no terms: sea battle or no sea battle; p or not-p. But no actual sea battle is just a sea battle – a featureless sea battle, a sea battle whose only feature is that it occurs or does not occur. All actual sea battles must have trillions of features – ranging from the precise time of its onset and its precise location to the position of Able Seaman Smith's arm at 10:15 a.m. on the day of the battle. No prospective description of finite length could, therefore, specify all the details of an actual battle. The yes/no dichotomy – battle or no battle – does not apply to the actual world only to the way a world of envisaged possibilities is framed, typically in relation to the interests of those who anticipate a battle. A sea battle as a prediction, while it may not be entirely featureless, will always be feature-poor compared with any sea battle that actually occurs. Its being, as with any merely possible event, will be identical with its mode of description. As such, it will fall short of what is necessary for it to be an actual happening. The future remains open.

This is true of any predicted event. We tend to overlook this because we retrofit predicted reality to anything that actually happens: the battle that happens becomes the one that was predicted, despite the unbridgeable gap between the latter and the former. No prediction can reach across to actuality; from which it follows that the prediction – that leaves out so much from, falls fatally short of, any actual event – does not have a truth value realized by the event and hence does not have a truth value before the event occurs. The prediction may predict an ill-defined class of event – "a sea battle" – but not an actual, individual event, which must be defined, when it occurs, right down to the precise location of every molecule that falls within the scope of the battle.

It might be thought that I am being unfair to those who predict the future. Surely the person who predicted that there would be a sea battle tomorrow would be closer to actuality than the person who predicted that no such battle would take place. Closer, yes; but the gap remains stubbornly open, however many features the predictor might correctly add to her prediction. What is predicted falls short of any actual event and the prediction – irrespective of whether it turns out to be true – does not predict what actually happens. The prediction that there will be a sea battle tomorrow could be equally satisfactorily fulfilled by a battle in which Able Seaman Smith raised his arm at 10:15 a.m., raised his arm at 10:30 a.m., or died before the battle began. Clearly, however, these events could not all be true of an actual battle.

9.3 POSSIBILITY AND PROBABILITY

When we articulate possibilities and are aware that they are "only" possibilities, we may assess the chance of the thoughts in which we articulate them turning out to be true. Possibilities have probabilities of being actualized ranging from 0 to 1, with 0 applying to possibilities that are deemed to be impossible and 1 applying to possibilities whose actualization is inevitable. Possibilities are usually assigned probabilities on the basis of previous observation of actual events – that are characterized as "outcomes" – or based on previously discovered quantitative laws connecting the succession of prior and subsequent events. As such, they are part of nature. That, at least, is the standard story. It does not, however, hold on closer inspection, a point that will become important to those who (mistakenly as I shall argue) believe that possibilities (and probabilities) are part of what-is.

The favourite example for exploring the idea of probability is the tossed coin. Coin-tossing seems to have only two possible outcomes – heads and tails – and, so long as no-one is cheating, and the coin has been properly minted, they have equal probability. It is expected that, as the sequence of tosses gets longer, heads and tails not only will but should occur with equal frequency. While it is possible to get a run of heads during a small series of tosses, there would be some suspicion if there were a straight run of 500 heads. We would suspect that the coin was weighted in favour of heads or, even, that it was double-headed.

It is examples like these that lead to the false belief that probabilities reflect something in the natural world, in the realm of what-is; that it is a fact of nature that the relative frequency of heads and tails should converge on 50:50. This would be misleading. First, it is part of the *specification* of the coin to land an equal number of times as heads or tails. Any serious variation from this sustained over many tosses would raise the suspicion already noted that something dodgy is going on.

This said, it is difficult not to rig the outcome – if only slightly and entirely innocently. A recent study of 350,000 or so tosses of coins of 46 denominations has found that the outcome slightly favours the side that is facing up when the coin is tossed: it is 51:49.[3] More precisely, it is *necessary* to rig the outcome – in the sense of keeping things controlled – to ensure that heads/tails is always converge to 50:50. Unless the conditions under which the coin is tossed – including the properties of the coin – are tightly regulated to an extraordinary degree, it would not be possible to guarantee that there are equal chances for heads and tails. We could go further. If there were genuinely equal chances of heads or tails, then it is not clear that the outcome of a toss could be one or the other. Only an imperfect toss – or an imperfect coin – can favour heads over tails or vice versa for a given toss. In short, there is no outcome without rigging – admittedly usually inadvertent and unconscious and hence entirely innocent. An unrigged toss would have the coin landing between heads and tails – on its side – with heads and tails effectively superimposed as unrealized possibilities.

While the outcome of interest for the coin-tossers is whether the coin lands heads or tails, that is not the whole story of what happens to the coin when it is tossed – just as a sea battle has more features than is captured by the fact that it does or does not take place. The result of coin-tossing has many other features, and there are consequently (many) other variables that acquire definite values when the coin comes to rest. Here are some: how far the coin travels from the umpire's hand; whether it wobbles before it lands and, if it does, how many times it wobbles, what is the duration of individual wobbles, and the character of the final wobble; which part of its edge it lands on; whether and how far it rolls after landing; how much time it takes to come to rest; how close it is to a particular white line on the cricket pitch; how many grains of soil it collects; how many blades of grass it disturbs; how much displacement it causes to the surface of a passing worm.

The inventory of possible variables, of possibilities subject to variation, is clearly endless. If we do not confine ourselves to macroscopic features, the number of variables – of possibilities to be realized – will explode exponentially, particularly if we can sharpen the definition of outcomes to the Planck limits (10^{-43} seconds of time and 10^{-45} of space). The range of possibilities extends even further if we define the outcome in terms of a *combination* of variables; for example, the precise number and duration of wobbles plus the distance from umpire plus the number of blades of grass disturbed by the coin.

While these other variables are irrelevant to the usual purpose of coin-tossing, they are no more or less natural than heads vs tails: they are equally relevant to the question of the *physical* outcome of tossing. If we did not artificially divide all outcomes into heads or tails (cf. sea battle vs no sea battle), there would be no way of characterizing the probabilities of the outcome. Our 50:50 heads/tails split is entirely dependent upon a *convention* stipulating – and greatly simplifying – what shall count as an outcome and legislating that all other sources of variation shall be ignored. This is why it appears that there are only two possible outcomes and that the probabilities of their occurring belong to the physical world. In reality, pure heads or pure tails as outcomes, outcomes lacking any other features – such as the ones we have specified – are impossible. The coin cannot land as heads or tails without *also* landing at a particular place, at a particular time, after having taken a particular trajectory, and having caused a variety of effects.

A coin cannot land both heads and tails – it must be one or the other and not both. They are incompatible. The outcome can, however, be "heads plus six wobbles before it settles down" or "tails plus coming to a halt six feet away from the umpire". It cannot, however, be "heads plus coming to a halt six feet away from the umpire and coming to a halt seven feet away from him". In short, it is only courtesy of the way the possibilities are framed, that compatibilities and incompatibilities are established. Impossibilities, like possibilities, are the child of explicitness.

What can we conclude from this seemingly simple example? Something not very surprising but often forgotten when the fundamental nature of probabilities is discussed. It should not be difficult to accept that possibilities, which exist only insofar as they are entertained by conscious subjects, are not part of nature. Nature consists solely of actualities; and actualities are not generic such as sea battles designated by the general terms "sea battle" and "tomorrow". The actual does not happen by a process of "actualization", as if it existed as a possibility before it happened. For an actual sea battle (there is no other kind) to have been the realization of a possibility – for an envisaged possibility to match what actually takes place – an infinite[4] number of details would have had to have been specified in advance. Predictions that humans make, based on finite descriptions using general terms, will for this reason be irreducibly open-textured, indefinite, and consequently always fall short of any actual event. Possibilities and the words that articulate them fall short of things – objects, processes, events: no finite number of yes's and no's will specify an actual sea battle.[5] While we might be able to make a reasonable estimate of the probability of "a sea battle" occurring, we cannot estimate the probability of a sea battle that actually takes place, given the limitless numbers of details (with their relative probabilities) that characterize an actual battle. And, as we have seen, the same applies with equal force, though rather less obviously, to the outcome of tossing a coin.

Even this critique concedes too much to possibilities. The unfolding of the universe is not a succession of transitions from possibilities to actualities. Rather, it is a transition from actuality to actuality. It takes place overwhelmingly in the absence of any conscious subject entertaining actualities in advance of their occurring or failing to occur and thus to occupy the slot allocated to them by the envisaging of possibilities.

This should be sufficient to discourage us from treating the relative probabilities of outcomes as if they were parts of nature; to warn us against reifying probabilities. Even the (seemingly!) simple example of coin-tossing demonstrates that they are not: there are no definite probabilities without an agreement as to the relevant aspects of the outcomes. Indeed, events are *outcomes* only when they are the subject of an uncertainty, when they are predicted, or are a surprise, all of which presuppose that they are of interest prior to their occurrence. They have this status only in the minds of subjects that give them a possible, a ghost, existence prior to their occurrence.

Indeed, for a state of the world to count as an outcome, at least two extra-natural determinants are required: the frame of reference that lays down the kind of state that shall count as the outcome; and the grain of attention that shall determine at what microscopic/macroscopic scale outcomes shall be defined.

9.4 NATURALIZING PROBABILITY

It might be argued that the non-natural character of possibilities applies only to situations governed by conventions such as coin-tossing while elsewhere in nature probabilities are a real part of what is "out there" independent of human interests. If that were the case, we could recover possibilities for the natural world in the form of probabilities. Natural science seems to offer two nature-based sources of probability.

The first are the so-called "laws of nature". If we know the limits of a closed system, its initial conditions, the specification of the input into the system, and the nature of the laws governing change in the system, we can predict with a good deal of certainty the consequences for that system of the impact. If, for example, we double the pressure applied to a fixed quantity of gas sealed off from any external interference, or other sources of change such as increased temperature, we can confidently expect the volume to halve. Now it may be objected that that 100 per cent confidence – inevitability – is not quite what we mean by probability. We can escape this by suggesting that imprecision in our measurements implies that there will be a range of measured inputs and measured outputs and consequently a probability distribution of possible outcomes within a certain range. This, however, places the source of the probability distribution within human observers and their limits. Nature does not make measurements – indeed measurements are not part of nature. As we discussed in Chapter 4, they are profoundly unnatural. There is consequently neither precision nor imprecision in nature. Nor are events distinguished within nature as outcomes. What is more, the "now" from which probabilities are judged is a *standpoint*, chosen or lived by a conscious subject, from which predictions are made such that what happens is an outcome, the result of an observation, a measurement. Probabilities require tensed time, most obviously a future to house states whose nature can be *pre*dicted. Tensed time is the child of conscious subjects though this is often concealed when it is sedimented in the collective consciousness into the post-tensed time of the diary and the calendar.[6]

There is a deeper point here, that we discussed in Chapter 4: there is nothing natural about the laws of nature; or, rather, about the laws of science passed off as "the laws of nature" . In nature, nothing separates the volume and pressure of a gas, specifies independent and dependent variables, the system to be protected and that from which it is protected by keeping things equal (*ceteris paribus*). Nothing in nature counts as "initial conditions" (or not after the Big Bang – which is the creation of all conditions) or (final) outcome (or not until the Big Crunch at which everything comes to an end). Nor is there a natural separation between the conditions surrounding a sequence of events and that sequence of events.

Any probabilities flagged up by the so-called laws of nature, therefore, require the artifices of complex, quality-controlled observation in order that they shall be extracted from a universe whose states are neither intrinsically probable nor improbable but simply, unfoldingly, are.[7]

This basic truth about probabilities is concealed when they are represented mathematically, and patterns of events are tidied up to a mathematical pattern, frozen to a wave, tokenized and given a local habitation and a name in, for example, a graph, as in quantum mechanics.

Once we understand that outcomes – happenings as answers to questions, as observations, measurements – are artefacts, we can see that even seemingly purely mathematical, even frequentist accounts of probability void of any reference to a conscious subject, do not reinstate them in the natural world. The conscious subject is still present in the frame of reference that circumscribes events as outcomes that are the bearers of frequencies. This is something to which we shall return.

9.5 THE BRAIN: NOT A BAYESIAN ENGINE

If we acknowledge that possibilities are generated by conscious subjects and that the probabilities that come with them are framework dependent, the increasingly popular notion that the brain (a physical object) is a "Bayesian engine" – making, checking, and revising, *pre*dictions – is clearly problematic. According to the Bayesian story, perception is probabilistic inference to which the brain responds by minimizing predictive error. What is wrong with this claim?

There is the point already made that there is no tensed time in the material world. The brain (like any other material object) is tense-free. Neurones at time t_1 are at time t_1; nothing that happens in them reaches back to time t_0 or forward to time t_2. That is one of many reasons why a theory of the brain as a Bayesian engine, which depends heavily on anthropomorphic talk of the brain "guessing", "signaling" and doing all sorts of things that require conscious subjects with a sense of future and of the past, should be resisted. The claim that the actual content of perception is the brain's best guess of the (past) causes of its (present) experiences and best guide to (future) experiences takes this to the extreme. It is no more appropriate to ascribe predictive processing to the brain than it is to ascribe it to a missile system that has built in feedback enabling it to hit a moving target.[8] It is not appropriate to see self-educated "guessware" arising out of ionic currents so that perception can be reduced to probabilistic inference.

Anil Seth and Tim Bayne's assertion that predictive processing provides "a framework in which phenomenological properties of consciousness can be addressed without attempting to account for the existence of phenomenology as such"[9] – that the contents of consciousness are generated by, or given in the brain's best guesses – is therefore puzzling. The idea that the brain somehow manages to build on, interpret, and correct the interpretations of, its experience without having to *have* those experiences, thus by-passing the so-called "hard" problem of consciousness, seems absurd. A missile with predictive processing built into it by its manufacturers is no closer to having phenomenal consciousness than a thrown stone.

As will be evident from the discussion in the previous section, there are no predictions, no errors, and no corrections in the absence of a (phenomenally) conscious subject. It is the latter that imports the distinction between a perception that is a prediction, a perception that checks a prediction, and a perception that is a fulfilment of a prediction. A piece of what-is – for example the brain seen as a material object – is not in itself a source either of errors or of corrections.

Nor, in the absence of a conscious subject, is there a contrast between the locality to which a prediction is directed and the relatively settled global background to which it belongs, the three-dimensional landscape that sediments into a world in which subjects pass their lives and locate themselves as self-present entities. The idea that that background is itself a neural construct seems to place an intolerable burden on the brain. Filling in a continuous background between local, episodic guesses remains also unexplained. Which is why sunbathing is not "sun-Bayesianing". There is nothing in the supposed Bayesian brain that delivers phenomenal consciousness.

There is an even more fundamental problem with the description of the brain as a Bayesian engine. It is that it draws on the idea that possibilities are part of the material world, by locating them in a (material) organ that is supposed to generate them. It is particularly attractive to those who, inspired by mathematical physics, come to believe that the physical world is fundamentally mathematical. Essentially, mathematical ranges of probabilities seem to be the very stuff of the material world. Which returns us to territory explored in Chapter 4.

9.6 CONCLUDING THOUGHTS

The crucial role of the mind in framing probabilities, as it frames possibilities, does not, of course, mean that the actualization (as opposed to the specification or definition) of possibilities lies within the gift of the conscious subject. It is nature, not Raymond Tallis, that determines that horses exist, and unicorns do not, even though the existence of each may be entertained with equal sincerity. And while (to refer back to something discussed in Chapter 4) the Schrödinger wave equation is a human artefact which cannot be understood in the absence of measurement, an activity of conscious subjects, it does not follow from this that the way physical systems unfold at the nanoscopic, or macroscopic level is controlled by the human minds.

Yes, the conscious subject is what makes possible the transition from what-is to that-it-is; but, no, it does not have the power to make a mind-independent what-is conform to expectations of how things are, except within the limited capacity of agency. Nor does the strength of someone's expectation of an event taking place dictate the probability of its actually happening – except perhaps in the context of human institutions such as the stock market where fears of a crash may make a crash more likely and optimism that the price of shares will go up will indeed make them go up. The connection in this case is through the influence of the behaviour of conscious human agents on the behaviour of other conscious human agents.

There are no probabilities or possibilities without expectations; no expectations without conscious subjects to host them and to provide the "now" from which the subject looks towards and anticipates, guesses, worries about, or has hopes for, the future; and no such gaze without what-is being made explicit and becoming the locus of a seething mass of possibilities. Possibilities and probabilities – an anticipation of what *might* happen tomorrow, or *might* turn out to be the case just now – import modality ("might be" or "likely to be" as opposed to "is") into a universe that lacks it, as modality belongs to articulated explicitness. Nature is what happens not what *might*, or will, happen any more than it is what did not happen. It does not anticipate, any more than it bets on, itself. And this remains true even if the piece of nature in question is the brain: as a material object, it cannot stand outside of itself, its present states, and assume the function of guessware.

An object either exists or does not exist and an event either occurs or does not occur but the probabilities or possibilities of its existence or of its occurrence are in the keeping of an observer who sees it as a realization of a type of possibility, a type drawn from its proposed menu of types. And of course, non-existence and non-occurrence require an anticipating mind making what-is explicit as what-is-the-case in order to flag up what isn't the case.

The naturalization or reification of possibility is connected with the reduction of the properties of what-is to dispositions – something that will be discussed in the next chapter. Whether or not all properties can be reduced to dispositions, expressed only under certain circumstances, thinking of natural properties in this way seems to offer a way of naturalizing probabilities. Seeing objects or events as clusters of dispositions or propensities which have a certain probability of being expressed under certain circumstances seem to make probabilities part of the natural world. There is, however, a high price for this: objects lose intrinsicality and evaporate to external relations. In the absence of the external circumstances that would trigger expression of their dispositions, they would be unexpressed potentials.

The failure to appreciate the mind-dependent nature of possibilities (and hence probabilities) – that they are free-floating explicitness to which nothing may correspond – has had an influence on philosophy that it would be difficult to exaggerate. To get some idea of this influence, it is sufficient to consider only the writing of Parmenides, possibly the most influential of all western philosophers.[10]

It will be recalled from the Overture that Parmenides denied the reality of change. That which is not – "What-is-not" – he says, is not. Since anything that comes into being would have to come *into* being out of what-is-not, things cannot come into being: what-is-not cannot generate what-is. Likewise, nothing that exists can pass away because, in order to do so, it would have to enter the non-existent realm of what-is-not. The notion of beings as *generated* or *perishing* is therefore literally unthinkable: it would require of us that we think at once of the same thing that it is and that it is not. The no-longer and the not-yet are modes of what-is-not. Consequently, the past and future do not exist. All of this points to one conclusion: there can be no change. The empty space necessary to separate one object from another is another mode of what-is-not, so a multiplicity of beings separated by non-being is ruled out.

What-is must therefore be continuous, a plenum. Since beings cannot *be* to a greater or lesser degree – this would require what-is to be commingled with the (non-existent) diluent of what-is-not – the universe must be fundamentally homogeneous.

By this means, we arrive at the conclusion that the sum total of things is a single, unchanging, timeless, undifferentiated unity. The universe does not, however, look to be unchanging. How, therefore, is this to be explained? According to Parmenides, we are misled by our senses. He instructed his followers not to be "guided by your dull eyes, nor by your resounding ears, but test all things with the power of thinking alone".[11]

He need not, however, have chosen this drastic way of squaring his seemingly robust conclusions about the nature of reality with everyday experience – however fruitful it has been for Western thought. All that was required was that he should have acknowledged that there are such things as possibilities, and they are neither things nor nothings. The what-is-not that seems to be required in order for a changing world to have a place to come from and a place to go to, is the past, where the possibilities realized in the present were once merely envisaged, and the future, where the possibilities not realized may yet come into being. What-is is actual; and what-is-not, insofar as it is articulated, specified, is a possibility that once was actual but is no longer so; or a possibility that was never actual and may never be. Their being articulated gives possibilities identity but we should not see that as a claim to their having being, as what-is. As we saw in the case of sea battles or Felix's location in the room next door, they fall short of sufficient definition to be existents.

A few final points. While what happens may be experienced by a conscious subject as the actualization of a possibility, it would be a mistake to think of the unfolding natural world as a succession of actualizations as opposed to merely events. In the absence of the anticipations entertained by a subject, what happens just happens. Events are not the singular realization or fulfilment of something that is conceived, however specifically, at a higher level of generality.

Likewise, to pick up on an earlier discussion, there are no *outcomes* in nature: what counts as an outcome – for example heads or tails – must be picked out by the interests of a conscious subject. What happens when I toss a coin is not captured by one of two descriptors or possible outcomes – "heads" or "tails". What's more, in the absence of a conscious subject with limited, focused interests, the consequences of tossing a coin do not reach a terminus with the coin coming to a halt on the floor. If you doubt this, ask the worm on the grass that has been hit on the head and is now fighting for its life.

Only conscious subjects project a future which is populated with possibilities; and only when there is the possibility of a future, and hence of possibilities, is there the possibility of probabilities, even though their mathematization in physics seems to conceal their tensed nature and their original provenance in conscious subjects, something that lies at the root of the measurement problem of quantum mechanics and, connected with this, the reification of the wave function.

In short, probabilities, for all that they seem more solid and subject to reification than possibilities, are no more part of what-is than are possibilities. It is this that justifies probability theorist Bruno de Finetti's assertion, which deserves being quoted again, that probability "if regarded as something endowed with some kind of objective existence ... is an illusory attempt to exteriorize or materialize actual probabilistic beliefs". Finetti compares conferring objective existence on probability with "superstitious beliefs in the existence of Phlogiston, the Cosmic Ether, Absolute Space and Time ... or Fairies and Witches". For this reason, we should resist the claim of natural science to be the last word on the intrinsic nature of what-is.

Meanwhile, we take away the thought that, while individual possibilities may be steeped in mystery, nothing is more mysterious than possibility itself and the conscious subjects who entertain them. Possibilities exist only insofar as they are made explicit. The space of possibilities is an entirely different dimension from physical space; it is opened up, developed, and maintained by conscious subjects individually or in a variety of groups and subject to explicit or implicit correctives.[12]

PART IV

Circling round what-is

CHAPTER 10
What-is, conceived as "reality" and "truth"

10.1 INTRODUCTION

Many of us are propelled into philosophy by the sense that things are not as they appear to us to be. This is not just a worry about our susceptibility to local error. While doubts may start near to home, philosophical doubt is global. Consequently, it cannot be addressed by scrutinizing more carefully what is before us or by travelling more widely in the world. Uncertainty is directed equally to my perception of the cup next to my word processor as to my knowledge of a volcano on a distant planet, the nature of an entity such as an electron, or the origin of the universe. Doubts as to the true nature of the stuff of the world or, indeed, of our own nature, may expand into profound uneasiness, as when we fear that we may be dreaming[1] or alternately into delight, when we sense that there may be more to the world, something more beautiful, unthreateningly mysterious and wonderful, than what seems to surround us as we go about our seemingly humdrum business in a seemingly humdrum world.[2]

The very process of questioning the reality of what seems to surround us, what we interact with in our waking lives, may seem itself to be questionable. The suggestion that everything revealed to us, everything we engage with on weekdays and weekends, may not be real seems to remove any standard against which *un*reality may be judged. There cannot be unreality without something – reality – as a benchmark. Radical doubts as to the nature of reality may therefore be haunted by meta-sceptical suspicion that systematic, universal, theoretical doubt about what is before, around, and within us may not be serious. It is not, after all, something that bothers us when we are running for a bus, trying to get a newborn baby to sleep, planning a career, fearing for our life, worrying about another's ill health, or just feeling annoyed with our boss. The shin-barking actuality, the push and shove and heave, of the material world, and the stresses and strains of the social realm, embarrassment and anxious responsibility, the unquestioned and unquestionable connectedness of the joys and sorrows of the daily round – in short, jam-packed quotidian life – may seem to discredit questions about "reality" and "truth". These questions seem to solicit answers that undermine the very ground on which we stand as we go about the business of busy days. Even those who question what they characterize as "naïve realism" are naïve realists in practice 99 per cent of the time.

For most of my waking hours I will not take the preoccupations of this chapter seriously, even when I am typing out its contents. All of the processes that lead from the tingle of doubt to the published book stand on assumptions that may be questionable in theory but are not questioned in practice. Pointing to things we have got egregiously wrong only highlights the bedrock of seemingly unquestionable truths on which our lives rest. If we can get some, or many, things wrong we do so against a background of generally getting them right. Or so it seems. Moment-to-moment consciousness is a fabric of usually fulfilled expectations lived out in a many-layered, seemingly transparent reality.

Questioning the nature of reality may therefore seem insincere[3] – even though fundamental science seems to provide a relatively respectable basis for a transformation of our understanding of the world in which we find ourselves. Formal philosophical inquiry is usually at a comfortable distance from the experience of being engulfed by the terrifying sense that what usually passes for reality may be unreal. There is only a tenuous connection between an uninvited and unwelcome fear that my waking life might be a dream and the voluntary cultivation of theoretical uncertainty, fueling an endlessly interrupted, endlessly resumed, line of investigation whose goal is to look beyond, even to dissipate, the "veil" of everyday appearance in search of a more coherent, and possibly more exciting, world picture arrived at as the point of convergence of different kinds of philosophical inquiry.

And yet, and yet ... an intellectual, even spiritual, conscience cannot help granting a profound validity to philosophical questions about the nature of reality and truth. At the heart of the multi-layered taken-for-granted there is an intermittent sense of a darkness, placing ordinary daylight in a different kind of light. At the very least, this is justified by the knowledge that, sooner or later we arrive at our last busy Wednesday, and, as we sink out of our own existence and its correlative world, we shall be forced to acknowledge that life, after all, is perhaps in some sense a dream and that reality, at least to the extent that its furniture and preoccupations are considered permanent, is an illusion.

What is more, the idea of objective reality systematically and permanently hidden from us is difficult to disentangle from that of a conscious subject. For this reason, among others, asking the question of what is ultimately real will be a way of engaging with, even illuminating, the fundamental preoccupation of this book: explicitness and the nature of the transition from what-is to that-it-is. It will address the challenge of trying to think, or imagine, what is revealed or concealed of what-is when it becomes explicit in the mind of conscious subjects; when what-is becomes "reality" – hidden or manifest – in or, for, conscious subjects who make it explicit.

Even if we have not completely armed ourselves against our second-order scepticism about the philosophical scepticism, we shall address the question of reality, noting a contradiction at the heart of the philosophical quest for reality: that it can seem to be both elusive and (existentially) inescapable. Our inquiry into the true nature of things will also encompass the nature of truth.

At the heart of what follows is the seemingly well-founded suspicion that the processes that underpin the transition from what-is to that-it-is, and thence to what-it-is-that-it-is, involves a process of revelation that transforms that which is made present; that appearance is inseparable from disappearance; that existential contamination goes to the very heart of the idea of reality.

10.2 APPEARANCE AS A VEIL

Many reasons have been advanced for believing that what is presented, what we see, engage with, inhabit, in our daily life as reality is in fact "mere" appearance. Appearance is sometimes characterized as a *veil*, hiding what-is, things as they are in themselves, from us – so that (perhaps contradictorily) revelation comes with concealment. We are invited to believe that the veil is, in the first instance, woven out of sense experience and thickened by the many-layered discourses which, while they try to unpeel things, may also come between us and things.

The most radical assaults in the Western philosophical tradition on the belief that what-is is anything like that which is revealed by our experiences have been driven by abstract arguments. We have already alluded to Parmenides' claim that the universe is an unchanging, undifferentiated, unified totality, a massive blob of Being in which nothing happens. This conclusion – which Aristotle judged as being "near to madness"[4] – has nevertheless been hugely influential.[5] Notwithstanding this seemingly absurd beginning, radically revisionary metaphysics, and the associated uneasy or excited feeling that our common-sense knowledge of what-is is fundamentally flawed, has been a crucial driver of philosophical thought in the Western tradition – and beyond – even though the wholesale dismissal of ordinary experience as a source of true knowledge has itself been challenged by many philosophers from Aristotle, via eighteenth-century Scottish common-sense realists, to some empiricists of recent centuries.

At any rate, faith in the senses has always been hedged about with caveats and uncertainties. The reasons for that suspicion have, however, changed since Parmenides urged his disciples, as we quoted earlier, not "to be guided by your dull eyes or by your resounding ears, but test all things with the power of your thinking alone". The empirical reasons for doubting empirical reality take various forms.

The most obvious reason to question the authority of sense perceptions as a guide to fundamental reality is that we perceive what-is from a sensory or, more broadly epistemic, *perspective*; from a standpoint that is independent of, extrinsic to, what-is. Phenomenal appearance will vary according to the circumstances in which we encounter an object and the state of our sense organs. An object assumed to be unchanging, will have a different appearance when it is observed from the back as opposed to the front, looked at from a distance rather than close up, seen from within or seen from without, in ordinary daylight, darkness, or artificial light. Only in the eyes of an observer is one part of an object concealed by its other parts. And the context in the broadest sense – the surroundings of an object, the prior experiences of the subject – may influence how it is experienced.

This is not in practice as worrying as some philosophers would like us to think. When I look at an object, I am aware not only of the object but also that I am looking at it from a certain angle, at a certain distance, in a certain light. Our sense experiences come with warnings: "You are not seeing object O period but object O from a certain angle, at a certain distance, in a certain light". Unlike the naïve curate Dougal in the comedy series *Father Ted*, we do not for a moment think that a person whose apparent size diminishes with distance is shrinking; or that a person a long way away is minute. We are pre-emptively undeceived before we are deceived by the phenomenal appearance of objects.

The allowances we make for the different conditions under which objects are perceived, generating different phenomenal appearances, are driven by many assumptions including what psychologists call "object constancy". This truly remarkable faculty requires a self-consciousness that incorporates into our sense experiences an awareness of the perspectives from which we perceive objects. We are conscious of the conditions under which we see things as well as the things themselves. We can see that we are seeing something in poor lighting, from the back, or half concealed by an object in front of it. To this extent, we are self-perceiving perceivers, which is just as well, since there could be no experience corresponding to seeing an object "in-itself" from *no* angle, no distance, in no particular light – in short in the absence of a see-er, with the conditions and constraints of the viewpoint that come with seeing. Importantly, this knowledge that we see items from perspectives that are external to them is central to the very idea of the object of our perception being more than any particular perception we have of it – to the full-blown intentionality of "weighty" object perception which knows, or thinks it knows, that the object exists when neither I nor anyone else is perceiving it – discussed in Chapter 5.

So, the table top appearing diamond-shaped and the penny looking elliptical are corrected for. We are not, for example, surprised when we pick up the penny that it feels circular or, indeed, looks circular when we hold it up to our eyes. And there are many resources for checking what we perceive. If seeing is tinged with doubt, other senses, notably touching, may resolve uncertainty. If doubt remains, we can trace out the roundness of the penny with a pencil and confirm the squareness of the table by measuring its sides. Consequently, we do not, in practice, first believe the table is diamond-shaped and then correct this impression. Armed with previous experience and knowledge, we see through what we see.

The most important reason that perception is self-correcting – that we are not uncritically impressed by our first or even subsequent impressions – is that, as has often been pointed out, perception is not entirely passive, a mere spectator sport. We engage actively with the world delivered to our senses. Looking is not passive gawping: we scrutinize, test, even measure what we see. More to the point, we handle, palpate, push against, manipulate, make use of, the objects revealed to our senses. The stick that appears bent as it is inserted into water is demonstrably straight when it is pulled out and, if there is any remaining doubt about its rectitude, it feels straight to the touch. Perception invites or obliges action. Our actions are *inter*actions with the stuff of the world around us and this is a continuous check on the nature of the objects with which we engage.

Behind these corrections there is a second-order explicitness. When I see an object, I am aware *that* I am seeing it and that I am seeing it from a viewpoint – the lived, active viewpoint of my lived body that tells me where it is relative to me, a realization that guides my reaching out for it. My walking towards or away from a tree spells out in lived experience in the changes in its distance from me and ascribes its changing appearance to those changing distances and not to the tree. Moreover, the tree's size relative to the items around it will not alter; for example, it will remain twice the apparent height of the wall next to it as it and the wall shrink in harmony, when I retreat from it; and the objects intervening between me and the tree will be registered accordingly.

Our ability to dispel illusions and hallucinations by testing them against other experiences is, therefore, a reminder of the extent to which our encounter with the world is not just a matter of isolated perceptions or sensations associated with single senses. Seeing, it is said, is believing; but touching, grasping, and handling an object that is available to be re-experienced may help to give a firmer foundation to our ordinary beliefs and weed out others that are ill-founded. The apple that we see, if it is a real apple, can be handled, weighed, and eaten. An hallucinated apple would not tilt the scales, and a diet of such apples would not be very nourishing. Additionally, we may see evidence of its interaction with other stuff in the world; for example, the branch springing back when the fruit is picked; or the crunching noise and the appreciative murmurs of someone with whom I am sharing the fruit. This capacity for self-correction is especially relevant to challenging the claim that perception is irremediably discredited by our propensity to illusions and hallucinations. Descartes' sceptical assertion that "it is prudent never to trust completely those who have deceived us even once"[6] does not consider the means by which we are undeceived – so that we discovered that we have been deceived, and assign some experiences to the categories of illusions, hallucinations, deceptions, and dreams.

Any residual tendency to take the phenomenal appearance of objects literally as revealing their intrinsic properties will be dissipated by our experience of the primordial object: namely, our own body – as discussed in Section 5.3. We are not even tempted to think of our feet getting smaller as we stretch our legs and they become more distant from our eyes. The existential ontology, according to which extra-corporeal objects are objects like our body, licenses the democratic generalization from the body we "am" to the objects that are around us.

The various strands in the foregoing argument can be gathered up in the idea, particularly associated with W. V. O. Quine, that experience is judged holistically and not atomically.[7] This is particularly clear in the case of the empirical sciences whose laws and principles – ultimately based on sense experience – face "the tribunal of experience" or "corporately". But it applies equally well in everyday life, where individual experiences are woven into, and indeed presuppose, a lived network of presences set out in the continuum of space and time. My judgement that this item I see before me is a table rather than merely a coloured surface is open to correction or to confirmation when I lean on it or place a cup on it, attempt to walk through it, see how others use it, note the shadows that it casts and the dints its legs leave in the carpet, or remember that I paid good money to buy it.

In short, perception is interactive, and it is of objects that interact with one another, including the object that is my body. My interactions with what I perceive has a history which is often – perhaps overwhelmingly – shared with others. I can arrange successfully to meet you in a café, which would not be possible if the café were a hallucination or an illusion, especially as we think of such deviant experiences as private. Sustained consensus about the enduring, solid objects that populate a world shared with friends and strangers alike seems to confirm the reality of what we see. Sensed objects are part of the personal or collective sense we make of a scene and of the world or worlds of which the scene is a part at and over time. "Let's meet up in London and go to the café where we met last year" implies a shared sense not only of the café but also of its wider location, and of a shared temporal depth underwritten by the continuing existence (in your absence and mine) of the café and of London and of the country which we both inhabit in the interval between our two meetings. Such objects are the staple of "objective reality".

These are the stringent tests to which ordinary life subjects our ordinary experience. That is why the examples of sensory illusions so often invoked by philosophers do not usually present any serious challenge to our seeing aright what is out there. Elliptical coins, square tabletops that seem to be diamond-shaped, sticks that are bent in the water, and bushes that look in poor light to be crouching predators, are exposed for what they are.

It hardly needs to be said that this does not satisfy many thinkers. For some, the answerability of perception to ourselves as agents servicing our own needs individually or collectively may be invoked as part of the case for the prosecution rather than the defence. Likewise, the interactive nature of perception. As we shall discuss presently, some neuroscientists and neurophilosophers see in the subordination of sense experience to the service of (ultimately) biological need as evidence that sense experience is far from being a means of accessing how things are in themselves. At a more homely level, the consensus about what-is that communities may arrive at as they cooperate in pursuing their common needs – as when I meet you by appointment in a café – may be regarded as a further distortion.

The activity of the mind – the imposition of a conceptual structure – in shaping experience of what is out there into a shared world picture seems to some philosophers to take us yet further from what-is. The top-down pressure of the mind to make practical sense of what is experienced may have profound consequences. For Kant, the synthetic activity of the mind was responsible for gathering up sense impressions into objects localized in space and time. More recently, similar ideas have come from rather surprising, tough-minded, sources: from thinkers of a naturalist or even physicalist persuasion.

This is what Quine, for whom natural science is our most potent guide to the nature of reality, such that even knowledge itself is to be understood naturalistically and investigated scientifically, had to say about physical objects:

> Physical objects are conceptually imported into the situation as convenient intermediaries not by definition in terms of experience, but simply as irreducible posits comparable, epistemologically, to the gods of Homer [...]. But in point of epistemological footing, the physical objects and the gods differ only in degree and not in kind. Both sorts of entities enter our conceptions only as cultural posits.[8]

For Quine, the objects that our experience reveals or seems to reveal to us – and to reveal them as, and in virtue of being, its cause – are in fact a construct of our minds which, for materialists like Quine, are our brains.

The notion that what we perceive is a construct of our brains has led to some rather interesting conclusions. The suggestion that material objects are "posits" ultimately generated by our brains – individually or working in conjunction with other brains or the collective of brains over history – raises questions about the status of brains themselves as material objects located in our material skulls. If, after all, material objects are constructs – rather than re-constructs – or posits triggered by sense experience, we are left with the problem of accounting for sense experience itself which, for a materialist, is the result of an interaction between a material object "out there" – such as a cup – and a material object "over here" – the human body with its eyes and its associated neural apparatus. Anyone who at this stage can hear the sound of a saw working its way through the epistemological branch on which the philosopher is sitting is not hallucinating.

These are issues to which we shall return later in this chapter, notably Section 10.9. For the present we may question the reassurance provided by the self-checking, or cross-checking, of perception. Do we really believe that the veil of appearance can be peeled off reality by further appearances? Or by the consistency and coherence of appearances?

10.3 PRIMARY AND SECONDARY QUALITIES

First, however, let us look at an aspect of the "veil of appearance" that may seem (even) less remediable: our propensity to perceive so-called "secondary" qualities and ascribe them to objects out there, though they are in fact generated by our minds.

The doctrine that the overwhelming majority of perceived qualities that we ascribe to objects in fact belong to our minds has a long and distinguished history. Democritus argued that it is only "by convention" that there are "sweet and bitter, hot and cold" while in reality there are only "atoms and the void".[9] Galileo asserted that "tastes, odours, colours, and so on are no more than mere names so far as the object in which we locate them are concerned, and that they reside only in the consciousness. Hence if the living creature were removed, all these qualities would be wiped away and annihilated".[10] Descartes' view sounds very modern:

> It must certainly be concluded regarding those things which, in external objects, the names of light, colour, odour, taste, heat, cold, and of other tactile qualities ... that we are not aware of there being any other than various arrangements of the size, figure, and motions of the parts of these objects which make it possible for our nerves to move in various ways, and to excite in our soul all the various feelings which they produce there.[11]

The downgrading of many of the properties ascribed to objects to the status of confections of the mind was most clearly articulated by John Locke in *An Essay Concerning Human Understanding*,[12] for whom they were secondary qualities, in contrast to size, shape, motion, number, and solidity, which he identified as primary qualities. While primary qualities "are utterly inseparable from the [observed] Body", secondary qualities are merely "powers to produce various sensations in us". In the absence of minds, objects are colourless, soundless, odourless, and tasteless. They may have a high or low temperature but there is nothing in them that amounts to the feeling of hot or cold. The leaf is not *green*. Its colour is an artefact of the mind, created, according to present-day thought, out of the interaction of the retina with a combination of different wavelengths of electromagnetic radiation. Even brightness seems to be a secondary quality: light becomes luminous and illuminating, only when harvested by a conscious subject. If God's first utterance had been "Let there be electromagnetic radiation" he would have had to wait billions of years for sentient entities to transform it into light – except in His own mind.

There are, of course, other qualities ascribed to objects – tertiary qualities such as being beautiful or valuable, ugly or valueless – that we can readily accept as being subjective and as having nothing to do with the object in itself. We may (perhaps) blame the cup for its fragility but not for the fact that it is rather ugly. But it is more difficult to think of secondary qualities as being as mind-dependent in the way that tertiary qualities are, if only because of the consensus that is built around them. A cloudless sky is blue for all people of all times and places, with the small exception of individuals who are blind or colour blind who will be aware of their blindness or colour blindness. A hot stove likewise for individuals who are not numb. And there is universal agreement that vinegar tastes sour. Nevertheless, blueness, hotness, and sourness, etc., do not seem to be features of a world void of conscious subjects.

The case against secondary qualities as being intrinsic features of objects in themselves is upheld. It is strengthened by the fact that there is limited read-across between the deliverances of the different senses. The tangerine that is orange to the eye is not orange to the ear. There is nothing in touch corresponding to colour; and little in sight corresponding to the tactile sensation of things. Of course, we *associate* the look of familiar things with how they feel to the touch, and this may lead us to think that we *see* how they would *feel* to the touch; and to our unsurprised fingers they feel as sight has led us to expect they would feel. That we do not have such a capacity, however, is brought home to us by the occasional surprise as when we touch an object unaware until that moment that it is hot. The connection between sight or touch on the one hand and sound, smell, and taste on the other is even more tenuous.

While it is obvious that the collaborative intelligence of senses focused on the same object may reveal more of the object than the deliverances of individual senses – as when I palpate something I can see – it remains true that what is revealed by my sensory acquaintance with the object is not something intrinsic to it. And it is, of course, possible to lose the ability to sense some secondary qualities and not others; to be blind and not numb or numb and not blind. It suggests that the intrinsic stuff of the object-in-itself is not revealed through the experiences into which it is translated. (We shall return to this when we address the contrast between intrinsic and extrinsic or relational properties).

While Lockean primary qualities such as size and shape seem to belong to an object more than do any secondary qualities – that is, we are inclined to think of them as qualities it would have in the absence of its being sensed by a conscious subject – the boundary between primary and secondary qualities is ill-defined. Despite initial impressions, it does not seem to correspond to the border between physical properties of matter and subjective experiences housed in the mind. Bishop Berkeley pointed out that shape is susceptible to variation according to the standpoint of the observer (we have discussed the elliptical coin), size likewise (the Dougal-*Father Ted* objection), and the state of being in motion or at rest will be relative to the dynamic state of the observer (as pointed out by Galileo). Number is no less problematic. Is that object over there, 1 sheep, ten trillion atoms, or one tenth of a flock of ten sheep? Does the view from a cliff reveal a single beach or a trillion grains of sand? Solidity is also questionable as an intrinsic property of an object. How or whether a macroscopic object is solid will depend on how roughly it is handled or how it is encountered. Its impenetrability – the most direct marker of its solidity – will depend on what is being used to penetrate it.

As will be recalled from the discussion of Eddington's Two Tables in Chapter 2, the solidity of the objects around us does not seem accurately to reflect their fundamental intrinsic nature, according to the physicist. For Eddington, objects are "nearly all empty space – space pervaded, it is true, by fields of force but these are assigned to the category of 'influences', not of 'things'".[13] Solidity, it seems, melts before the scientific gaze. Heaviness and lightness – as opposed to weight recorded on a machine or mass – are, of course, relative to the size and strength of the conscious subject. There is nothing in a feather that makes it "light" as opposed to "heavy" or in a boulder that makes it heavy rather than light. Of course, a feather at 10 grams is lighter than a boulder at 100 kilograms – but "lighter than" does not mean "light". The feather is massively heavy compared with the molecules of which it is composed. Non-relative weight boils down to numbers of units and this seems quite a distance from an object. A pebble and a bag of feathers can have the same weight.

Kant challenged the distinction between object-dependent primary and mind-dependent secondary qualities:

> That one could, without detracting from the actual existence of outer things, say of a great many of their predicates: they belong not to these things in themselves, but only to their appearance and have no existence of their own outside of representation ... [has been] generally accepted and acknowledged ... To these predicates belong warmth, colour, taste, etc. ... I, however, even beyond these, include ... also among mere appearances the remaining qualities of bodies ... extension, place, and more generally space along with everything else that depends on it ... [The] existence of the thing that appears is not thereby nullified ... but it is only that through the senses we cannot cognize it all as it is in itself.[14]

The paragraph is riddled with problems that have kept philosophers busy for a quarter of a millenium. The idea that what-is as experienced through our senses is nothing like what it is in itself raises questions as to what, if any, constraints apply to how we experience things and what it is that we experience, and why there should be such constraints. If, for example, space is something imposed on experience, ordering it into the seeming presence of spatially extended and spatially located objects, then there seem to be no grounds for the body of an embodied subject (a spatially located object) to be located in one quarter of space rather than another – so that I should be in my study in Chester as I am writing this – and to be aware of and interact with some objects and not others – my laptop rather than a crater on Mars.

The idea that the phenomenal world should be a manifestation of what-is in itself and yet have nothing in common with it, nor (more importantly) be determined or constrained by it, seems difficult to understand. That experience A should be a revelation of object B and yet have nothing in common with it is deeply puzzling. This difficulty is signalled in the very opening sentences of the main text of *The Critique of Pure Reason*:

> There can be no doubt that all our knowledge begins with experience. *For how should our faculty of knowledge be awakened into action did not objects affecting our senses partly of themselves produce representations*, partly arouse the activity of our understanding to compare these representations, and, by combining or separating them, work up the raw material of sensible impressions into that knowledge of objects which is entitled experience? (emphasis added)[15]

Objects are described as "*affecting* our senses" as a result of which those very objects "*produce* representations". Sensible impressions are "raw" material, to be cooked into a world by the mind. It would be a fundamental mistake to interpret Kant as thinking of objects being present to my mind in virtue of their causal efficacy. Kant's talk of "*affecting* senses" (having an effect on senses) and "*producing* representations" (generating representations) sounds like this but, for Kant, causation is not a property of the realm of things in themselves but rather one of the categories of the understanding, of the mind. The "causal law is an *a priori* principle of human understanding rather than an empirically discoverable fact about the world".[16]

It therefore remains unclear as to how individual objects affect our senses, given that spatial and temporal relations are not properties of the extra-mental world but "forms of sensible intuition" and causation is a category of the understanding, imposed on experience by the mind. It is equally difficult to know what is meant by the *representations* produced by objects, when, we are to believe, there is no similarity or even correspondence between objects in themselves and sense impressions.

In short, Kant's things in themselves, stripped of secondary and primary qualities, do not seem to have the power to ground, or provide justification for, any actual appearance, even if they do exist independently of appearances. On this matter, scholars differ. Michael Oberst has identified the deepest division of opinion: "In the interpretation of Kant's transcendental idealism, a textual stalemate between two camps has evolved: two-world interpretations regard things in themselves and appearances as two numerical distinctive entities, whereas two-aspect interpretations take this distinction as one between two aspects of the same thing".[17]

Lucy Allais has also distinguished one-world and two-world interpretations of Kant's transcendentalism. According to the one-world interpretation, the realm of appearances is that of the *appearance* of things in themselves. According to the two-world interpretation, "Kant's appearances and his things as they are in themselves [are] different kinds of entities that are in some kind of (unknown) relation to each other".[18]

We can to some extent connect the one-world versus two-world interpretations of Kant with two interpretations of the nature of the noumenon. One-world Kantianism defines the noumenon as "a thing insofar as it is not an object of experience" and two-world Kantianism defines the noumenon positively as "a thing-in-itself", that is an object of non-experiential, of intellectual intuition. Alas, it remains unclear in what sense the thing-in-itself is an *object* – whether it has any features (including unity or plurality), whether it is an immaterial entity, whether it is an unknown and unknowable entity beyond the reach of sensible intuition, or whether it exists at all. Most importantly, however, it seems that, for Kant, the noumenon has no identifiable role in justifying, explaining, or causing the particular appearance of particular objects in the phenomenal realm.

Perhaps this is an inevitable outcome of his inquiry. The search for something that justifies, by underlying, appearances, necessarily leads to entities or stuffs that have none of the features of the phenomenal world. The trouble is, it is difficult to think what these featureless features could be. Nor – and this is the crucial point – what role they could play in constraining what is experienced, in shaping appearances. If noumena (according to the two-world interpretation) are things-in-themselves but nevertheless are not localized in space and time and lack causal connectedness, it is difficult to know what work they might do in shaping, constraining, and justifying appearances, underpinning a particular world in which a particular conscious subject is located. If, on the other hand, we embrace the one-world, two aspects, interpretation, so there is no distinct noumenal realm populated by noumenal entities, it is difficult to understand how the inhabitants of that one world can be the substrate of the rich, spatio-temporally extended, causally connected, phenomenal world and yet lack those features in their other aspect.

Thus, the puzzling consequences of Kant's "Copernican revolution", according to which the common-sense view that objects generate perception as a result of interacting with the senses and perception underpins knowledge is replaced by the far from common-sense view that the objects we perceive are mental constructs out of sense experiences. As already noted, Kant's transcendental idealism leaves the role of the body of the ambodied subject entirely unclear. The interpretation of space as one of the forms of sensible intuition, something imposed by the mind on sense experience, undermines understanding of how it is we find ourselves in a particular space co-occupied by our body and by the material objects that constitute and populate the surroundings of our body; how I locate myself in a space shared by my body and the cup I reach for or the chair I am sitting on as I reach for the cup.

It is difficult to understand how, if space is the child of mind, the body-as-object located in space could have the capacity to define what counts as "here", as "in my vicinity" and what counts as "elsewhere". In short, it is not clear why mind should attach itself to this embodied subject who is located in this place – and thus has one set of experiences – rather than that other place, populated by these objects rather than others, and consequently having other experiences.[19]

While we may not be willing to go so far as Kant in assimilating even primary qualities to the mind of the subject, it seems that the closer we examine those qualities, the less they appear to qualify as intrinsic properties of the objects to which they are ascribed. The endeavour to unpeel the veil of appearance and discover what lies beneath it, removing phenomenal qualities, can slim what-is to mere quantities, retaining none of their appearance, as we discussed in Chapter 4. What-is becomes how-much. Primary qualities are "de-phenomenalized" as pure quantities, as numbers whose nature is unrelated to any qualities. As we saw in Chapter 4, this will lead to the conclusion that behind the veil of appearance there is only a set of relations best seen as a mathematical structure. The aspiration to achieve a "view from nowhere" to borrow Thomas Nagel's deliberately paradoxical phrase,[20] to escape the contaminant of viewpoint, of the sensorium of the observer, ultimately delivers a view of nothing.

All of this leaves open the question of whether unpeeling the veil of appearance will result in something close to *dis*appearance.

10.4 RELATIONAL VERSUS INTRINSIC PROPERTIES 1

The contrast between primary and secondary qualities to some extent overlaps with that between intrinsic and relational properties of an object. Primary qualities, at least insofar as they resist total dissolution into quantities, are intrinsic; while secondary properties are relational, with the *relata* being respectively the perceived object and the subject conscious of it – the leaf and the sighted person who sees it as green. The question then arises as to what content we can give to the idea of intrinsic properties belonging to objects, independently of, uncontaminated by, their interaction with the subject. Examples of relational properties seem easier to find than examples of intrinsic ones. The harder we look, the more supposedly intrinsic properties seem relational, since their expression appears to be conditional on external factors.

Here are some classic, indisputably relational, properties of an entity A: (1) being a few miles from another entity B; (2) being bigger than another entity B; (3) being B's brother-in-law. In each of these cases, the property of the object A depends on the properties, including location, of other entities such as B. Object A could lose the external properties listed above by changes occurring in the other object B. If object B is moved so that it is contiguous to object A, the latter would cease to be a few miles from it. If B grew sufficiently, A would no longer be bigger than B. If person A's sister got divorced from B, then A would no longer be the brother-in-law of B. None of these changes of status would require any change within object A. The physicist John Bell gives a tongue-in-cheek example of how the social properties of entities can be changed in the absence of changes in the entities in question. When Queen Elizabeth died, Prince Charles became king and this would be instantaneous, irrespective of the physical distance between the two entities. It would not be constrained by the speed of light.[21] Relational properties above a certain level of specification can have two values simultaneously without being contradictory. Object A can be "two miles away from" and "ten miles away from" at the same time or both "taller than" and "smaller than" so long as the relationship is to a different object in the two cases.

So much for relational properties. What of intrinsic properties? They are not to be found by looking at Locke's primary qualities. As we saw, stripping objects of their mind-dependent phenomenal appearances seems to leave only numbers. Such numbers, being numbers of units, are not intrinsic to the object in question. Objects are not self-numbering. The beach does not dictate whether it is one beach, 2 square miles of sand, a trillion sand grains, or a trillion, trillion molecules of silica.

Have we looked too deeply too quickly? Less fundamental characteristics of objects temporarily seem more promising as candidates qualifying for the status of intrinsic properties. This promise, however, often evaporates on closer inspection. Take the weight of an object. It depends on the presence of other objects creating a gravitational field. A pebble would be heavier on Pluto than on the moon. Buoyancy will depend on the specific gravity of the fluid into which the object is dropped. My body is more buoyant in the saline-rich Dead Sea than in fresh water.

Even so, it may seem premature to jettison the contrast between relational and intrinsic properties. After all, there is a contrast between an object's being 5 metres from another object and an object being six feet long or my weighing 10 stone: the latter properties seem *relatively* intrinsic compared with the former. The object can take at least some credit for these properties, whereas it could not take credit for being 100 miles from another object, notwithstanding that this separation has to do with the histories of the two objects. Indeed, the very idea of an object is inseparable from that of something that is something in-itself such that it brings something of its own to the party. If we abandon this contrast, we drift – via the position that all properties are extrinsic – to the idea that there are no entities that qualify as things in themselves with more or less defined or constrained capabilities. It is surely absurd to deny that there are fundamental differences between a bubble, a human body, and a planet, or to suggest that these differences have nothing to do with the intrinsic properties of three items.

What is more, there is something dubious about giving external properties a status equal to that of intrinsic properties. There need be nothing in common between "all the objects 5 miles away from a particular object" though they share the same external property in relation to the reference object. If it is reality we are after, such a relational property seems less real than the most secondary of secondary properties. I cannot tell by examining an object – however carefully – all, or even some, external relations it has. Besides, it will have to accommodate that external property alongside its property of being other distances from other properties – of not only being 5 inches away from another cup but also 240,000 miles away from the moon. An object can therefore be both "5 inches away from" and "240,000 miles away from" at the same time. But it cannot be both 5 feet and 10 feet high at the same time. There is, what is more, such a thing as an intrinsic position. An object cannot be both at position A and position B at the same time. And if "away from" is defined with respect to another individual object, then the index object A cannot be both 5 inches and 5 miles away from object B at the same time.

The status of location is, it seems, particularly complex. It seems both relational and intrinsic. *Where* an object is is clearly an external relation if "where" is defined in relation to other objects (5 miles from Cambridge) or a frame of reference (4,3,6, where these figures are derived from a range made visible by graphical means). While the relation between location and the object is contingent, *that* it has a location is, however, necessary – at least in the case of a material object. It has to have some location, just as it has to have some size.

There lingers a suspicion that since observation – irrespective of whether it is sensory perception or mediated by sophisticated scientific instrumentation – is *interactive*, intrinsic properties, even if they are real, must nevertheless lie beyond scrutiny. For this reason, those intrinsic properties cannot contribute to a world-picture understood as something corresponding to what is experienced by a conscious subject. This is, of course, the Kantian position for whom the phenomenal world, made possible by the receptive capability of the mind ordered by its active capacity, is an impenetrable veil hiding things-in-themselves – whatever sense we may ascribe to their existence.

There is a more radical expression of the assertion that all properties are relational. It is a position we mentioned in Chapter 4: there are no such entities as things understood as self-individuating objects. This is the view that Ladyman and Ross expressed in *Every Thing Must Go*: that there are only relations and that these relations do not require discrete, self-individuating relata to uphold them.[22] This view, which for Ladyman and Ross is rooted in quantum mechanics, is gaining wide traction in that field, in the form of relational quantum mechanics. This is how it is expressed by Carlo Rovelli: "When an electron does not interact with anything, it has no physical properties. It has no position; it has no momentum".[23] It is, however, reasonable to question whether the picture that emerges at the microphysical level should be taken as the last word on, the fundamental truth about, reality – given that it is experienced quite differently in everyday life, including that part of everyday life that is lived by physicists doing their physics. Rovelli is confident that a relational ontology applies throughout the universe which is "A world in which, rather than independent entities, with definite properties, there are entities that have properties and characteristics only with regard to others, and only when they interact".[24] Relations, we are to believe, are ontologically more fundamental than any putative self-subsistent, self-individuating entities. The latter are derivative, emerging only out of interactions.

This is clearly problematic. If objects exist only when they interact with other objects, what do the individual participants bring to the party? If every object depends for its existence and its properties on other objects, and those in turn depend on other objects, it is difficult to see whence any object acquires its existence or properties. What is the nature of the "inter" in the interaction? Surely this implies objects with distinct existences, separate histories, that have to close or cross any spaces between them in order to interact? Even greater problems would arise in making sense of the claim that one of the objects was a subject, an observer, and the other an object that is observed. What could be the basis of this hierarchy? If, at any rate, "everything is what it is with respect to something else", it is not clear what the basis of "else-hood" would be.

We should perhaps consider the possibility that relationality and intrinsicality are not dichotomously separated and that properties lie along a spectrum between relationality and intrinsicality. This will enable us to retain the idea of the *relative* intrinsicality or *relative* externality of the properties of objects – and indeed to retain the idea of objects as their locus – while accepting that no properties are 100 per cent intrinsic or 100 per cent relational. After all, not only do properties, once classified as intrinsic, seem to be relational (for example, weight), but relational properties also seem to have, or to require, an element of intrinsicality. For example, in order for me to be a brother-in-law, I have to be a male human being, even though I can be a brother-in-law without being aware of, of bearing any mark of, my status as a brother-in-law. A stone could not be anyone's brother-in-law. Likewise, an entity has to have certain properties – that of occupying a certain portion of space – in order to be at a certain distance from another object. Having a location and indeed a size seems to be something for which an individual object can take credit. A thought could not be externally related to, in the sense of being a few miles from, a boulder or a city. And a boulder could not be logically connected with another boulder, in the way that one thought is logically connected with another. And while we may get to know stones and photons only through what they do when they interact with other entities, this does not demonstrate that they have no intrinsic differences. In short, an entity has to have stand-alone properties in order to bring something specific to the party.

While this may help us to deal with the fact that even a (relatively) intrinsic property such as weight requires certain conditions – namely, a gravitational field provided by other objects – to be expressed, it is not entirely reassuring. The idea of a spectrum between intrinsicality at one end and relationality at the other end of a spectrum may seem to undermine something central to the idea of the properties of an object: that they are standing states or (as we shall discuss) stable dispositions rather than existing only insofar as they are (intermittently) expressed. To say this is to invite further reflection on the nature of a property.

There are other properties that seem to be continuants of potentiality rather than occurrents. When the mass of a rock translates into weight courtesy of being in a gravitational field created by the masses of other objects, and thence into a pressure that deforms the material on which it stands, this has sustained effects. A mass attached to a spring stretches it and keeps it stretched. While certain primary qualities, shorn of phenomenological appearance, may also seem to be genuine standing properties, this is an illusion. Size and shape do not, of themselves, cause anything to happen. The squareness of the table does not actively square. It is a standing state of the table but not a continually acting property, unless one characterizes the appearance of squareness as an activity, which is rather stretching things.

So long as the expression of any property of an object is interactive, therefore, all properties are to a greater or lesser degree relational. Their expression is conditional, and the conditions are created by other objects or other aspects of surrounding circumstances. This does not, however, undermine a fundamental asymmetry between objects and their properties: objects can exist without their properties being expressed but properties cannot exist without objects to instantiate them. Objects possess properties but properties do not possess objects. Moreover, an object has several distinct properties, which are highlighted under different circumstances. The relational nature of properties makes it possible not only for an object to have more than one property but to have properties that are incompatible if they are envisaged as being expressed at the same time. An entity may be soluble in one solute and insoluble in another.

External circumstances, including the direction and grain of attention of a subject perceiving the object, tease out different properties from an entity identified as a single thing.[25] The ascription of several properties to an object, making the latter the bearer of the former, creates a kind of hierarchy of substantiveness, with the material object being more substantial than any one of its properties. This may be because the individual properties cannot be stand-alone, in the way that the object is – or seems to be. An object cannot have a weight without size, shape, location, and so on but none of these properties can exist without an object as their substrate. What is more, properties are necessarily general – they can be realized in an indefinite number of particular instances. This is a consequence of the fact that they are classified and, by definition, are capable of being realized in a population of entities. But, unless one is a Platonist ascribing ontological superiority to classes or types or universals over instances or tokens, then properties depend on the existence of singulars in order to be instantiated while the reverse is not true.

Bare (propertyless) existence, however, is not possible. At the macroscopic level, whatever exists must have at the very least a location, a size, a state of motion in order that it can be the bearer of properties irrespective of whether those properties are (relatively) intrinsic or (relatively) relational. This is the necessary condition of the convergence of a multiplicity of properties: a *site* of convergence, realized in an object, is required.

It might be argued that there are entities, not the least those that starred in the previous two chapters – namely, thoughts and possibilities – that have a discrete identity and have distinctive features and yet are not located in space. Nor do they have a size or a state of motion. This, however, is possible because they are had or entertained by conscious, embodied subjects. It is courtesy of such subjects that examples of pure explicitness have a connection with space and time. A thought may occur at Stockport station because I, who had the thought, was at Stockport station when the thought occurred. And it occurred at 1:30 p.m., at the same time as my train was drawing in, because it occurred to me at 1:30 p.m. Beyond this borrowed source of spatio-temporal location conferring singularity upon the thought, securing its descent from thought-type to actual token, the thought has no intrinsic properties. As an instance of pure that-it-is, any what-is the thought has is lost to referential relations. Such properties as it seems to have – its phenomenal surface, the imagined words, tone of voice, accent, in which it is thought – are hollowed out by its referential function.

10.5 A GLANCE AT THE GIVENNESS OF THE GIVEN

We are trying to understand what it means to unpeel the veil of appearance that seems to coat what-is when it is made explicit and what we might expect to discover if we did. Although genuinely or purely primary or intrinsic properties, as the mark of what is intrinsic to what-is, seem to elude us, it is difficult to resist the intuition that certain properties such as mass are more primary than others such as colour, which explicitly require a conscious subject to elicit them. Given that consciousness is something of a *parvenu* in the order of things, we may assume that properties such as colour, unlike mass, would not have been features of the universe for the greater part of its history. Objects will have been having their impact on each other courtesy of their mass long before they operated on their surroundings via any colour experienced by conscious beings. We tend to think of (metaphysical) reality as the what-is that lies underneath those properties (colours and the like) that require newcomers such as conscious subjects to bring them out. The attempt to separate the intrinsic properties of what-is from relational properties is adjacent to another problem, which we might characterize as a question of "the givenness of the given".

Givenness does not seem to be intrinsic to what is given. Most obviously nothing is given without a receiver – something that is conspicuously overlooked by the paninformationalists. Hence the state of affairs we explored in our discussion of the veil of appearance in Section 10.2: that revelation is perspectival, incomplete, relative to the conditions in which it is registered, and coloured by the properties and interests of the receiver – of the properties of the sensory system and the interests of the conscious subject. An object can be revealed only insofar as it affects a *recipient* such as ourselves; and how it affects us depends not only on its, but also the receiving subject's, properties and circumstances. It is, in many senses of the word, perspective-dependent and perspectives are self-evidently not intrinsic to perceived objects. Moreover, what affects us is not the entirety of the object. Most obviously, we see, and feel, surfaces. Even when an object is broken open, this simply reveals different, additional surfaces. The surfaces of an object do not amount to an object and the surfaces as a totality are not intrinsic to an object. An object is neither its surfaces nor (just) its depths but what we might call its "through and throughness".

For these reasons, it seems naïve to imagine that there could be a mode of givenness of things in which what is received is entirely faithful to the intrinsic reality of what is out there; that being could transition to presence without fundamental change; that what-it-is-that-is could be an uncontaminated revelation of what-is, particularly where what-is is experienced as an instance of an entity with a general sense.

As we have seen, the question of how true the given is to the intrinsic nature of what it is that is given in the absence of its being given is not made any easier by postulating, as does Kant (at least according to the two-world interpretation of the *Critique*) a distinct noumenal realm which underpins, justifies, the phenomenal realm and which, if it does not cause it (because causation is a relation *within* the phenomenal realm) at least guarantees that it is not a free-floating illusion. It is unclear how a putative featureless noumenal given could provide the basis for, could *ground*, the experiences of particular subjects with their rich and varied worlds. It seems as if the featurelessness of the noumenal world both disqualifies it from the job of justifying features and is necessary for it to perform this job. What, in short, a universe of purely intrinsic properties – given without all the messy business by which the given is taken, or what-is becomes the given – would look like. How would it appear when it is not being processed by a sensing subject – not appearing to anyone?

Why, you may think, does this matter? How is this connected with our concern with access to reality? Because there is a background assumption that to access reality is to access its intrinsic properties and yet those properties that are accessed or indeed accessible are relational rather than intrinsic given that experience and knowledge, however abstract or sophisticated, are necessarily relational. If our idea of access to reality is of access to purely intrinsic properties of what is out there, the very process of accessing what is out there seems to erect epistemic barriers between ourselves and that which we would access. That the process of "taking" necessary for what-is to become "the given" will change the nature of what is taken, since taking involves sensory receptors that have certain properties, are influenced by the direction and scale of attention, and personal and communal interest, and the language in which what is taken is articulated. And this would be true even if the process of taking did not act upon that which is taken.

Thus, one path to the problem that, since *being given* is not an intrinsic property of an entity (it requires a recipient), what is given is not intrinsic to that which is given.

10.6 RELATIONAL VERSUS INTRINSIC PROPERTIES 2

It will be evident from the story so far that the distinction between intrinsic and extrinsic (or relational) properties is neither hard-edged in fact or straightforward in theory. Yes, certain properties of a thing would not be manifested in the absence of other things but at the same time there would be nothing to be expressed in the absence of intrinsic properties. Yes, solubility requires the presence of a solvent for it to be expressed but nevertheless it is an expression of intrinsic properties that, for example, differentiate sugar from flint. We may think of the capacity to express such properties as being what intrinsic properties are. If the thing does not bring anything of its own to the party, it would have no *extrinsic* properties, notably those expressed in specific uses to which it is put. The fact that we can use a brick but not a bubble to hammer in a nail, and that we cannot use a leopard as a notebook, surely gives us some clue as to their respective, different intrinsic properties. What is more, the properties expressed in interactions, while they are powerless in the absence of the appropriate circumstance (as the attractive power of a magnet is not exerted on a wooden box), must nevertheless have some basis in the intrinsic nature of the thing. Even so, what seems an entirely sensible (Lockean) claim that "Some properties of things are entirely *intrinsic*, or *internal*, to the things that have them: shape, charge, internal structure".[26] is more vulnerable than appears at first sight. Which is why so many have accepted the relational ontology of quantum mechanics that we discussed earlier and have bowed before the mathematized gaze of the quantum physicist, in which (to reiterate the edict of Ladyman and Ross) "every thing must go". Self-individuating objects with intrinsic properties dissolve into a sea of relations.

Upholding the distinction between intrinsic and relational properties may sometimes be problematic at the homely, macroscopic level, as we have seen. This is particularly clear in the case of entities that have multiple parts which are required to interact with each other for them to express their properties. There is an internal–external relation provided by the parts; while their common environment supplies the material for a third party to supply external relational interactions. For example, a match cannot burn unless it interacts with oxygen (external relations) *and* its head is rubbed against a rough surface (another external relation). However, the burning is also the result of the interaction between the phosphorus at the head of the match setting fire to the stalk (an internal relation).

Perhaps we are making too much of the problem of differentiating intrinsic and relational, internal and external, properties as a result of starting with the idea that reality is made of discrete objects so that we can clearly distinguish between intrinsic or internal properties on the one hand and relational or extrinsic properties on the other, a distinction whose paradigm is that offered by a bounded physical thing and the spatial contrast between the properties of internal contents and the external relations that come from the items in its neighbourhood. Even if, as it seems, the real world that surrounds us is indeed populated by discrete objects, those objects are not encountered or experienced or engaged with one by one. The leaf is part of the tree; the tree is part of a landscape; the landscape is something that is encountered as evolving over time; and it is animated by attention, sense-making, and need. Its presence (as part of a foreground or mere blurred taken-for-granted background) is shaped by the recent and less recent, individual and shared, history of the conscious subject. When, in short, we think of what-is as the given, as presence, the borders of its individual elements that provide a primordial basis for separating intrinsic and relational, internal and external, properties become blurred, even arbitrary or conventional.

This counter-argument is undermined by the fact that, even in the case of a landscape unified by a gaze, that which is unified is also justifiably separated into elements. The trees on the mountain are distinct from the mountain itself; the blackbird on the tree from the tree; and one leaf on the tree from another. If this were not the case, there would be no basis for the heterogeneous, differentiated world with which we engage. As we saw in Chapter 2, some have argued that there is no such basis: everything boils down to homogeneous fundamental particles. That, however, delivers an insoluble problem of explaining how it boils up to a heterogeneous world in which there are entities that have different properties.

10.7 THE INTERACTIVE NATURE OF ANY GIVEN

Access to "reality", the transformation of what-is into that-it-is, into presence, is inescapably interactive: the capacity to be "given" relies on the collaboration of the (conscious) receiver, the taker of the given. It is this necessary interaction that makes the causal theory of perception – and of consciousness more broadly – so tenacious, notwithstanding that it is (as we saw in Chapter 3) deeply problematic. If what I perceive is not caused by the object or event that I perceive, then my perception seems suspect – indeed, an hallucination.

If there is a "myth of the given" – again to appropriate Wilfrid Sellars' phrase, if to not strictly Sellarsian purposes – it is the myth that givenness is an intrinsic property of what-is; that givenness can be built into an entity because of its intrinsic properties; that Being has the wherewithal to make itself present such that what is received is uncontaminated and in no sense constructed by the recipient. In truth, nothing can be given unless there is something – at the very least a sensorium – to receive it.

The world of objects we face is an *interface* that has two, admittedly unequal, parties. Of course, the entity has a say in what is given, what is revealed, on whether it is manifest as a pebble, a snail, or a volcano; and we cannot reduce matter and material objects to the mere "Permanent Possibility of Sensation"[27] as John Stuart Mill described it. Or at least we cannot do so without ignoring the seemingly solid fact that the material world precedes conscious beings and hence the possibility of sensation. If material objects carry the possibility of (future) sensation, a possibility realized in sentient beings, it is because they are something more than, other than, actual sensation – and precede the sensations in which they are received.

The irreducible opacity – the cognitive impermeability – and intransigence of the material world highlights its fundamental otherness to mind. Nevertheless, there is an irreducible "beholder's share" (to borrow the phrase from art historian Alois Riegl), additional to what-is, which at the very least is necessary to prompt any intrinsic properties into expression, so that those intrinsic properties, insofar as they are experienced, are changed into something other than they are in themselves.

While this transformation, even misrepresentation, is obvious in the case of classical secondary qualities – where, for example the interaction of an object with light is harvested as a colour which is then ascribed to the object in itself – we have seen that it can seem to be true also of primary qualities, to the point where the pursuit of pure givenness, of the thing in itself, through supposedly objective quantification, at best delivers *numbers* which betray little, perhaps nothing, of the nature of any individual object and ends with the dissolution of the thing in itself.

The suspicion that that in virtue of which we are open to what-is, and make it the world in which we live, conceals what-is is justified by the fact that we receptive subjects are ourselves entities, beings in that world, which have properties of their own. The fear that the given is inescapably altered by the processes of reception seems well-founded. It is not just that the process of revelation "coats", and so partially conceals, that which is revealed. It is a more radical concern than that which acknowledges that an item is necessarily altered by the clumsy touch of our fingers or the lighter touch of our gaze or that cheese is not smelly in a noseless universe. It goes deeper than the judgement that all mediation results in contamination; that we cannot sense things without translating them into something they, in themselves, are not; that when I feel sunlight on my arm, what is revealed is sunlight-on-my-arm not sunlight, the sun, or even (possibly) my arm. That-it-is cannot be entirely faithful to what-is, especially as the latter will have been harvested by something very specific, profoundly unusual, namely our bodies, a slice of what-is – not to speak of our society, our language, our thoughts. The very idea of access to the given as an uncontaminated revelation of a reality identical with what-is is problematic.

The fundamental issue is expressed by G. P. Hodes: "The difficulty is not, as with Plato, about knowing change or things that change; it is about *how we can know anything by changing it or being changed by it*".[28] Or indeed both: the exchange that generates the transition from what-is to that-it-is changes both parties involved – the subject that makes of a portion of what-is its object and the relevant portion of what-is.

Thus, we don't know what are the fundamental constituents of what-is, only how they behave in response to the provocation of the encounter with a conscious subject, even when the latter takes the form of a scientist, or scientific team, performing experiments under strictly controlled conditions. In short, inertial mass – and other entirely respectable physical properties – are dispositional.[29]

It is difficult, therefore, to shake off the suspicion that any kind of interaction necessary to reveal the nature of an object – irrespective of whether the interaction delivers ordinary sensations or observations mediated through the wizardry of CERN – will weave another veil of appearance over reality. Lifting observation above the vagaries of individual observers and the internal and external circumstances under which they make observations – the necessary condition of arriving at generalizable laws and descriptions – does not entirely remove the element of interaction.

The necessity for there to be a conscious subject to bring about the transformation of what-is into "that-it-is", into how-it-is when it becomes something "which is the case", raises the suspicion that the interaction necessary to reveal what-is must conceal or cloud its intrinsic nature; that any such interaction – even the light touch interaction of, say, visual perception – is a kind of *contamination*. Our endeavours to find out what something "really" is may involve a good deal of pushing it around, a testing, sometimes to destruction. And behind the light touch of the glance is the heavier touch of the embodied sensory system, not to speak of the conceptual grids to which the seer subscribes.

This is particularly clear in the case of scientific investigation where the encounter between conscious subject and an object of its attention is considerably more brutal than a glance or a touch – as in particle physics where entities are subjected to high energy poking and prodding. Nevertheless, these kinds of interactions are still thought to be revelatory rather than veil-casting. We see how the object *responds* to our poking and prodding and that as a result we discover what kind of thing it is. We think that this reveals its intrinsic nature – a belief that can be justified only if its very being is seen to be active or interactive.[30]

The attempt to shake off or minimize the interactive element in perception and knowledge is repeatedly frustrated. The privileging of quality-controlled *measurement* as the source of empirical truth, in the endeavour to burn through the veil of appearance, does not deliver things in themselves. There may be some consolation in the fact that things in themselves – lacking perceptible qualities and seemingly inescapably uniform – may be near to being featureless, as we discussed in Chapter 2. This is true not only of the universe rewritten as an equation, as an endless series of cash transactions of mass-energy, but also of the world seen through the eyes of pre-scientific atomistic metaphysics or the totality of things as an undifferentiated, unchanging Parmenidean unity. The latter is a consequence of condemning sense perception – the "dull" eyes and the "resounding" ears – and relying entirely on the power of thought.

It may be possible to accept that there is a fundamental gap between what-is and that-it-is or what-it-is-that-it-is without succumbing to a full-blown Kantian transcendental idealism and embracing the idea that objects "out there", set out in space and time and causally interacting with other objects, are not things in themselves but mere appearances. What is more difficult to resist, however, is the possibility that the "taking" transforms the "given" so that it lies on the far side of an epistemic barrier created by the processes by which what-is becomes what-is-given.

If we accept that all those properties of which we are aware are necessarily relational, being the product of interactions, we may feel that we are pushed to ascribe a greater role to the mind in shaping any possible world than is entirely comfortable. It is, after all, not easy to dismiss the common-sense view that the room we walk into is something like the room before we walked into it. We are equally reluctant to accept that, if all the properties of which we are aware are necessarily relational, that there is not much to be found behind the veil of appearance; indeed that unpeeling appearance leads to disappearance. This seems to be the case with Kant's view where the thing-in-itself, if it exists as a thing, is remarkably lacking in features.

The aspiration to what Nagel ironically characterized as "the view from nowhere" – the epistemic asymptote of objective, in particular scientific, inquiry – starts to look self-contradictory. Notwithstanding the dephenomenalization that comes with abstract, quantitative science, it is, ultimately, a dream of seeing what reality looks like when it is not being looked at.

The very idea of uncontaminated access to, a pure revelation of, reality understood as the intrinsic nature of what-is is therefore deeply problematic. It assumes that what-is can be entirely translated, in the individual or collective minds of subjects, into its own appearance, into that-it-is without remainder. There are many barriers to this, but four will be especially obvious from the discussion so far:

1. There is no canonical appearance: close up is no more faithful than far away. A tree is no more correctly seen when one is climbing it than when it is part of a view of a distant wood. There are no privileged scales in the material world. What-is, that is to say, has no look of its own. If it did, different appearances – close up and from a distance – would be in conflict.
2. Knowing a piece of what-is is knowing what it is *like*: cognition is among other things *re*cognition, classification. The object will fall under one of many possible general categories none of which does full justice to its individual nature.

   These first two objections may be met by pointing to the dephenomenalized view from science (though this view from nowhere approximates to a view of nothing). But two others are not so easily dealt with.
3. We cannot simultaneously reconcile direct realism – the belief that we could at least in theory get to know the objects of knowledge as they are in themselves – with the belief that the presence of those objects

in consciousness are ultimately rooted in the *effects* of their interaction with our bodies, in particular our sensoria; that perception is mediated by inner representations of what-is.

4. And, finally, revelation will necessarily be of something other than what is revealed. Knowledge is inevitably a *relation* between a subject and an object.

It is important not to over-state "the beholder's share". It is not up to me, the perceiver, whether something I see is revealed as a snail, a puff of smoke, or a mountain. There are grounds within the object for my seeing the object before me as a snail. If there was not a significant share from that which is beheld, the world around me would be either totally chaotic or perceived as undifferentiated.

10.8 HOW WE OVERLOOK THE INTERACTIVE NATURE OF WHAT WE COUNT AS THE GIVEN

Our overlooking the necessarily interactional nature of that to which we gain cognitive access has much to do with the strange, indeed unnatural, nature of knowledge. When experience is transformed into a matter of fact, it seems as if, through its ascent to factual knowledge, my consciousness is liberated from the status of being a "mere" viewpoint such that that which it knows is uncontaminated with the processes involved in knowing. I do not seem to get in the way of my access to reality.

Let me illustrate this with a humble example. When, this evening, I see that there are daffodils in my garden, I see them from a certain angle and distance, in a certain light, and through eyes that are in a particular condition. When, however, I learn from you that "There were daffodils in the garden at 5 Valley Road, on the evening of 25 February 2022" the knowledge I have acquired does not seem to be tethered to a certain angle from which the daffodils were viewed, or the conditions under which they were seen; nor is it accessed through senses in a certain condition. The daffodils in your assertion are not blurred by my poor vision or half-hidden in the twilight. What is more, if you said that the daffodils were half-hidden in twilight, their half-hiddenness would itself be revealed by what you said.

The fact you have shared with me is, as Quine pointed out,[31] timeless: it is true at any time. Nor is it a prisoner of any spatial location. Every fact that is not context-dependent has a splinter of eternity at its heart. While "There were daffodils in the garden at 5 Valley Road, on the evening of 25 February

2022" is eternal, "There are daffodils in the garden" may or may not be true. It may be true on 25 February and false on 25 September. Statements that are implicitly or explicitly indexical have a limited shelf life which they often make plain.

Reality presented through propositional, factual truths looks therefore to be liberated from the sensory system of any person, even though the latter may be deployed to check a fact – as when I look at a photograph labelled "Garden, 5 Valley Road, 25 February 2022" in my picture gallery on this computer. We can think of "objective" realities free from the connection with conscious subjects. Or imagine we can. Ultimately, of course, the audit trail leads back to sense experience.

Granted that token assertions of this fact are localized in time and – given that they are materialized as utterances coming out of someone's mouth – in space; but their location in time or space does not alter their revelatory power. It is not, for example, dependent on the current presence of daffodils. What is more, the statement shares none of the sensory or other qualities of the state of affairs it asserts. To reiterate an earlier point, the word daffodil is not "yellow" and any colour its token might have is irrelevant to its meaning: it can have any colour in writing – the blue of ink used to inscribe the word "daffodil" does not clash with the hue of its referent – and no colour if spoken. It doesn't matter when or where you or anyone else declared that "There were daffodils in the garden at 5 Valley Road, on the evening of 25 February 2022", it remains true forever if it was true in the first instance. And its truth is independent of who is asserting it, most obviously of the state of their sensorium and their location in space and time.

This independence of space, time, and person is connected with the property of factual statements: that they may connect classes or categories of objects (such as daffodils) with singulars – a particular garden in a particular place on a particular evening. (Of course, the evening identified as 25 February is not my evening or your evening. It belongs exclusively to no-one's biography.)

As knowledge has increasing generality there is a gradual replacement of primary intentionality of sense experience anchored in a particular embodied subject who has a partial, incomplete, perspectival, view on what-is, by joined intentionality mediated by systems of discourse that are not tethered in this way. Although encounter with even these facts is mediated by experience – of spoken words, written words, images and other symbols – we look past or through those experiences and their objects to what it is they mean. In such cases, they seem uncompromised by being partners in an interaction: they are untethered to the physical now of a conscious subject. Facts, most notably quantitative and general ones – graphs, formulae, equations, and such-like

– are not owned in the way that experiences are; I have an experience, but I do not have a fact, though I may be said to "grasp" it. Facts, therefore, seem to be detached from those who formulate and communicate them and from that which, the reality, that is disclosed in them. Facts seem as if they have always been in the realm of that-it-is. The messy business of the transition from what-is to that-it-is, that in virtue of which the given is taken, is out of sight.

The interaction between the embodied subject and her surrounding world may be off-stage. Nevertheless, even in quite abstract discourse, what-is does not disclose itself in a manner that would correspond to givenness without interference from the mediation necessary for reception. What-is has no language of its own. Contrary to what Galileo said of nature – and what many contemporary scientists and philosophers believe – it does not speak in the language of mathematics. There is no discourse, however objective, that is not tethered to some notional cultural, historical, or disciplinary viewpoint. Facts may be *facta* – things that are made, artefacts – but, unlike counterfactuals, they are not, as we noted in the Overture, merely *made up.*

It may reasonably be objected that the example I have just examined of a trivial fact does not do justice to the length of the journey from manifestly interactive, contaminated sense experience to our, seemingly uncontaminated, best knowledge of reality – or to the endeavour to bore down on, to get back to, what-is from that-it-is and its many-layered elaboration. Cue for a brief return to science.

We have already noted that science seems to get closer to things-in-themselves, not the least because it sheds secondary qualities (and even primary qualities) by reducing what is there to quantities that are expressed in units. While measurements that generate these data are performed at a particular place or time, they are liberated from that particular place or time in virtue of (a) the exclusion from the result most of what is experienced when the measurement is made; (b) a consensus-based control of the conditions under which the measurements take place; (c) treating the measurements as examples which are of sufficiently high quality to be generalizable to an indefinite number of circumstances. Even so, measurement is not an aperspectival revelation of a particular entity but captures aspects of the world – parameters such as gravitational field, velocity, pressure, charge and so on – that are instantiated in the entity.

While they ultimately depend on the sense experience of individual observers – *someone*, after all, has to make an observation – observations-as-data standing for a boundless class of data are remote from the sense experience of a particular individual. Nevertheless, remoteness does not entirely sever the relationship between the observation and an observer. Leaving aside the special circumstances of the actual observer – which should in theory be rendered irrelevant by appropriate methodology – observation is refracted through the lenses of the evolving history, concepts, theories, instrumentation of the relevant sciences and the skills and understanding of the individual scientist or team of scientists. The angle from which we approach that which is being studied is that of a "collective expert" consciousness that has evolved and been refined, checked, and corrected over a long period of time.

So, while the interactive nature of the relationship between the knowing subject and the known object is concealed, the connection between a knower (often a representative of a collective of knowers) and a known (admittedly an exemplary known) remains. The incorporation of the knowing subject into a boundless community of minds, and the distribution of the task of confirmation of knowledge across the experiences of an equally boundless congregation of subjects, does not finally sever the connection and allow what-is to say what-it-is-that-it-is in its own dialect.

The pessimistic conclusion that the interactive nature of our access to what-is, concealed when it is transformed into factual knowledge, means that reality will always be hidden from us may, however, be challenged. Perhaps, after all, it is the subject, far from distorting reality, who makes what-is into reality.

This is something to which we shall return later in this chapter. First, however, we should look at some of the more radical views that have arisen out of the fact that the passage from what-is to that-it-is is not merely interactive but that it is complexly mediated and that the ultimate mediator is the body and, so we are assured, more specifically the brain, of the conscious subject.

10.9 RADICAL DOUBT: MEDIATION IMPLIES HALLUCINATION

> [We] are never in any direct epistemic contact with the world surrounding us even while phenomenally experiencing direct contact.
>
> Thomas Metzinger[32]

> My Perception Is not of the World but of My Brain's Model of the World.
>
> Chris Frith[33]

The grounds for looking with beady-eyed suspicion upon our beady eyes are sometimes quite superficial. They do not cite the observer-dependence of the appearance of what-is (perspective, secondary qualities, etc.) but simply note that we are prone to making mistakes. If we can sometimes be misled by our senses, so the argument goes, it is always possible that we might be misled by our senses, and even that we *are* always misled by our senses. The classical early modern statement of this belief is in Descartes' *Meditation on First Philosophy*: "Whatever I have up till now accepted as most true and assured I have gotten either from the senses or through the senses. *But from time to time I have found that the senses deceive and it is prudent never to trust completely those who have deceived us even once*" (emphasis added).[34]

There are seeds of self-contradiction in this passage. In order to support the conclusion that I ought not to trust my senses, I have to trust my memory of being deceived by my senses and I have also to be confident in the validity of the sense experiences by which the deception was uncovered. Yes, I was deceived in thinking that the stick in the water was bent. But subsequent experience – the stick looking straight when it is pulled out of the water, feeling straight to the touch, passing without obstruction through a straight pipe – is also sense experience.

Descartes' reason for extrapolating from particular deceptions to universal deceit, from local error to global mistakenness, looks to be self-undermining. His taking the trouble to write down these reasons for universal doubt, in order that he might share them with others, and his expectation that those others might be persuaded by what he has to say, suggests that he may not wholeheartedly embrace his scepticism. And this, of course, is true; he is, so he tells us, confident that a benign God will ensure that he is not deceived.

The appeal to a deity may seem a rather expensive way to buy one's way out of the possibility of being deceived by our senses and the suspicion that that reality lies forever beyond our reach. A less expensive strategy is to look more closely at the self-contradiction built into Descartes' argument. Gilbert Ryle pointed out that the very idea of falsehood and deception would have no meaning where falsehood and deception were universal: "There can be false coins only where there are coins made of proper materials by the proper authorities".[35] The temporary authority of dreams, hallucinations, and illusions is borrowed from the justified authority of our waking hours and their overwhelmingly veridical conscious content, and the part they play in enabling us to act on the world and leave enduring marks of our actions. You can get things wrong – and know that you have done so – only if there is the possibility of getting things right, against the background of generally getting things right.

The argument that our proneness to mistakes justifies a global scepticism is less popular with philosophers than it used to be. However, the idea that the world in which we feel ourselves immersed does not correspond to reality still has considerable attraction among neuroscientists of a philosophical persuasion and philosophers of a neuroscientific persuasion. The suspicion, or sometimes confident claim, that we are excluded from reality begins with the acknowledgement of something we have just discussed, and which appears beyond dispute: that sense experience, as the ultimate source of our knowledge, is *mediated.* As a consequence, so it is argued, we do not have direct access to things out there, to things they are in themselves, but only to intermediaries, generated by the interaction between potential objects of perception and the sensory apparatus of the perceiving subject. The most proximate of those intermediaries are the contents of consciousness.

In the first half of the twentieth century, before the philosophy of mind was engulfed by neuroscience, the intermediaries were items of uncertain status. They were usually called "sense data". They reveal themselves when what we see is different from what subsequently – or at least on other occasions – we discover to be out there. When I see an elliptical penny I am seeing something different from the (actually circular) penny out there before me. The divergence between the object that is sensed and the sense impression of it highlights what is always the case in sense experience – or so it is argued. Even when, for example, I examine a penny from above, so that the visible appearance of its shape corresponds to its actual shape there is still an intermediate entity between the object out there and my experience. The sense datum in this instance happens to coincide with reality.

Sense data have fallen out of fashion in recent decades. This does not, however, entirely abolish philosophical preoccupation with mediation between perceived object and the perceiving subject and the feeling that this mediation gets in the way of direct access to the object. Perception, that is to say, is still seen to be indirect – and this is betrayed by hallucinations.

What hallucinations and veridical perceptions have in common can be presented diagrammatically:

Veridical perception: Object → Sense impression → Perceptual experience
Hallucination: Sense impression → Perceptual experience

If the hallucination of (say) a cat is entirely compelling, it would suggest that it is due to a mental content (sense impression) identical to that which is formed in a veridical perception of a real cat out there. The hallucination and the veridical perception of a cat are of "A Common Kind", being intrinsically the same, though having different origins.[36] The fact that I can experience a cat without encountering a cat shows, so it is argued, that experiences are independent of their objects and that even veridical experiences are separated from both the object (though, according to the causal theory of perception, they may be *effects* of the object) and the subject by the intermediary that connects them.

The grounds for this claim are set out succinctly by Tim Crane and Craig French:[37]

1. In an illusory/hallucinatory experience, a subject is not directly presented with an ordinary object.
2. The same account of experience must apply to veridical experiences.

Therefore

3. Subjects are never directly presented with ordinary objects.

The most vulnerable step in this argument is (2) – the claim that veridical experiences are fundamentally the same as illusory or hallucinatory ones. That might seem to be true if it were already accepted that there are intermediate entities between perceived objects and perceiving subjects. This, however, is precisely what the Common Kind argument must prove. Even if this claim were waved through, there would remain an unignorable difference in the status of the intermediate mental entities between the two situations. In the case of veridical perception, the intermediary is occasioned by the external entity that is its intentional object. There is no such entity in the case of hallucinations.

The intermediaries which both hallucinations and veridical sense perceptions are thought to have in common may seem rather tenuous. We shall come to what has made the idea of intermediaries more respectable presently. Let us, however, re-examine the term "sense data" which is particularly revealing in this regard.

Sense data are envisaged as the items of direct mental acquaintance. We cannot be wrong about them. They are what they seem to be: their being is their seeming; their seeming is their being. This, of course, guarantees nothing about the nature of any entity that we believe that we perceive: when we infer from the sense datum of a green patch that there is a leaf "out there" it is possible that we are wrong (though we typically are not). Sense data are, however, distinct from any perceived object. They intervene, that is, between any object and the subject. They are, so it seems, appearances separated from the object that appears. As such, it is possible that they could be part of a "veil" of appearance somehow concealing that which they reveal.

The tenuousness of the idea of "sense data" becomes more apparent the more closely we inspect the term. A "datum" is that which is given; and sense data correspond to the givenness of what-is. But this gift relies on the senses. Even so, to grant separate existence to the givenness of what is given – identifying the sense datum as a (mental) entity with which we have direct acquaintance – seems a classic example of multiplying entities unnecessarily. It is rather like making the presence of an object an additional something which, unlike the object, is "directly" perceived.

Nevertheless, it is not easy to lay the ghost of intermediary entities creating the fabric of the veil of appearance hiding what's actually out there, when the mediator is identified as something as respectable as events taking place in a material object, when the events are – wait for it – neural discharges and the relevant object is the brain. Envisaging the veil of perception as neural activity seems to license an even more profound scepticism than does the notion that the properties of the sensorium get in the way of, as well as making possible, sensing; that, for example, my view is blocked by my retina and visual pathways. The events in the brain that supposedly transform what-is into that-it-is seem less ethereal than sense data, and so (we are told) obtrude more decisively between conscious subjects and what-is. Sense data as neural activity become suddenly respectable.

Hence the view that the brain – according to certain philosophically inclined neuroscientists and neuroscientifically inclined philosophers – conceals what is out there by the very processes by which it reveals enough of what is out there to increase the chances of the survival of the owner of the brain. If "the various properties of consciousness depend on, and relate to, the operations of the neural wetware inside our heads"[38] then we seem entitled to draw the conclusion expressed by Bryan Magee: "This is what experience is: it is something in our brains and central nervous systems ... The world of our experience must in some fundamental, categorial way be radically different from independent reality".[39] What we experience is not what is out there but the neural activity in here. There is, of course, a connection – usually portrayed as a causal connection – between the former and the latter but the connection is not such as to make the conscious subject a transparent window on what-is.

As we discussed in the previous chapter many contemporary psychologists and neuroscientists characterize brains as Bayesian engines, making predictions as to what is "out there" on the basis of current neural activity. This is how brains manage to be "complex control systems that process sensory signals and use the information to regulate behaviour".[40] Notwithstanding that they are material objects, these brains reach into the future with guesses, hypotheses and inferences: "what we see, hear, and feel, is nothing more than the brain's 'best guess' of the causes of its sensory inputs".[41] The wetware of the brain is "guessware" and any knowledge it makes possible must fall short of what is in fact out there in extra-cerebral space.

The use of terms such as "prediction" and "error" are – to put it politely – personifications. The very notion of "prediction" requires a sense of that which is not-yet; in short, of tensed time, the co-presence of the present at time t_1 and a future at time t_2, which is not part of the material world.[42] While we often apply the notion of "prediction" to insentient machines, this is only because machines are prostheses serving the purposes of conscious subjects who *do* live in tensed time. The prediction is, of course, ultimately, harvested by the users of the machines, not by the machines themselves. Physical processes outside of this context are no more *pre*dictions than they are reminiscences.

The reduction of perception by the Bayesian brain to internally generated, presumably localized, expectations and comparisons between that which is predicted and that which actually takes place seems to empty it of the kind of wall-to-wall content – charged with secondary qualities and set in a continuous arena only patchily relevant to action – that constitutes our moment-to-moment surroundings. "Guessware" would deliver a world of fragments.

Even as experience, what-is does not consist narrowly of conformity to, or deviation from, predictions: this would not add up to the continuous multi-layered presence that it has. I might erroneously think that a bush ahead of me in the twilight is a person and it will be corrected as I get closer to it. This error and its correction, however, will take place against a continuous background woven out of a veridical perception of the road, the traffic, the street lamps, the houses, the pavement, the person walking next to me, the conversation we are having, the birdsong, the feel of the air; in short a 360-degree experience of a *world*, a world, what is more, with which we are purposefully and effortfully engaged, most obviously through our bodily movement or, more generally, our embodied realization of the necessary responses to what is going on around us and our immediate, medium-term, and long-term intentions. We live in a bath of largely justified certainty, though of course we may cultivate uncertainty, especially when, as philosophers we endeavour to wake out of wakefulness – or, as scientists, we test our experiences to, alas, destruction.

None of this prevents "neurobayesians" such as Anil Seth from drawing a conclusion, radically at odds with ordinary understanding, that the experienced world is "a neuronal fantasy". It is the product of "top-down predictions" generated by cerebral tissue rather than "bottom-up" sensory input. Where, however, guesses prove wrong – as when I reach out for a pink elephant, and my hand passes through its hide – there is a correction. While every perception is an hallucination, so Seth claims, it is a *controlled* hallucination.

It is difficult to know how seriously this is meant to be taken. If consciousness really were identical with the activity of a brain generating a top-down neuronal fantasy, it would be difficult to see how we could get to know this supposedly universal truth about our experience of the world. By what means would we get past the neuronal fantasy to call it out as a fantasy?

"Science!" is not the answer because ultimately science depends on perceptions. If perception were the epistemological prisoner of the brain there would be no means by which scientists could uncover what the brain is up to – never mind share the news that daily experience is a controlled hallucination. The implicit claim that science is epistemologically, even metaphysically, privileged – so that it enables us to nip round the back of the veil that is cast over reality by the brain – undermines itself if perception is activity in said brains. If Seth's account of perception as hallucination were true, what would we make of our knowledge of the brain and indeed of the status of the brain? Is it, too, a hallucination? If so, on what grounds could we ascribe to it a central role in perception and other conscious experiences?

One might suspect that Seth's "neuronal fantasy" is so completely reined in – that the hallucination is so tightly "controlled" by "reality" – that it is not a fantasy at all, and sense experience is no kind of hallucination, controlled or otherwise. In which case, aren't things just what we thought they were all along? An hallucination subservient to reality – and, what's more, the kind of reality that we can all see if we are correctly positioned – seems indistinguishable from veridical perception. If I have the experience of seeing a cup in front of me and (as is usually the case) there *is* a cup in front of me, this is not a question of experiencing any kind of hallucination, controlled or not.

It is easy to see, however, what has led thinkers to this strange position: the neural *mediators* of the subject's access to an external world seem to slip between the perceiver and the perceived, as is expressed in Seth's further claim that our conscious perceptions are "indirect reflections of hidden causes that we can never directly encounter".[43] This is another claim that seems to undermine itself. By what indirect means did we discover those causes and observe that they are hidden? How did the predictions discover that they are "tied in useful ways to their causes" if the causes are hidden? We shall return to this presently.

The theory that the brain is not only in the business of prediction but that it is "an organ for prediction error minimization" leads to rather daunting conclusions, according to Jakob Hohwy.[44] It implies that the brain is essentially "*self-evidencing*" creating an "evidentiary boundary":

> All perceptual and active inference happens in an interplay between the evidence to the system, that is activity at the sensory epithelia, and the predictions generated under the overall model in the brain.

> This creates a sensory blanket – the evidentiary boundary – that is permeable only in the sense that *inferences* can be made about the causes of sensory input hidden beyond the boundary. The brain doing the inference is secluded at least in the sense that certain kinds of doubt about the occurrence of the evidence are unanswerable without further, independent evidence. Of course, once we average over the entire sensory input, there is no possibility of independent evidence, which would require us to crawl outside of our own brains.[45]

Perceptions are not of the extracerebral world but of the brain's model of that world.[46] As Hohwy points out, "this becomes an affirmation of a simple Cartesian skepticism". Whether or not Descartes would recognize this, it certainly has the scope and depth of Descartes' systematic doubt, with (neuro) science replacing God as the guarantor of some (desperately limited) access to reality.

Of the many problems associated with the fashionable notion that the brain is an organ for predictive error minimization and that its mediations come between us and what-is, to the point of concealing the latter from us, the most fundamental is one that we have already hinted at: its proponents saw off the branch on which they are sitting. The brain as an object in its own right should, according to the theory, lie beyond the impermeable evidentiary boundary it is supposed to draw between the inside and the outside world. According to Seth, Hohwy and others, the brain as inference machine could not make itself evident as a collection of neurones, located in space, wired up in a certain way, engaged in a vast number of physico-chemical processes supporting its own life and its functions, and connected with the spinal cord, and through the sensory nerves and their relevant organs, the motor nerves and their associated effector structures, with the remainder of the body. Despite this, it seems that scientists believe they can get past the evidentiary boundary after all – in order to see what it is that is making evidence possible and discovering the limits of that evidence.

If, to use Hohwy's phrase, we cannot "crawl outside of our own brains" how can so much be known about those brains, and how can it be ascribed to such a central position in the theory? If perception is not fundamentally different from hallucination – with the outside world being a construct of the brain – how can the brain find its true or actual place in the outside world? How can this object – supposedly inside my skull, five feet away from my toes, two feet away from the screen I am looking at as I write this – be revealed if the other objects it is apparently spatially related to are beyond the evidentiary boundary?

The idea of the evidentiary boundary is not entirely new. The nineteenth-century polymath Hermann von Helmholtz placed the boundary further out:

> The result of [scientific] examination, as at present understood, is that the organs of sense do indeed give us information about external effects produced on them but convey those effects to our consciousness in a totally different form, so that the character of a sensuous perception depends not so much on the properties of the object perceived as on those of the organ by which we receive the information.[47]

The neural solipsism that follows from the claim that our only access to what is out there is mediated through an organ that is sealed up inside itself, should make us question how we arrive at a shared, indeed collective, knowledge and understanding as to what brains do, how they do it, and what they are made of; how we discovered brains and discovered that they are the basis for the world we are supposed to construct out of brain activity; or, more precisely, how the brain constructs the knowledge of the neural sources of its own knowledge out of those neural sources. This question returns with added bite when neuro-epistemology is replaced by what we might call neural idealism. Some thinkers fully embrace the consequences of the idea that what we know is pure hallucination.

A particularly striking example of this mode of thought is the radical Darwinism set out by Donald Hoffman.[48] As we shall discover. Hoffman takes Seth's argument to a place where Seth himself, a neuroscientist, would not presumably wish to go. Hoffman argues that material objects, including brains, do not exist: they, too, are artefacts of perception.

Hoffman's argument that perception does not reveal a real world out there rests on two premises: the fitness-beats-truth theorem (FBT) and the interface theory of perception (ITP). FBT theorem claims that our perceptions are entirely determined by "fitness payoff": our senses "forage" for fitness not truth. We evolved to act in such a way as to maximize fitness – increasing our chances of surviving to reproduce – and this does not require us to perceive the true structure of objective reality. From this, Hoffman draws astonishing conclusions. While the FTP theorem allows variation, selection, and heredity to be real aspects of the real world, it does not allow material entities such as DNA, RNA, chromosomes, organisms, and resources, to be real: "Universal Darwinism ... is an abstract algorithm with no commitment to substrates that implement it".[49] Hoffman rejects the very idea that objects located in space-time are real.

How can we be so deceived? After all, believing in objects that are located in spacetime is not a matter of being susceptible to local mistakes, illusions, and hallucinations, that may be corrected by veridical perceptions. That pink elephant evaporates when I try to touch or feed it but no imaginable corrective experience could expose all objects, including our own bodies, as unreal.

Which brings us to Hoffman's other theorem. According to the innocent-sounding interface theory of perception (ITP), evolution shaped our senses to be a *user interface*, tailored to our needs. Those needs are best served not by knowing what is really going on either in the world or in our brains – which is anyway unmanageably complicated – but by having our experiences encoded in something similar to the icon on a computer screen that reveals nothing of what is happening in the machine. Far from being "out there", independent of our perceptions, spacetime is the "desktop" of this interface and physical objects are icons which need not resemble anything of an objective reality. While the interface "hides reality", it "helps us raise kids". Spacetime is "in the eye of the beholder".[50] It is a virtual reality headset.

Hoffman is serious. Rocks and trees and spoons, being "messages about fitness", mere data structures, do not exist when no-one is observing them. Hoffman declares that even his own body (and the brain that is supposed to be in it) is an icon. This is an idealism even more radical than that of Bishop Berkeley. While Berkeley, too, claimed that objects exist only insofar as they are objects of perception, he did believe, after all, that even when they were not being perceived by us, they continued to exist in the mind of God.[51]

Hoffman does not seem to offer any basis for, or constraints on, the phenomenal "reality" that is the theatre of our lives as organisms or as persons. If objects existed only as cerebrally-generated icons, it is not clear why I should generate this icon rather than that; in short, given that my body-as-object would itself be an icon, why I should believe myself to be situated in one place rather than another, even less gain survival advantage from doing so? What needs would these beliefs serve? How could I have any needs? If the brain-mind brings space and time to the party, it is difficult to see what constraints there are on what the mind is exposed to and hence what it should experience in pursuit of truth-beating fitness.

Hoffman tries to make his idealism more respectable by appealing to physics and that happy hunting ground for radical thinkers: quantum mechanics, in particular the relational interpretation discussed earlier in this chapter. Quantum mechanics, he argues, undermines our belief in local realism: elementary particles are entangled; a measurement ascribing a definite property such as spin to one of a pair will instantaneously confer a definite property on the other. From this he concludes that there are no discrete objects.

Measurements, he further points out, do not reveal pre-existing properties of the entity that is measured; rather they are *interventions* that confer definition upon otherwise indeterminate reality. More recently, he has supported his denial of the reality of spacetime by claiming that "there is no operational meaning to space-time beyond the Planck scale" (10^{-33} centimetres and 10^{-43} seconds).[52] He also recruits some physicists of high repute, such as the Nobel prize-winner Gerard 't Hooft and Leonard Susskind (one of the fathers of string theory), who have suggested that the three-dimensional world of everyday experience, populated with galaxies, stars, houses, and people, is a *hologram*, an image of reality encoded on a two-dimensional surface. This image is, we are to believe, generated by the brain. Or, more particularly, particular parts of the brain that deliver the transition from the quantum to the macroscopic world. It is something that lies beyond the capability of most of the nervous system (cerebellum, spinal cord, etc.) while the cerebral cortex and certain salient connections not only manage this but also do it for themselves so that living brains can see living brains as macroscopic objects lacking superimposed states.

There are, of course, serious doubts about scaling up from the microphysical world of quantum mechanics to macroscopic objects such as flowers, frogs, blue tits, human beings, and other members of the cast of characters in Darwinian evolution. What's more, the observations that, we are to believe, have demonstrated the unreality of space and time, or locations in space and time, appear to have taken place in space and time and to have involved macroscopic entities (such as oscilloscopes and CERN) and the macroscopic physicists who employ them. If we can be deceived as to the very existence of space and time, why should we be confident of the results of esoteric experiments in physics? And what sense should we make of the fact that it seems to be necessary to travel to CERN in order to make the observations that demonstrate the non-existence of space.

But there is a particularly pertinent difficulty with Hoffman's "case against reality", even more challenging than the dubiousness of his appeal to nanophysics to uphold his case against spacetime, and against perceptual experience as a source of truth about objects independent of it. It is the question of how much of *Darwinism* would survive Hoffmanesque radicalization.

The very idea of evolution – and of fitness payoffs that make it advantageous to be good at fleeing from tigers and moving towards food – relies on the idea of real organisms as discrete, spatially bounded objects, with capacities and vulnerabilities determined by their intrinsic nature, rather than as perceptual constructs. It assumes that organisms are separated from one another in space, such that there is a distance between predator and prey, between mouth and food, and the beast and its environment. Without space and time, there is no individuation – or not at least of the kind we think of in the world of living creatures. And organisms grow, move, and decay in spacetime. It is difficult to imagine entities that do not occupy space having the kinds of needs that we ascribe to organisms – and experience ourselves. The four f's – fighting, fleeing, feeding, and mating – are all activities that take place in spacetime.

Evolutionary theory also presupposes a temporal ordering of the emergence of species. The standard story tells us that unicellular life preceded large mammals, paramecia preceded exotic megafauna such as you and me, and that the path from one to the other was marked out by countless mutations and the operation of natural selection upon them resulting in gradual changes *over time*. The theory also envisages successive generations of organisms, with offspring (presumably) arriving later than their parents and of organisms lasting or not lasting sufficient time to reproduce. As Hoffman himself says, "Human vision is shaped by eons of natural selection".[53] "Eons" sound like time – lots of time. So do the two billion years that, he reminds us, cyanobacteria have been on the earth emitting the oxygen that has made the planet habitable for organisms like us.[54]

In short, Hoffman's appeal to Darwinism to justify his revisionary metaphysics ends up dismantling the world picture presupposed in Darwin's theory. The very idea of "fitness" cannot make sense in the absence of organisms – objects located in spacetime and existing independently of perception. As for those perceptions, if there is nothing "out there", there would be nothing to justify any perception of the world. Or, if there is no such spatially discrete object as a brain, it is not clear how organisms perceive anything – or anything in particular. My present location accounts for the fact that I see my laptop screen and that I could not see my copy of Hoffman's book the other day, the dark side of the moon, or the Battle of Hastings that ended a millennium ago. How therefore can Hoffman reconcile his embrace of evolutionary science with his rejection of the conditions in which it makes sense, the very ground upon which it stands?

There is of course an additional question as to why an organism would benefit from fundamentally misreading the physical world. How could "fitness" "beat truth"? It is difficult to see how there could be survival benefits arising from seeing the world through classical eyes if reality were as it appears in quantum mechanics. As already noted, there would be no basis for the item whose survival is at issue. The well-adapted prey would not be safer from the predator nor the predator more successful in predation if there were no classical entities. It would hardly account for the evolutionary journey, supposedly driven by the self-replication of organisms, from micro-organisms to higher mammals or from an insentient material world described by physicists to the physicists describing that world

More broadly, it is difficult to reconcile Hoffman's commitment to science with his belief that the universe has no history apart from observers, so that "[t]here was no sun before there were creatures to perceive it".[55] What should we make of the standard scientific story according to which the Big Bang took place before there were planets, the creation of planets preceded organisms, unconscious organisms arrived before conscious ones, and conscious organisms moved over the surface of the planet long before the conscious organism C. Darwin stalked the earth and formulated his theory? More fundamentally, how can Hoffman give credence to the claims of microphysics and evolutionary biology while rejecting the truth of the perceptions upon which they ultimately stand? If CERN and the bodies of the physicists who work there are illusions what is the status of physics?

Hoffman's thesis places the brain front and central. If, however, my brain were not an entity located in space and time what warrant would it have for locating me just now in my study rather than on the moon? Hoffman is aware of this problem and even goes so far as to accept that the brain is not something that exists in itself: it is a perceptual construct. And so, too, is the organism Donald Hoffman. The sound of sawing gets louder.

Hoffman's two theorems privileging evolutionary fitness over veridical perception, far from being consequences of Darwinism, are therefore incompatible with it. As such, his "case against reality" is a particularly egregious example of pragmatic self-refutation in which an argument put forward to support a position undermines its own premises and even the possibility of it being asserted. There are many aspects of Hoffman's self-refutation. He requires us to assume that there is such a thing as a well-founded theory of evolution which is already in place as illustrated by his claim that "Our minds evolved by natural selection to solve problems that were life-and-death matters to our ancestors not to commune with correctness".[56] Nevertheless, since time does not exist, he must deny that the theory and the processes it describes have been in place before his argument. And he also requires us to believe that Donald Hoffman – like other apparent physical objects – does not occupy space and endure over a definite period, yet this must be the condition of his writing *The Case Against Reality* at a particular place and a particular time. And what is the truth status of his claim that evolution shows that "fitness beats truth" if we have no access to truth to see that it is beaten by fitness? With what eyes did we discover that "evolution hid the truth from our eyes".

There is another problem with Hoffman's FBT theorem, beyond the concern as to how it can be to our advantage to get things wrong. As we noted in Chapter 3, it is not clear why we should "get" things at all? It is difficult to pinpoint the evolutionary advantage accruing to organisms from having phenomenal consciousness of the environment with which they interact. This becomes even more difficult when we are told that survival requires that the organism gets everything wrong about the environment in which it is endeavouring to survive.

Something like Hoffman's founding theorem that "fitness beats truth" is often invoked by thinkers who want to pull humanity down a peg or two and claim that we believe things to be true because they are biologically useful rather than that they are useful because they are true. John Gray, argued in *Straw Dogs: Thoughts on Humans and Other Animals* that "the faith that through science humankind can know the truth" has been exposed as groundless because Darwin's theory of natural selection has shown that "[t]he human mind serves evolutionary success not truth. To think otherwise is to resurrect the pre-Darwinian error that humans are different from all other animals".[57] We have to believe the truth of Darwinism – and the validity of its displacement of "pre-Darwinian error" – in order to arrive at the conclusion that we do not have access to the truth about our nature or anything else. The sound of sawing – of the branch on which Gray is sitting – is deafening. Hoffman, however, goes further, sawing not only the branch but the tree; indeed, the forest in which the tree is situated.

In *Darwin's Dangerous Idea* – a classic manifesto of universal Darwinism – Daniel Dennett described the theory of evolution as a "universal acid ... that eats through just about every traditional concept and leaves in its wake a revolutionized world view".[58] In Hoffman's hands, Darwinism ends up eating itself. Thus, The Origin of the Specious.

10.10 BEYOND DOUBT: REALITY AS SOMETHING LIVED

> Reality is that which, when you stop believing in it, does not go away.
>
> Philip K. Dick[59]

The discussion so far has been framed by the intuition that there is something called "reality" that is "in itself", pure what-is, defined to a significant degree by intrinsic properties, approached, explored by conscious subjects from without, but never fully revealed. If we think of reality in this way, the interactions necessary to access it will seem like sources of contamination, of distortion, even of concealment: appearance will be a "veil". Because there can never be a pure revelation of what-is, so the often-implicit argument goes, we shall be denied full, or perhaps any, access to what is real, understood, as how things are in themselves, how they are in the absence of consciousness.

Notwithstanding the seemingly inescapable conclusion that reality is cognitively elusive there is the day-to-day, moment-to-moment, certainty that it is in practice far from elusive: it is inescapable. The reality of reality is to be found in what directly or indirectly impacts on conscious subjects. Instead of being some fundamental stuff, defined solely in terms of inaccessible intrinsic properties, reality is (for example) material objects and events that inescapably *matter* to us, in part because of the nature of the material of which we are ourselves made. The reality of reality lies in its value, meaning, or significance for human subjects and these are determined at least in part by what counts as an issue for those subjects.

A radical development of this suggestion is that the very term "reality" makes sense only in the context of its opposite: *un*reality – events or entities that are experienced as, discovered to be, or exposed as, unreal. This reflects the ambiguity of the question "What is reality?". It could equally mean "What do we mean by reality?", "What counts as real?", "What is the ultimate grounding of reality?".

J. L. Austin argued that, of the contrasting pair reality and unreality, it was unreality that "wore the trousers": "A definite sense attaches to the assertion that something is real, a real such-and-such, only in the light of the way in which might be, or might have been, *not* real".[60] We may assert that a duck is real against the counterclaim that it was "just" a dummy, a toy duck, a decoy, or a picture of a duck. Reality is granted to an entity that meets certain applicable criteria, and its reality is asserted where there is the possibility that someone might be making a (local) mistake. It is this sense that gives the idea of reality its *borrowed* force – on loan from the idea of unreality.

This is a challenge to the traditional way in which the contrast between reality and unreality is deployed in philosophy and the way we have so far discussed it in this chapter. An assertion that things universally are not as they appear demotes how things appear to "mere" appearance: I might be dreaming; I could be mistaken as to the nature of everything I am currently experiencing, or have experienced, even where it seems to concur with the experience of others. Against this it might be argued that what-is, understood as the totality of material things, forces, and energy, falls short of reality because the very idea of "reality" is inseparable from our (individual and collective) lives as conscious subjects. At the most basic level, reality is what what-is "is like" when it is made explicit and enters our lives.

The case against our having access to reality is based on the notion that what-is made explicit is nothing "like" itself. And the contrary case is that what counts as real – or indeed what could count as, or be, real – is what confronts us as living creatures consciously making their way in the world. It is what we must take account of, and this is not stand-alone stuff defined by its intrinsic properties but things as they matter to us. Reality is something lived or lived with – suffered, enjoyed – and as such defined existentially. It is inseparable from a partnership between living subjects – with their needs and interests – and what-is. Getting things practically right is seen as sufficient evidence that we have drilled down to the level of reality. Interaction, far from being contamination, is the very essence of what is real. The capital of this reality is the body that has needs and an agency that serves them, and which we suffer, enjoy, and, most fundamentally, "am".

This will help us to break out of the circularity that we encountered as when, in Section 10.2, we tried reassuring ourselves, on the basis of their coherence, that the deliverances of perception are on the whole true. I know, for example, that a stick which appears bent in water is in fact – in itself – straight. It not only appears straight when it is not in water, but I can feel its straightness as I run my fingers down it. Moreover, I can insert it into a straight cylinder close to it in diameter without it getting stuck. Approaching an object from different directions, with different sensory modalities, at different times, can seem therefore to enable us to sort out false from correct appearances of it. It can also enable us to distinguish hallucinations from veridical perceptions. Attempting to mount or to feed a pink elephant or sell it to a zoo, will cure me of the belief that the creature is a real beast rather than a coinage of my mind.

Consistency with other perceptions and overall coherence, however, does not address the concern, unpacked in the previous sections, that everything that is given is contaminated by the processes by which it is taken, by which it is received. Trying to break out of the bubble of perception by multiplying perceptions is (to borrow a humorous analogy from Wittgenstein) to be like a person buying "several copies of the morning paper to assure himself that what it said was true".[61]

We can address this apparent circularity by acknowledging that we are already on the far side of any barrier between ourselves and reality, in virtue of the fact that reality is something that we *live* – live in, live with, live by. In this regard, the transition from what it is to that-it-is (as an object of perception, of knowledge, of thought) is not the spinning of a veil over what-is, by conscious subjects who contaminate that of which they are conscious. On the contrary, reality is to be found in that which is real *for* conscious subjects. Access to reality is existentially guaranteed by our ambodiment, which places us both sides of any barrier between ourselves and what-is, between what I am and experience and what is fundamentally out there. What-(really)-is is primordially revealed in my being what I am.

In this context, profoundly revisionary accounts of reality – whether they are offered by physicists or philosophers – seem lacking in fundamental seriousness. No-one doubts the reality of enduring material objects located in space and time when they are running for a bus, going downstairs to make a coffee, keeping an appointment, or planning a career. It does not seem right to think of bus-chasers being deluded and living their lives in an epistemologically fallen state. At the very least, it seems implausible to think of reality as being available only to experts – if it is available to anyone; that one should, for example, not be able to know what reality is without a sufficient training in advanced mathematics to be able to understand the Schrödinger wave equation.

While pain and pleasure are ranked as "mere" secondary qualities – because they vanish as we advance upon the what-is with the instruments of science – there could be no more profound, inescapable, invasive, reality than is captured by the difference between toothache and the pleasure of sunbathing. And there are many uncomfortable truths – prompting fear, shame, and frustration – that cannot be ignored as unreal. The way we are seen in the eyes of others who know more about us than we would like them to, and who may judge us adversely, is something we cannot wriggle out of, notwithstanding that their judgements seem remote from the kind of reality supposedly unpeeled by science. There is no doubt what reality is when we have been discovered doing something wrong. Caught with our hand in the till, we cannot seriously entertain the possibility that hands, tills, and the visual field of the person who has caught us, are constructs out of our own sense data. Reality must include back doors, a welcome cup of tea, uphill slopes, a spider, and an enigmatic smile – things that we can all understand.

Rather than being something beyond our reach, therefore, reality construed existentially seems something we cannot escape. Instead of eluding our most sophisticated inquiries, reality becomes something we are exposed to and know through being exposed to it. Far from being something that has little or even nothing to do with us it concerns us intimately because we live it. Existentialized reality places tertiary qualities – values, and the pain and pleasure caused by states of affairs – centre stage.

This is evident at a homely level when we recruit perceived objects to assist in the exercise of our agency or otherwise interact with them; when they are encountered as opportunities for action; or, indeed, items that demand action.[62] While this is most obviously true of artefacts such as cups and saucers and motor cars and phones, which have been manufactured with action in mind, it can be equally true of items such as stones (things to throw, to use for building, to avoid tripping over, to pose as obstacles) or threats (such as predators). The reality of a cup is revealed when it is used as something you can drink out of. You cannot doubt that a cup from which you are presently drinking is "really" solid, "really" localized, "really" out there. The purposes that artefacts serve overshadow that in them which is unrelated to their purpose. A cup has many properties – such as the capacity to cast shadows or to leave a mark on a table – that are only distantly related to its function.

Our encounter with the material world is not disinterested and no less authentic for that. It seems highly unlikely that even Donald Hoffman, notwithstanding his argument that objects are artefacts of our perception, is really capable of embracing his own conclusion and doubting the existence of his own body or even of the airplane that will take him to the conference at which he will present his ideas. There is, in other words, something fundamentally insincere in putting forward "the case against reality".

There is another dimension to the existential understanding of reality. We have hitherto approached reality as something that is disclosed, uncovered, revealed by the conscious subject. Thus seen, the traffic is largely one way: I expose the pebble and make it present to me, but the pebble does not reciprocate by making me present to it. But, as we discussed in Chapter 5, the traffic is not entirely one way. I, too, am an entity in the world, rather than a free-floating consciousness, lighting up this, that, and the other. I am to some degree disclosed, in the sense of being in part defined, made real, by the world that I disclose, though that world is for the most part unaware of me. At any given time, I can see or otherwise sense parts of my body standing in relation to material entities. While my location determines what objects are evident to me, those objects define my location. The cup is two feet from my hand; but my hand is two feet from the cup.

This kind of reciprocity is unpacked in a multitude of ways as I actively engage with the world around me – peering and palpating and listening and sniffing and tasting – and express an agency shaped by immediate, medium-term, and long-term aims specified with varying degrees of generality or abstraction. The pushback against, the resistance of, the world in relation to our agency is a mode of disclosure of ourselves by the world we have disclosed.

As physically located entities with physical properties, as needy creatures with limited powers, therefore, we are exposed by the world we have exposed. Reality, far from being hidden, tucked away in a transcendental realm beyond the reach of our senses is our mode of being there. Reality is revealed not as the imagined object of an epistemically asymptotic view from nowhere but as the object of a view from someone (individual or collective), or somewhere (physical or discursive). Reality as it is lived is that which is encountered, engaged with, suffered, exploited, enjoyed by the embodied subject who is immersed in it.

There are problems, however, with the existential domestication of reality. First, it by-passes the awkward fact that the sum of things has a history that has lasted much longer, and a geography that is much wider, than anything revealed to conscious subjects, to individuals for whom what-is matters, who care for, and take care of, themselves, to knowing subjects who can get things right or wrong. Conscious subjects are confined to relatively recent history and, so far as we know, to a minute part of what-is. Secondly, while we may think of reality as something in which, like it or not, we are immersed, it remains reasonable to think that we do not have complete knowledge of that in which we are immersed or even the nature of the "we" who are immersed in it and of the relationship between us and it. After all, our understanding of the kind of object that we are is manifestly incomplete. There is no embodied subject who, so far as I know, understands how her body relates to her subjectivity. Thirdly, to demote the idea of "reality" in this way seems to ignore advances we have made towards uncovering truths about what-is, notably those achieved by science. The replacement of a geocentric by a heliocentric view of the solar system or of Aristotelian by Einsteinian mechanics, would seem to suggest that we are in at least one sense closing in on a reality that still remains beyond our cognitive grasp. Or, if that is epistemically a little too optimistic, that reality is in some important sense *not* how it seems to us as we go about our everyday lives. And if we accept the scientific image as being closer to the reality of what-is, then the veil of "mere" appearance that intervenes between our moment-to-moment experience and what is out there is thick indeed.

It hardly needs repeating that there are reasons for not regarding the everyday and the scientific images of what-is as being of equal standing, with the former being close to the existential reality of the lived world and the latter to a reality that goes beyond, even though in some respects it may underpin, daily life. The most obvious reason is that seeing what-is through the lens of science has enhanced our individual and collective agency. The landscape of technology in which we live seems a spectacular demonstration of the superior, even fundamental, truth of the world picture of science. Indeed, it offers a coherent account of the universe before and after, above and beyond, the emergence of conscious subjects.

There is a further reason for questioning whether what counts as "reality" must be confined to that which is lived or liveable. We are conscious, at every moment of our lives, of multiple horizons which place limits on what we are aware of, and hence what can, even in principle, be lived. I do not live all of the reality I know, or know of, though I cannot sincerely (existentially) doubt its existence. In short, the sense of reality is just as much present in our intuition of things that are absent or hidden as in those things we are engaged with, and which are revealed to us.

We cannot, therefore, conclude that what should be taken to count as reality is exclusively either: (a) the epistemic asymptote, quite unlike what we take for granted in everyday life, of an as yet unended (and possibly endless) scientific or more generally intellectual project in which ever more precise and wide-ranging observation and increasingly sophisticated conceptual developments interact iteratively; or (b) the existential truths we take account of, take for granted, suffer and enjoy, in our daily lives.

It would be satisfying to think that there would be a point where reality as the ultimate object of knowledge and reality as something lived (suffered, enjoyed, negotiated) converge. After all, they have a common point of origin – in the conscious, embodied subject, in the "I" that suffers itself and knows the world and knows itself as part of the world, in the "am" that merges knowing and being. But that is only the starting point and "amming" does not deliver Cartesian-strength certainty regarding what-is.

Perhaps, after all, the epistemic and existential versions of reality are not rival accounts of the same thing. It would, after all, be unsurprising that such a hard-worked word as "reality" has a multitude of meanings. I hope nevertheless that it has been worth exploring some of its ambiguities, as a way of discovering ourselves, and the profound cognitive ambitions and expectations that are gathered up in the notion of "the search for reality" or the search for the what-is that is thought to lie behind, to be the hidden source of, the that-it-is.

10.11 REALITY AND TRUTH

Our discussion of the nature and idea of reality has so far made little use of a word closely associated with it: "truth". The search for reality overlaps with the search for fundamental, enduring truth. We might connect "reality" and "truth" by thinking of the latter as the explicitness of what-is, experienced as reality, made more explicit – typically in the form of propositions: that-it-is in propositional form. A brief treatment unloading some of the burden carried by this loaded term seems appropriate.

The most intuitively attractive account of truth is the "correspondence theory" according to which truth is a *relationship* – of correspondence between a proposition or assertion and a state of affairs. My statement that it is snowing in our garden is true if (and only if) it is snowing "out there" in our garden: there is a correspondence between the claim that it is snowing in the garden and the falling of snow in the garden; between the belief that I wish to instill in you by stating that it is snowing and what is happening in the garden. Falsity is the absence of the relevant correspondence. The nature of this correspondence – of the relata and how they relate – remains unclear as will have been evident from our discussion of thought. What is clear, however, is that the correspondence theory seems to apply most straightforwardly to discrete, factual truths – hence the example chosen of snow falling in the garden.

There are many truths where there is no simple, direct correspondence between what is asserted and a state of affairs that could be perceived or pointed to. They are often general but not necessarily so. Rather, they are embedded in a collective cognitive soil. Consider our much-used example – "Paris is the capital of France". The correspondence between the statement and that which makes it true is remote from the kind of correspondence exemplified by "Felix the cat is black".

In many cases, the truth claims of an assertion are upheld in virtue of their being consistent with, or following from, or instantiating, other truths. This, the *coherence* dimension of truth, is particularly applicable to science, where underlying theories may not only be confirmed by observations but also by their compatibility with already well-established bodies of knowledge and understanding. As Quine said in a passage quoted earlier, "our statements about the external world face the tribunal of experience corporately". The coherence theory of truth is most typically manifested in the prediction of the consequences that would follow if certain assertions were true and in looking for evidence of those consequences. Any assertion deemed to be true is required to be consistent with other assertions that are known to be true.

The most striking manifestations of coherence are predictions of particular states of affairs on the basis of more general patterns of events. The validation of truths through coherence is, however, at work in our everyday life. We might say, for example, that Jane's assertion that she was in New York at noon on 20 November 2023 cannot possibly be true because I saw her in Chester at noon on 20 November 2023 and people cannot be in two places at once.

The clearest distinction between correspondence and coherence theories is that, in the former, the immediate truth conditions of propositions are objective states of the world and in the latter the immediate truth conditions of propositions are those of other propositions that are directly or indirectly true of certain states of the world. The distinction may not, however, be as clear cut as this suggests. No proposition directly assesses what-is, as will be evident from the discussion in the previous sections of this chapter. Propositions are about aspects of what-is that have been made explicit as that-which-is-the-case. Given that assertions *en route* to formulation have passed through the lens of subjective consciousness coming upon itself through the lens of collective consciousness reflected in the mirror of language it is a gross simplification to think of any asserted truth as a mirror of what-is.

A third theory of truth – the pragmatic theory – asserts that what counts as truth is "what works". True statements work for us in the sense of assisting us in our passage through life, while false ones do not. A statement or belief is true if embracing it directly or indirectly promotes our ends or, more ultimately, supports our survival. It is this that is "the cash value" of a belief. In William James' classic formulation of the pragmatic account of truth "The ultimate test … of what a truth means … is the conduct it dictates or inspires".[63] This is more likely to apply to higher-order beliefs and claims – such as religious beliefs, notwithstanding that those beliefs would not acknowledge utility as the sole or even the major, grounds, for believing them. More importantly, it takes the idea of truth beyond that of a narrow correctness – of a string or web of propositions.

Nevertheless, it overlooks the fact that many truths do not inspire conduct. For example, it is not clear how we would differentiate the conduct inspired by the discovery that the universe is 13.8 rather than 5 billion years old. And it can be argued that assertions are not usually true because they are useful but useful because they are true. Yes, it is useful to know that Paris is in France if I want to visit Paris. But it is useful because Paris (truly) is in France. And, even in those (many) cases where there are behavioural consequences, the truth of the claim that there is a missile heading towards me does not lie in its prompting me to duck and hence save my life. Rather it lies in the fact that there is a missile heading towards me which makes ducking a good idea rather than a pointless act.

One of the most far-reaching examples of a pragmatic interpretation of truth is Quine's assertion, already referred to, that Homer's gods and physical objects are on the same "epistemological footing". If the idea of physical objects seems closer to truth than that of Homer's gods it is that "the myth of physical objects is epistemologically superior to most in that it has proved more efficacious than other myths as a device for working a manageable structure into the flux of experience".[64] The question that Quine ignores is why the myth of physical objects should be so efficacious. The reason for this is that it is not a mere "posit". As we discussed in Chapter 5, it lies in our embodiment and the ontological democracy between our bodies and other objects.

So pragmatism can be turned on its head: beliefs work because they are true rather than being true because they work. As we saw in the case of Hoffman's "fitness-beats-truth" theorem, extreme pragmatism that excludes other aspects of truth leads to self-refutation. There are exceptions, of course – most notably religious or other very general beliefs. Even so, they may be open to testing on the basis of facts. Does religious belief "work"? Well, let us look at the consequences, insofar as they can be judged. If we cannot judge them – the calculus is too complex and the criteria for success too contested – then we may conclude that we cannot say whether they are (pragmatically) true or false. A belief that brings comfort to some and justifies the persecution of others cannot be seen as corresponding to an eternal truth.

Perhaps the most fundamental weakness of pragmatism is inadvertently revealed in its appropriation by those such as Hoffman who would wish to biologize truth. The biologized versions of pragmatic theories – according to which seeing what is truly there boils down to being attuned to the environment in a life-supporting way – do not withstand the observation that there is no "there" or "out there" or "reality" without intentionality. Biology does not secrete propositions with truth values along with saliva, urine, and path-smoothing slime.

Besides, the judgement that a belief "works" is based on consequences that are themselves subject to factual judgement – and exposed to the test of correspondence. And, as we have already noted, theories defended on the grounds of coherence also have to be upheld ultimately by other putative truths to which correspondence tests may apply. It is not necessary therefore to adjudicate between these three theories and identify the one that tells the whole truth about truth. Rather, we should acknowledge that they highlight different aspects of truth.

Even so, the suggestion that the three rival theories are not competing accounts of the nature of truth but different aspects of a single complex concept does not seem quite right. Correspondence seems to be closer to the fundamental truth about truth. Truths could not subsist simply in being consistent, or coherent, with other truths. Foundational truths need to be correspondently true.

Nevertheless, the differences between correspondent, coherent, and pragmatic modes of truth are less important than what they have in common: explicitness. That explicitness is central to different aspects of truth and fundamental to its nature is revealed, by default, in a fourth theory of truth: the theory that truth doesn't amount to anything. I am referring to the redundancy (or deflationary or disquotational) theory of truth.

Frank Ramsay (and before him Gottlob Frege[65]) advanced a redundancy theory of truth by arguing as follows.[66] To assert that "p is true" is to say no more than to assert that "p". They have the same truth conditions. "Is true" therefore adds nothing to the proposition p, any more than saying "Yes" or "I agree" when someone else says it. Does this argument, however, empty the notion of truth? It would do so only if truth were, or were required to be, a predicate or a property of all true statements (and "is false" of all false ones). In fact, it is neither a predicate nor a property nor, indeed, any kind of stuff. The absence of difference between "p" and "p is true" does not demonstrate that truth is empty but rather that *truth is equally present* in both "p" and "p is true". "Is true" does not add to the content of a proposition, though it may be asserted in response to a challenge. Truth should not be sought in the *difference* between these two propositions. Truth most typically lies in the *relationship* between an assertion and that which (typically a state of affairs) it makes explicit; it lies in the intentional relation of a propositional thought: the "that" and "is the case" in "that p is the case". Both "p" and "p is true" lie on the far side of the transition from what-is to that-it-is and their truth lies in the relationship between the explicitness and that what-is they have made explicit.

A variant of the deflationary theory of truth is the Polish logician Alfred Tarski's semantic theory according to which truth is a property of sentences.[67] To use his standard example, "'Snow is white' is true if and only if snow is white" sets this out clearly. By gathering up truth into sentences – or the relationship between first-order sentences and higher-order sentences about those sentences – it misses the connection with extra-linguistic reality and hence loses all the layers of intentionality and explicitness that underpin true statements and the very nature of truth – truths by which we live.

The endeavour to deflate truth does not therefore deliver what the deflators may have hoped: to demystify it. On the contrary, it highlights its nature; more specifically, the *explicitness* behind the surface manifestations of truth such as correspondence, coherence, and pragmatic cash value. They are all downstream of explicitness. Truth is not a property of what-is; indeed, it is not a property. (What-is is neither true nor false.) Nevertheless, the assertion that something "is true" does serious work: it draws attention to a claim that has been made and evaluates this claim: it is a second-order explicitness – that makes explicitness explicit.

There is no truth in the material world; nor, we may guess, in the *umwelt* of primitive organisms that do not share claims and counterclaims that refer, via complex intentional relationships, to states of affairs present and absent, actual or merely possible. While a pebble is neither true nor false, correct or incorrect, valid or invalid, justified or groundless, the *assertion* that something is a pebble or that there is a pebble over there, can be any of these things. Indeed, it *has* to be either true or false. For there to be truth and falsehood there need to be conscious subjects facing the world and sharing their experiences with fellow subjects.[68]

Deflationist theories of truth shrink, even collapse, the cognitive space of the human world which resides in the difference between what-is and *that* it is. The distance between a cat's being black and its being the referent of the true assertion that "Felix is black" is the fundamental dimension of the distance between mankind and nature. It is because explicitness is ubiquitous that it is easy to overlook it and to empty the truths that express it. The universe is not in itself, however, a vast bag of frozen p's.

It is obvious that falsity is not part of what-is. By definition, it belongs to what-is-not, and is therefore not part of what-is. But it should be equally obvious that *truth* is outside what it is true of. This is a non-trivial point as it reminds us that, as with perception, factual truth (and likewise factual falsehood) is not part of the network of material events and objects. Truth begins with the inchoate "that" of intentional experience. There is then a long non-causal journey between revelatory perceptions and the truths of homely assertions and, ultimately, of scientific theories. At the very basic level, the claim that a causal relationship between that which is perceived and the perception, which is the condition of the latter becoming a truth, overlooks the fact that material connectivity between physical elements in what-is does not make later events into truths about the earlier.[69]

The important point is that propositions subject to the test of truth are not merely *effects* of that which they are about in the way that thunder is the effect of lightning. This is captured in the striking passage in the final pages of Wittgenstein's final notebook, *Of Certainty* quoted in Chapter 6: "Someone who, dreaming, says 'I am dreaming', even if he speaks audibly in doing so, is no more right than if he said in his dream 'It is raining'. *Even if his dream were actually connected with the noise of the rain*" (emphasis added). Causal ancestry does not deliver truth values because causation does not deliver truths – any more than it delivers falsehoods. Behind this is the fundamental truth that what-is does not, of itself, cause that-it-is, that-it-is-the-case, the-fact-that-it-is – any more than what-is-not causes that-it-is-not, that-it-is-not-the-case, the-fact-that-it-is-not-the-case. Hence the invalidity of the causal theory of truth – which is betrayed by the way in which it is typically expressed as a causal relationship between a belief and the fact that is believed.[70] To think that "the fact that such-and-such is the case" could have causal powers is to conflate a state of affairs with the state of affairs being made explicit. Of course, my knowing the fact that such-and-such is the case – for example that there are bears in the woods where I am walking – may influence my behaviour. But my *knowing that* bears are at large is not an aspect of the material world in the way that bears in the woods are. In the absence of a knowledgeable, conscious subject, bears in the woods would not generate the fact that there are bears in the wood.

The inappropriateness of causal theories of knowledge and belief is particularly evident in those cases where knowledge is (actively) sought out, even when what is sought is as homely as that which becomes available to me when I pop next door to test your claim that my cat is asleep in my neighbour's garden. While I check that truth by my positioning my sense organs to see the garden and its contents, it is not generated, caused in you or us, by the cat's being in the garden – any more than the cat's being in somebody else's garden generates the falsehood that the cat is in your garden or the fact that it is not in New York. The causal theory of truth would make assertions that are false or explicitly about non-existent objects impossible to explain and counterfactuals impossible to account for.

The very ideas of reality and truth - or unreality and untruth – could not arise in the absence of conscious subjects because what-is does not make itself explicit. There is no space in the realm of what-is for truths, falsehoods, about itself, even less *world pictures* including, as we have noted, that of natural science, with its ideas of principles, laws, causes, and raw materials with fixed properties. Truth is a relationship between what-is and a proposition while what-is generates only one partner in the relationship. That is why, as we discussed in Chapter 4, the scientific world picture cannot accommodate the processes that generated that world picture, and its being entertained, by the part of the universe that is pictured in it.

Ramsay's deflationary account of truth is one of the most spectacular examples of point-missing in philosophy and the clearest example of how explicitness – or making explicit – is overlooked. Far from eliminating the idea of truth, the deflationary argument highlights its extraordinary nature. Explicitness transforms portions of what-is into truthmakers or truth conditions and objects and their properties into respectively the subjects and predicates of sentences expressing propositions. While truth itself is not a property of the world of which it speaks, the very idea of truth is not empty. Every aspect of our lives is guided by spoken and written, particular and general, truths; but those truths are not identical with what-is. At best, they express a mode of (directly or indirectly) *facing* what-is rather than being part of it, though the audible spoken and visible written token words and non-verbal signs in which they are expressed will also be part of what-is – things that can be sensed (seen or heard).

The distance between truth (that what-is is) and what-is is underlined by the fact of, and importance of, negative truths. While there is clearly nothing in the realm of what-is corresponding to the assertion that "proposition p is not true", negative propositions may express truths about the world. They transform what-is, where it does not correspond to what-is-claimed, into a falsity-maker. The rotation of the Earth round the Sun would not constitute a truthmaker for the assertion that the Earth goes round the Sun or as a falsity-maker for the assertion that the Sun goes round the Earth before the emergence of sentient creatures. It was billions of years before "The Earth goes round the Sun" became a truth, long after the Earth had started orbiting the Sun. It was an equal number of years before "The Sun goes round the Earth" became a falsehood. What-is exists irrespective of whether or not it is made explicit while what-is-not exists only insofar as it is made explicit, entertained in a negative proposition – usually in contradiction to a proposition that has been entertained by oneself or asserted by someone else.

While truth and falsehood seem to be equal and opposite – and they emerge together – they are not on a par. "The cat Felix is in the room next door" (p) tells us more than "The cat Felix is not in the room next door" (not-p) tells us. Proposition p, while ruling *in* one possibility rules *out* an indefinite number of others. Not-p, on the other hand, rules out only one possibility and does not rule in any other. Being ruled in is, of course, no more a feature of what-is than being ruled out. While the cat being in the next room rules out its being in the garden, in New York or Greenland, or dead 200 years ago, these not-p's exist only insofar as they are made explicit.

There is an irreducible vagueness in propositions that are true. As with thoughts and possibilities – discussed in the previous two chapters – the grain of precision of what is proposed in factual assertions will always fall short of that of what-is. This vagueness is often flagged up in what we say – as when, for example, the assertion that the cat is "somewhere" in the garden. But there is an unacknowledged blurring in whatever factual assertions we make. "The cat is on the sofa" does not deliver the precision of our seeing the cat on the sofa – which reveals exactly where it is on the sofa and what space it occupies. The humblest, most pedantic proposition is, in this respect, fundamentally different from the what-is that it makes explicit. Any true statement falls short of precise truth and, with respect to that of which it is true, it has a halo of imprecision.

This is yet another reason for resisting the belief that the truth or falsity of propositional that-it-is is underwritten by a causal relationship between the relevant what-is and the proposition. A further reason is that the relevant what-is is not episodic in the way that the propositional that-it-is is episodic.

I can think or assert that "Paris is the capital of France" at a time which is not related to any occurrence of the relevant what-is, which is anyway, a continuant. The occurrents of thought do not track the standing or episodic realities that they are about. The status of Paris as the capital of France does not account for the scattered, random occurrences of the thought *that* Paris is the capital of France, not even if that status is reduced to a disposition to increase the likelihood of such token thoughts occurring and being expressed as propositions.

Take the relatively straightforward (true) assertion that "London is a busy place". Does the busyness of London cause the truth of this statement? It is difficult to grasp the idea of the causal powers of something as abstract as the busyness. How was the busyness separated and abstracted from an already abstract concept (London defined by municipal boundaries) in order to gather itself up and exert causal powers? If it had causal powers it would be permanently causing appropriately tuned people to say or think that it is busy as the effect of its busyness. Likewise, the truth of the less abstract fact that "The population of London is less than 10,000,000 people". Does having a population less than 10,000,000 cause this truth? Would having a population of 1,000 people or being entirely depopulated also cause it?

The fundamental point is this: explicitness transforms what-is into truthmakers, but the latter do not act as causes of true statements; nor are they congruent with them. Notwithstanding their name "truthmakers" do not *make* anything; rather they justify certain assertions and the implicit claim that they are true.

The relevance of this to the central preoccupation of this book will be apparent from what has gone before. The transition from what-is to what-is-the-case – from what-is to that-it-is – is not an effect of the causal powers of what-is. The natural world does not have the power to put itself in inverted commas. It is in finding, asserting, and recognizing truths that we most clearly *face* the natural and human worlds and engage with them from a virtual outside rather than being wired into them. As we have already highlighted, this lies at the heart of our capacity to act within and upon a universe, of which we are a part, as agents with a significant margin of freedom.

10.12 INCONCLUSIVE CONCLUSIONS

What-is simply is: it is not unreal, nor is it real; it is neither false nor true. If reality seems elusive, it is because the transition from what-is to that-it-is has to be seen through a multitude of lenses, en route to generating the general and particular truths in which it is articulated.

Our exploration of reality has been to some extent confused by conflicting senses of the word, such that it is on the one hand beyond our reach and on the other existentially engulfing; elusive and at the same time inescapable. Disinterested inquiry – or inquiry in which interest is at the very least postponed – may lead ultimately to an account of reality as seen in natural science. But this portrait of reality may seem to be at widening distance from reality as it is lived and the scientific enterprise points to an unlived and unlivable reality beyond the horizon of our knowledge. The gap between the reality beyond the edge of knowledge and the existential reality that engulfs us may be narrowed when knowledge of abstract reality gives rise to technologies that serve ordinary everyday interests such as eating, keeping warm, communicating, and so on. But it is never closed.

Notwithstanding the distinction between objective (elusive) and existential (inescapable) reality, it is important to see what they have in common. The elusive reality that lies beyond the edge of our knowledge and the inescapable reality that we suffer and enjoy are both aspects of what-is made explicit. To be real, what-is has in some sense to be actually or possibly present – either experienced directly or via the mediation of discourse. Hence the interactive, or contaminated, nature of reality. And hence also the tension in trying to understand reality as something-in-itself. The transition from what-is to (for example) "matter" or "mass-energy", intuited as the asymptote of our scientific understanding, and from what-is to the mattering that engulfs us from the day when we first become conscious converge in the preoccupation of this book.

We are aware of having limited knowledge of what is around us. Without having hidden depths, that which is revealed to us does not have reality understood as independence of that which is perceived, detected, known. The visible is always of something that is otherwise invisible. If the entire universe – inside and out, over here, over there, and elsewhere – were visible its contents would have to be transparent so that nothing was hidden by anything else. But a totally transparent universe would be invisible: we would see *through* everything and hence see nothing. The gaze would have no point of arrival: it would have no *object*. The very fact that perception is *of* such-and-such implies that it falls short of complete revelation. Likewise complete epistemic transparency would leave the mind no point of arrival, nowhere to alight. Knowledge, passing through everything, would not know anything. In short, that to which our knowledge is applied is always more than what is revealed by the means by which we acquire knowledge.

It will be evident that the idea of mind-independent what-is must not be confused with that of mind-independent *reality*. Reality is a feature of a *presence* that can never be utterly mind-independent, howsoever remote this presence may be from direct sense experience. It is ultimately *for* and *to* someone; it is what-is manifested. The scientific portrait of what-is, displaying it in, for example, the form of equations may seem to be closer to a mind-independent reality, since it is not mediated by individual sensoria or minds. It is, however, mediated by an intersubjective, collective consciousness expressed in an evolving discipline.

Sense-making has to have an *object* that is intuited as other than, as a constraint upon, the sense that is made. The knowledge relation – like the perceptual relation – requires two partners: the (embodied) I and the other that is presented first, and last, in the form of material bodies and events that happen in them. Material bodies must remain to some degree senseless. To perceive X, to know X, to make sense of X, is to perceive, know, or make sense of something that is other than perception, knowledge, or sense-making. The "of" in perception of etc. marks the necessary distance between the conscious, knowing subject, and that of which it is conscious, the object of its knowledge.[71]

It may seem paradoxical that we can be aware of that of which we are unaware, entertain known unknowns; that we perceive, of our objects of perception, that they lie beyond our perception. It seems less paradoxical when we recall that we are aware, even at the epistemic ground floor, of limitations to our awareness: our perceptions of an object are mediated through different senses and each sense knows that it gets only part of the story. I see something that I can also feel, and I am consequently aware that what I see is not the whole story: seeing falls short of touch and vice versa. And it is true that our sensory field has an explicit horizon, most obviously in the case of vision. This may be the surface of a deeper contradiction at work throughout the hunt for reality in this chapter: that it is the endeavour to transcend the processes by which what-is becomes explicit to us, make those processes explicit, and somehow get behind them.

We have touched on the idea of points of convergence between metaphysically elusive and existentially inescapable reality. As we discussed in Chapter 5, it is our ambivalent relationship with our own body – something that we are forced to be (or to "am"), that we embrace as ourselves, and yet as something that is other than us – that confers the sense of the objective reality of our body, and that of the countless other objects with which we engage in our moment-to-moment lives. The intuition that the experienced object transcends, is more than, is independent of, my experience of it is central to my sense of its reality, itself drawing on my sense of the reality of my body – its lived, experienced, existential reality. And, less intimately, the incomplete scientific and the existential sense of reality converge in the science-informed artefactscape in which we pass our daily lives, illustrated in the quantum-based smartphones which we use to advance our daily relationships with each other.

The intuition of the independent reality of what-is (out there) has many connections with other intuitions; that, for example, an object will be there when I return to the location where I have found it, and that others will see it and, if necessary, report it as present. Foremost among such objects are those that knit together into familiar territory (landscapes, townscapes, and artefactscapes) and those that we possess and rely upon to support our lives – objects whose disappearance would require an explanation and, if necessary, a criminal investigation. This is not to deny the fact that they have an intermittent existence in our lives. Close to 100 per cent of the city I am currently writing in is not present to me. I am assuming, however, that it could be present to me, that much of it is present to others who are differently located, and that other parts of it would be present to me if I changed my location. The justified robustness of these and other assumptions of the folk metaphysics of daily life make philosophical scepticism seem insincere.

At the heart of the discussion of "reality" there is not only the problematic relationship between "what-really-is" and what is lived, but also the equally problematic relationship between the epistemological question of what could be known, in the light of what we know about our means of acquiring knowledge, and the ontological question of what there is in itself, beyond its status as an object of knowledge. A failure to think at sufficient depth about this relationship underlies some major ambiguities in philosophy.

Perhaps the most spectacular example of such an ambiguity is to be found in Kant's first *Critique*. His acknowledgement of the fact that what we know is mediated through our means of knowing leads him to the conclusion that our knowledge is bounded by insuperable limitations. He then endeavours to set out in what respect our knowledge is limited. It is constrained by the forms of sensible intuition (space and time) and the categories of the understanding (such as causation) imposed by our minds on sense experience. That he is able to spell these out suggests that he has somehow found a means of getting outside of the limits of knowledge to see what those limits are – which seems, perhaps, contradictory.

He then, at least according to one interpretation, goes a little further, drawing ontological conclusions from his epistemological inquiry, arguing that there is a realm of things in themselves, the noumenal realm, which is ultimately responsible for the experiences that we have. And he back-pedals by asserting that, concerning the noumenal realm, there is nothing to be known or indeed said. It is intelligible but not sensible. This does not, however, cancel the claims he has made which seem to exceed the limits he has set himself. To go beyond the position that our knowledge is constrained by mind-structured experiences to the claim that there is something in-itself that accounts for those experiences, which is quite unlike them, is to wobble between epistemology and ontology that has left scholars quarrelling – exemplified in the stalemate, discussed earlier, between advocates of two-aspect and two-world interpretations of Kant's vision as set out in the first *Critique*.

Our profound sense that there is more to the world than we sense, know, or think – to what-is than what is evident in that-it-is, to the things of which we are aware beyond our awareness of them – has fueled the various journeys outlined in this chapter. The intuition of the elusiveness of what-is is at bottom the result of our having to try to reach back across the transition to that-it-is, using instruments forged in that realm – most notably language.

This notwithstanding, for some there is a vehicle to take us closer to the end of the journey back to what-is: natural science, specifically fundamental physics. This is a view, an attitude, semi-seriously summarized in the claim by physicist and philosopher Sean Carroll that "reality is a vector in Hilbert space evolving according to the Schrödinger equation".[72] This is an extreme manifestation of the belief that what-is somehow boils down to (almost featureless) "how much" and that measurement gets the measure of all things, of what reality really is. And we have seen where this leads us: to the evaporation of what-is.

Coda: (In) conclusion

In the Preface to this book, I described its aim in two ways: to make explicitness explicit; or, less ambitiously, to make explicitness more explicit. I am confident that I have not fully succeeded in realizing the first ambition. Whether, or to what extent, I have delivered on the second the reader who has gone this far can now judge.

I began with a direct attempt to capture explicitness, teasing out its many aspects and dimensions. While I hope this was helpful, I fear that it may not have been very illuminating, even less inspiring. Unpacking "explicitness" into its semantic neighbours may have served only to demonstrate how slippery it is and how it forces anyone attempting to dive into its depths to float towards the surface. This may be a question of like forces repelling one another: explicitness resisting making itself explicit. Or, less fancifully, of the impossibility of using words to take us to a place beyond words – and indeed beyond perceptual awareness – whence we can adopt an outside view on explicitness. Hence the indirect approach adopted in the chapters that followed, where I have examined how contemporary approaches to philosophical problems have so often overlooked or sidelined explicitness and have for this reason been seriously point-missing.

Chapter 1 unpacked the seemingly mundane revelation of an individual looking at himself in a mirror. The physical reflections prompted philosophical reflections on what he sees: the unholy trinity of the thing or the Blob, of the organism or the Beast, and of the person or the Bloke who makes the other two, and himself, explicit. This three-in-one highlighted the gap between the material world and conscious subject, a gap that becomes an explanatory gap. The gap becomes even more problematic in the light of the fact that, by one account considered by many to be the most authoritative – that of contemporary physics – the Blob or the thing is emptied of its common-sense properties, even solidity.

Chapter 2 examined unsuccessful endeavours to explain, explain away, or narrow, the gap between the thing and the person, by invoking the concept of "emergence". "Emergence" acknowledges that something new has arisen and tries to account for it in terms of the laws and forces evident in the old. It fails in both its "weak" and "strong" forms to close or enclose the explanatory gap between what-is, understood as the physical world seen through the eyes of physics, and the conscious subject. While weak emergence does not require any breach of or addition to the laws of physical nature, it fails signally to give a satisfying story of the passage from material stuff to conscious subjects. Strong emergence effectively says that new things, with new powers, turn up – "just like that". In summary, emergence offers nothing to account for the arrival of sentient beings able to transform what-is into a scene, as when, after a several billion-year pause following the switching on of electromagnetic radiation, there is an awakening of a consciousness for whom that energy becomes luminosity making what-is visible.

Chapter 3 focused on the brain, an organ whose distinctive activity is supposed to make possible, even to explain, the conscious subject, notwithstanding that it is a material entity subject to the laws of physical nature. The attempt to see how the brain might be "the brewery of explicitness" fails because neuroscience cannot provide a naturalistic or physicalist account of the transition from what-is to that-it-is, more specifically the origin of consciousness with its intentional relation to that of which it is conscious. It disguises this failure with "pantasies" such as paninformationalism and panpsychism. Evolution by natural selection explains neither the "Why?" or "How" of the awakening of the material world to consciousness.

Chapter 4 reflected on the most spectacular manifestation of man the creature who makes things explicit: natural science. The chapter began with reasons for challenging the claim that physics is the royal road to metaphysics. Among these reasons, four are of particular importance. Firstly, there are crises within physics, notably the failure to reconcile the most powerful theories – general relativity and quantum mechanics. Secondly, the world picture of physics appears to be incompatible with the mundane reality in which physicists live and physical experiments are performed. This is linked with the third reason: the unintelligibility of many aspects of quantum mechanics. And, fourthly, and most importantly, the reduction of what-is to how-much-it-is, in turn reduced to numbers, empties it of content and of the capacity to explain the rich variety of the experienced world. This was most clearly reflected in the proximity of fundamental physics to a portrait of the universe as a featureless mathematical structure.

In the next three chapters, the focus was on the capital of explicitness, the self. First-person being in the form of "ambodiment" – in which an entity makes itself explicit *as* itself, as "I", and its surroundings explicit as its world – was explored in Chapter 5. The central role of the incomplete and variable colonization of the "it" of the body by the "I" of the self in giving a sense of objects – including the body – as transcending our experiences and existing in themselves was highlighted. Chapter 6 examined the self in more detail, defending it against "autocides" who claim the self is an illusion, mere time slices or trickles of consciousness, a theoretical construct comparable to a "centre of gravity", or a process amounting to no-one. The connectedness of the self within and over time, and its inner and outer scaffolding that enables it to transcend change and interruption, were discussed, as was the active contribution of the self and other selves in making and curating itself. Chapter 7 discussed agency, addressing the role of the "ambodied" agent in introducing "doings" in a universe overwhelmingly composed of mere happenings. In actions, happenings are not only made explicit but requisitioned as the realization of possibilities (that exist only insofar as they are entertained by conscious subjects). The capacity of humanity to amplify its individual and collective agency through natural science is examined as a particularly spectacular expression of how human agents act upon nature from a virtual outside.

Chapters 8 and 9 engaged with (almost) pure explicitness. Chapter 8 discussed the nature of thoughts and their status as (relatively) "free-floating" explicitness, seen most strikingly in articulate propositional thoughts which assert *that* something is or (even more strikingly) is *not* the case. Chapter 9 focused on possibilities and probabilities. They point forward in time, are of a general nature, and are not part of the natural world, which is necessarily confined to actualities. The unnatural nature of possibilities and probabilities was highlighted by emphasizing the extent to which they are dependent on frameworks introduced by conscious subjects and shaped by their interests.

Chapter 10 circled round the idea of what-is conceived as "reality". It began with the traditional philosophical preoccupation with the "veil of appearance" supposedly concealing things in themselves. Much of the discussion addressed the concern that the transition to explicitness seems to require contamination of what-is in the consciousness of the observer. This opened on to questions as to whether what-is has intrinsic properties or whether they are all relational and reflected on the idea of the Given. The view of reality as necessarily concealed behind an hallucinatory wall (possibly fabricated by the brain), and hence systematically elusive, was contrasted with the acknowledgement of reality as something lived and inescapable – an existential as opposed to an epistemological or ontological understanding or reality. The chapter also investigated the connection between the idea of reality and the concepts and judgements of truth and falsehood. Assertions of truth or falsehood were identified as the most striking expressions of explicitness. Deflationary or redundancy theories of truth were exposed as particularly egregious examples of how explicitness can be overlooked.

Thus, a journey in which explicitness has been explored and, I hope, made visible, not the least by reminders of how it is so often overlooked. Explicitness has not, however, been *explained*; nor would this be expected, given that the entire journey inescapably took place on the far side of the transition from what-is to that-it-is. Notwithstanding the sometimes tetchy polemics addressed to those who overlook, marginalize, or try to explain away explicitness, the present volume has also been a celebration of ourselves as beings who make things, including themselves, explicit. To re-enchant the world, we need only to remember those selves.

It would be nice to say that I, indeed we, have reached a destination, the end of a journey that began before the beginning of the present volume. It is time to make myself explicit and leap to the front side of the page. If this *isn't* the time, then I ask forgiveness in advance from those readers who would rather wish that I would stay hidden behind the *verso.*

The voice you have heard in these pages is that of an individual in his late seventies. The explicitness I have been circling around has preoccupied me for over half a century. That preoccupation was already 20 or more years' old by the time it surfaced in print in *The Explicit Animal* in 1990.[1] Much of my writing has been explicitly or implicitly directed against accounts of our nature that would make explicitness invisible or assimilate it into various forms of what-is, most typically matter as seen through the lens of natural science.

Some of the most direct assaults on "the explicit animal" are variations on the eighteenth-century theme of *l'homme machine,* updated in the idea of man as a cerebral machine, a material object stitched into the material world. In the last century, modes of behaviourism – from the crude mechanization of human behaviour by John Watson, I. P. Pavlow, and B. F. Skinner, to more subtle functionalism, where the brain-mind has the fake interiority of a computer – endeavoured to collapse the distance between agents and the worlds in which they act and to squeeze out explicitness.

This, however, has not been the only way in which the connected, fundamental intuitions, *that* "I am", "*that* thou art" and "*that* it is" have been, by a bitter irony, suppressed by theorists of humanity. I squandered a decade or more, writing several books[2] and numerous articles, defending the self, the "I", the conscious subject, against the theorists of so-called Theory, for whom belief in the subject was a naïve (and sometimes "bourgeois-indoctrinated") failure to acknowledge that the subject was dissolved in various solvents, including the forces of history, of class-consciousness, and of The Unconscious unearthed by depth psychology. Even speech, the most powerful and versatile generator and curator of explicitness, was dissolved in language which was understood as a structure that belonged to no-one and whose manifestations were meant by no-one. Speakers and writers were no more the source of what they said, even in the sustained, articulated, output of a written text: it is, so it was claimed, language which does the talking. Hence Barthes' "The Death of the Author" according to which the author – like all subjects – is merely a conduit for the passage of the collective (un)consciousness owned by no-one.

While this endeavour to marginalize the conscious subject grabbed the headlines in the humanities for several decades, it drew upon and embraced other endeavours to deny the explicitness of the explicit animal. Late Marxism, bearing the message of the historical unconscious, Durkheim preaching the social unconscious, and Freud drowning the ego in the ocean of the instinctual unconscious (set out in the 24 volumes of his Collected Works that are products of a very high level, if misguided, consciousness), joined the scrum of the already mentioned theorists of Theory in exposing the author absent from works of literature, the speaker absent from speech, and the self absent from her world. All were discussed and taken out to be shot in my final engagement with Theory.[3]

Much of the writing that endeavoured to marginalize consciousness was a sustained act of pragmatic self-refutation. We are not conscious but, so we are reassured, we can nevertheless, courtesy of certain *maîtres à penser*, become (mysteriously) conscious of our lack of consciousness, of the reasons for this, and of the reasons we should accept those reasons.

Within philosophy as conventionally understood, there have been numerous, less radical, endeavours to marginalize consciousness. Much of what has been criticized in this book has had the (not always conscious) consequence of belittling consciousness. Paninformationalists and panpsychists seeing "information" and "consciousness" everywhere – in ships, shoes, ceiling wax, and electrons – are motivated by the desire to make consciousness less extraordinary, not the least by avoiding the difficulty of understanding how it could be particularly associated with a small subset of material entities, notably the bodies of human beings.

By this means it was divorced from its special association with *H. sapiens*, The Explicit Animal. The assault on agency, undermining the difference between happenings taking place throughout the material world, and doings that seem to happen only courtesy of the intentions of individual subjects, also marginalized the role of explicitness – notably in the form of possibilities entertained by conscious subjects, and our capacity to shape our own lives by fulfilling them.

Thus the background to the central preoccupation of this book: the need to restore explicitness to its central place in our thinking about philosophical problems and, indeed, our humanity. Arriving at, or recovering, "that" – that-it-is, that-it-is-the-case – which begins with perception, and is spectacularly developed in propositional awareness, may hardly seem to amount to a conclusion, even less adequate reward for getting to the end of a book. It does not even suggest a beginning. A beginning at least has ideas as to where it might go, what direction it might take, what destination it might seek, set out for, or point in the direction of. The prepositions stranded in post-position in the preceding sentence betray that the beginning only sends out feelers towards a goal of travel but does not have sight, even less grasp, of it.

The failure to arrive at an end should perhaps have been anticipated because we did not really begin at the beginning. Philosophy, contrary to what philosophers sometimes think, always begins *in medias res.* The astonishment or anguish or puzzlement with which they address the mysteries that demand their attention have been framed in advance – often explicitly by quotable philosophical predecessors or implicitly in the assumptions of what has become a discipline and more broadly in the discourse which they employ, and which employs them, when they descend from wonder to inquiry, from mystery to puzzles. This is of particular relevance when our concern is explicitness. At the most fundamental level, our thoughts begin at a point many iterations of the transition away from what-is to that-it-is to wondering what-is is from the standpoint of that-it-is. "Thatter" is many-layered and the "thatosphere" has many mansions.

Often an ending can pass itself off as a point of arrival, the completion of a journey. There is the satisfaction of completing a work, even of adding to the *oeuvre.* The consolations of philosophy may not, however, be as profound as one might hope. Discovering a new argument, defeating rival theorists, solving problems of largely technical interest may deliver disproportionate satisfaction. All of which make it rather too easy for the author of *Circling Round Explicitness* to experience an insufficient level of disappointment when he abandons the hope of understanding the transition from what-is to that-it-is.

There is something even more subversive of philosophical seriousness than the consolations just identified. It is a dynamism of consciousness that keeps moving on to the next thought, the next topic, the next project – even when life does not intervene with "events", with things rightly demanding non-philosophical attention, or the habit of yielding to, or seeking out, distractions. Ideas and counter-ideas, questions and the struggle to answer them, too often crystallize into token thoughts that take their place in a lightly curated stream of consciousness and are swept aside by calls from within and from without. "The long-legged fly" referred to in the Preface rarely if ever stands still, its mind on silence. Thoughts are inevitably part of the current of noise, the unstoppable torrent, that endlessly interrupts the silence it seeks in which to make explicitness, and hence itself, explicit. This onward propulsion explains why thoughts often seem deeper when they are in the process of being discovered rather than when they are being reported, connected with other ideas, gathered into a Topic.

Just now I mentioned the joy of completion as one of the all too available sources of consolation for failure to reach a real conclusion to a question, true understanding or, more narrowly, satisfactory explanation. En route to that (in)conclusion there are many paths already beaten by others. With the scaffolding of a certain form, of chapter headings, of footnotes, of all the habits of the scholar entering a conversation with other scholars, it is too easy to be fluent, even when what one has to say is a long way from the place where the question turns the tables and questions the questioner. Instead of thinking about the world, feeling frustration at the failure to enclose the totality by which she is enclosed, the philosopher is absorbed in the work-in-progress, that takes up and develops a position on a menu of problems formulated by philosophy-as-a-discipline remote from the joyful, or anguished, astonishment appropriate for philosophy as a calling. Philosophers – who are not philosophers for most of the time, even when they are writing their books – are in this respect no different from religious believers who are only intermittently engaged with the deity even when they are in church.

And as one book grows towards completion, and the spirit of enquiry is displaced by the commitment to finishing off and tidying up, there are stirrings of infidelity. Unfinished business becomes the seed of the next book that will attempt to finish it. Completion proves to be an awakening from a dream of something conclusive, however ill-defined that something was. The dream is illuminated by the hope of escaping from a parish of consciousness – defined by accidents of birth, of space and time, by language, by history, by culture – from the many layers that seal oneself off from the totality and, incidentally, make one's continuous importance to one's self both absurd and inescapable. And so, the dream wakes up again and the journey begins towards another line in a bibliography, a flickering vision barnacled with footnotes.

And thoughts can awaken of a different kind of philosophical work that is not an argumentative journey to a conclusion in some respects as featureless as the world picture of fundamental science. It might correspond to Francis Bacon's vision of an aphoristic work that offered "a doctrine of scattered occasions". Bacon, so R. H. Stephenson tells us, "called on his fellow scholars to adopt aphoristic thought in order to promote active, dynamic cogitation in an endeavour to counteract the effects of monolithic systematizations of Scholasticism".[4] Bacon thought "that knowledge is uttered to men as if everything were finished ... whereas antiquity used to deliver the knowledge which the mind of man had gathered in observations, aphorisms, or short and dispersed sentences ... which did invite men, both to ponder that which was invented, and to add and supply further".[5]

So much for the form of the work. As for content, the emphasis might not be on universal explanation, description, or even revision, but tilt further towards celebration through occasional essays or verses unpacking all that is explicit and implicit, all that is "holy", in certain, "minute particulars"[6] and intensifying the wattage of awareness, perhaps in the hope that, notwithstanding their scattered occasions, they would add up to at least sketches towards a portrait of what-is.

And so, we enter the genre wars and the dream of The Book in which the possibilities of philosophy, poetry, fiction, and occasional essays, converge. But that is a story for another day – or another book. Instead, let me return to something nearer to the preoccupation of the present book.

The slithering of the Overture may suggest that the advice, supposedly given to a traveller asking for directions, that "I wouldn't start from here" may have a certain wisdom. Similar wisdom is dispensed by Wittgenstein: "We have got on to slippery ice where there is no friction, and so in a certain sense the conditions are ideal, but also, just because of that, we are unable to walk. We want to walk so we need friction. Back to the rough ground!"[7] To some extent, I have in the body of this book found the necessary rough ground, outside the skating rink of the Overture, in arguments with other philosophers who I believe have overlooked explicitness. But after a long pilgrimage of disagreement, I have still not found anything to fill the space created by the ejection of those philosophers whom I deem to have missed the point of all points. The ground, it seems, needs to be rougher still – but not, perhaps, with the roughness that is entirely defined by furrowed brows.

Part of, perhaps the root of, the problem is that words allow us to be ahead of, beyond, wider than, ourselves. We can enclose totalities that we cannot encircle with our imagination, even less our lives. Our thoughts use a currency remote from the gold standard of lived, or liveable, experience.[8] The epistemic *chutzpah* of making cities, countries, planets, the subjects of sentences, items to which we attach predicates is extraordinary. We can talk about the sum total of what-is, having named it "the universe", and we can even talk about something that, by definition, is more than whatever is, namely infinity. We require more rough ground so we can try to catch up with what we can say, what our words allow us to articulate, or to refer to too easily. Hence the need to dismount from philosophy and embrace loving descriptions of minute and not-so-minute particulars in poetry and essayistic tours of the surfaces and depths of things.

This becomes more pressing when we endeavour to grasp our location in the vast realm dimly illuminated by what we can say. The mundane experience – looking into a mirror – with which this book began is anything but mundane. The mirrored light by which your author was enabled to see himself as an unholy trinity had travelled trillions of miles over billions of years while those miles unpacked themselves from a near-dimensionless point to enable that unimaginable small part of the universe (literally) to face itself. That Raymond Tallis should have begun at a particular time is difficult for Raymond Tallis to grasp except as a mere matter of objective fact. I can say it but I cannot *realize* it, cannot truly encircle the truth that I entered the universe at a point in time after so much unchronicled history has passed.

While I know objectively that the universe existed long before we humans were aware of it, before a minute part of it became "our world", I cannot imagine the endless centuries – lit by the retrospective virtual gaze of factual knowledge – of our absence. Yes, it is easy to say that the universe has managed without us for all but the thinnest sliver of its existence. And it is easy to acknowledge that the stuff of life into which we were knitted into the possibility of post-natal lives (with the vital assistance of our organic landladies, our mothers) waited until there had passed 13.7 billion years after Nothing exploded into Something, 4.5 billion years after the Earth peeled off from the rest of the material world, around 4 billion years after life began in the modest form of single cells, and 200,000 years after so-called modern humanity first walked the earth. I can rehearse these facts but, because everything prior to my flash of sentience – made available to me courtesy of vast stretches of time being curated by the mouths, pens, and relics created by others – I cannot truly think them.

One thing is sure. We are not just pieces of nature – which for the most part does not know that it exists. Nor are we supernatural creatures, protected from being swept away by the flow of the natural world that swept me into being. As explicit animals, we have distanced ourselves ever further from nature; but the Bloke cannot, ultimately, separate himself from the Beast or the Blob; nor can he avoid the fate that befalls all beasts and all blobs. We cannot deny that the purposes of our many-purposed lives will end in extinction rather than fulfilment. Such fulfilment as there is will not, alas, reach a place where all our purposes converge.

If this sounds a bit glum, it is in part because I am unsatisfied at having failed to explain explicitness and its many manifestations in consciousness, self-consciousness, a coherent self, free will, the community of minds and most features of our human world. Naturalism unpacked by the natural sciences falls before the first hurdle: it cannot account for the very ground on which it stands. Supernatural explanations, however, simply parcel up our incomprehension, often in the notion of an entity – God – that is not only unexplained but riddled with contradictions and often unintelligible.

As we have circled round explicitness, *it* has been circling round us in the form of the collective consciousness gathered in our language. And if we have been denied any sense of arrival, it is (perhaps) because we had landed on our destination before we had set off on our journey, so had nowhere to go.

Notes

PREFACE

1. See, for example, Krauss, *A Universe from Nothing*. Less controversially, for some, is the claim that there are examples of low-energy quantum states which can be combined to produce an energy total greater than the sum of their parts; see Podovic-Callaghan, "Quantum 'super behaviour' could create energy seemingly from nothing".
2. Cited in Goff, *Why? The Purpose of the Universe*. This figure has been challenged by several equally well qualified authorities; Adams, "The degree of fine-tuning in our universe – and others" has suggested that "the universe is surprisingly resilient in its fundamental parameters" and "it is not fully optimized for the emergence of life". The chances of a biophilic universe are still however, vanishingly small.
3. Laing, *The Politics of Experience*.
4. Yeats, "Long-Legged Fly".
5. The distinction between mysteries and problems, highlighted by Gabriel Marcel, was central to his thought.
6. Heidegger, "What is metaphysics?", 336.
7. I am grateful to the author of *The Explicit Animal* for permission to borrow this epithet.

OVERTURE

1. Online etymological dictionary.
2. A formulation particularly associated with Heidegger, in whose writings it is denoted by the term *Dasein* or *Da-sein*.
3. For a discussion of "full-blown" and shared or communal intentionality, see the "Overture" to Tallis, *Freedom: An Impossible Reality*.

4. Thagard, "When did consciousness begin?".
5. For a discussion of other reasons why the transition is incompletable, see Tallis, *Logos: The Mystery of How We Make Sense of the World*, ch. 8.
6. See Heidegger, *Being and Time*. Explicitness encompasses what Heidegger designates in both "the present-at-hand" and "the ready-to-hand".
7. Miriam-Webster Dictionary. This seems to be the most salient of the half-a-dozen definitions of identity that it offers.
8. Sartre, *Being and Nothingness: An Essay on Phenomenal Ontology*.
9. There are circumstances in which objects can seem to have merely predicative status. In the indicative sentence "This is a cat", the indexical "this" displaces its referent-of-that-moment to make that referent a predicate. That, however, is a matter of grammar not of ontology.
10. Kant, *Critique of Pure Reason*, A 599/B627.
11. *Ibid.*
12. See Meillassoux, *After Finitude: An Essay on the Necessity of Contingency*. Meillassoux's discussion of what he calls "ancestrals" – what-is prior to the emergence of human or any other kind of consciousness – is illuminating.
13. The central role of the act of pointing in building a human world distant from the mere *umwelt* of non-human animals is explored in Tallis, *Michelangelo's Finger: An Exploration of Everyday Transcendence*.
14. A formulation borrowed from Dreben, "Putnam, Quine – and the facts".
15. Hoffman, *The Case Against Reality: How Evolution Hid the Truth from Our Eyes*.
16. See Tallis, *Logos*, ch. 3.
17. Sellars, "Empiricism and the Philosophy of Mind".
18. Watkins, "Kant, Sellars, and the myth of the given", 312.
19. *Ibid.*, 313.
20. There are living creatures who are not persons and yet seem to make other entities explicit as the contents of their environment. How much "that-it-is" is available to non-human living animals, how world-rich or "world-poor" they are (to use Heidegger's phrase) are matters for debate. Equally debatable is how satisfactorily the question can be resolved by observation of behaviour. This notwithstanding, there seem to be good reasons for concluding from their mode of living that non-human animals – even our nearest primate kin – do not swim in as deep and wide a realm of explicitness as we do. I have addressed this most recently in *Seeing Ourselves: Reclaiming Humanity from God and Science*; see especially chs 2 and 3. For a discussion of the status of animals in Heidegger's philosophy, see Tonner, "Are animals poor in the world? A critique of Heidegger's anthropocentrism".
21. A journey nicely captured in Quine's title for his final book: *From Stimulus to Science*.
22. See Tallis, *Seeing Ourselves*, ch. 4, §4.4.
23. *Ibid.*
24. George Musser has talked about "the hard problem of matter" – with a sideways glance at David Chalmers' "the hard problem of consciousness"; see Musser, *Putting Ourselves Back in the Equations*, 21.
25. For a discussion of post-tensed time, see Tallis, *Of Time and Lamentation*.
26. Wittgenstein, *Tractatus Logico-Philosophicus*, 7e.

27. *Ibid.*
28. *Ibid.* This paves the way to the now universally rejected picture theory of propositions, according to which an array of terms depicts a state of affairs understood as an arrangement of things. The most radical rejection was by Wittgenstein himself.
29. It is addressed in many places in the text that follows but especially in Section 3.5. We shall also discuss relationship between facts and thought – propositional awareness – in Chapter 8. A similar slither between what-is and that-it-is is seen in this famous passage from David Lewis: "[A]ll there is to the world is a vast mosaic of local matters of fact, just one little thing and then another" (Lewis, *Philosophical Papers Volume 2*, x).
30. Wittgenstein, *Tractatus*, 5.6.
31. Russell, "Introduction", *Tractatus*, xi.
32. See Searle, *The Construction of Social Reality.*
33. Wheeler, "Law without law".
34. Frege, "The thought: a logical inquiry", 37.
35. *Ibid.*, 38.
36. See, for example, Tallis, *Freedom* and the discussion by David Scott of the treatment in this book of de re necessity: "Disarming causation in the service of agency: Tallis on Hume".
37. See Tallis, *The Enduring Significance of Parmenides.*
38. See Reicher, "Non-existent objects".
39. As was intended by Thomas Nagel, who introduced the notion in his classic *The View from Nowhere.*
40. Rovelli, *Helgoland: Making Sense of the Quantum Revolution*, 69.
41. Borges' pessimistic description of the aesthetic experience is given in "The wall and the books".
42. Sartre, *Being and Nothingness*. Heidegger dallied with this idea or intuition when, in "What is metaphysics?" he pronounced that "The Nothing itself noths". For a brave attempt to make sense of this, see Inwood, "Does the Nothing Noth?".

CHAPTER 1 LOOKING IN THE MIRROR: A VISION OF THE UNHOLY TRINITY

1. See Tallis, *Of Time and Lamentation: Reflections on Transience*, §6.2.1.2.
2. Sellars, "The myth of the given".
3. Eddington, *The Nature of the Physical World*, xii.
4. Stebbing, *Philosophy and the Physicists.*
5. See, for example, Ladyman & Ross, *Every Thing Must Go: Metaphysics Naturalized.*
6. Wigner, "Remarks on the mind–body question".
7. Cited in Davies & Brown, *The Ghost in the Atom*, 26.
8. Suddendorf, *The Gap: The Science of What Separates Us from Other Animals*, 1. I have quoted this passage from Suddendorf before – in *Seeing Ourselves* – but, as I cannot improve on it, I quote it again. It beautifully traces hundreds of millions of the years of the unconscious, involuntary journey that led from the Blob to the Beast in the mirror and is a reminder of the complexity of the kit incorporated into its make-up.

9. Varela & Maturana, *Autopoiesis and Cognition: The Realization of the Living.*
10. Houseman, "A Shropshire Lad".
11. Lewis Wolpert quoted in Hopwood, "Not birth, marriage or death, but gastrulation: the life of a quotation in biology", 1.
12. Quoted in Godfrey-Smith, *Other Minds: The Octopus and the Evolution of Intelligent Life*, 37.
13. Most recently in *Seeing Ourselves*, ch. 2 and most extensively in Tallis, *Aping Mankind: Neuromania, Darwinitis, and the Misrepresentation of Humanity*. As for how we came to be so different, a trilogy beginning with Tallis, *The Hand: A Philosophical Inquiry into Human Being* offers a Just-So story. The story draws on the unique features of the human hand, liberated by the upright position, and the hegemony of the visual sense, to account biologically for our partial escape from biology.
14. The ascription to Augustine has been challenged. It is not to be found in any of his collected works.
15. This may set alarm bells ringing. It sounds as if I believe in a human essence. Well, I do believe in this; but I am aware of the dangers this may present. If we identify, or think we identify, fundamental characteristics of what it takes to be human, then some human beings may be judged to have failed to make the cut with the hideous consequences this can have for the way they are treated by those who consider themselves to be both human and able to define humanity. History has made us all too aware what happens when humanism asserts the rights of man and then defines "man" in such a way that some are left out, are denied those rights, and left only with duties.

    This can be avoided very simply by defining a human being as someone both of whose parents are biological humans. As for the characteristics that I shall mention as being characteristics of humans, they are common to and in all human groups. However, some individuals may not have these characteristics, but they are still entitled to be treated with humanity, and respect, and in all the ways those who do possess them would expect to be treated. It is central to our own dignity and standing as human beings that we treat all human beings with dignity. What is offered in these pages is a metaphysics, not a politics, a sociology or even an anthropology, of humanity. (I have discussed this in more detail in the opening chapter of *Seeing Ourselves*).
16. Gallup, "Self-recognition in primates: a comparative approach to the bidirectional properties of consciousness".
17. See Tallis, *The Hand*.
18. For a start on this subject, see, for example, Tallis, *Aping Mankind* and *Seeing Ourselves*.
19. I say this notwithstanding a recent paper suggesting that scans may be able to read thoughts. You may want to consult Tang *et al.*, "Semantic reconstruction of continuous language from non-invasive brain recordings". I hope you are unpersuaded that scans make thoughts visible.
20. Grosseteste, *On Light or the Beginning of Forms*, 10.

CHAPTER 2 FROM THINGS TO PERSONS: "EMERGENCE" AS A NON-EXPLANATION

1. For a luminous discussion of this issue, see Wasserman, "Material constitution".
2. Levine, "Materialism and qualia: the explanatory gap".
3. Crane, "Cosmic hermeneutics vs emergence: the challenge of the explanatory gap", 22.
4. As opposed to "ontological surprise". This is how Hans Jonas characterized life in *The Phenomenon of Life: Towards a Philosophical Biology*. (I am grateful to Simon Bowes for drawing my attention to this phrase.)
5. Wilson, *Metaphysical Emergence*.
6. O'Connor, "Emergent properties".
7. Chalmers, "Strong and weak emergence", 251.
8. James, *The Principles of Psychology*, quoted as the epigraph to Godfrey-Smith, *Other Minds*, 37.
9. Democritus 689B (Diels & Kranz).
10. Feynman, "Atoms in motion" (emphasis in original).
11. Paul Dirac quoted in Simoes, "Dirac's Claim and the Chemists", 255.
12. Ladyman & Ross, *Every Thing Must Go*, 20.
13. Close, *Elusive: How Peter Higgs Solved the Mystery of Mass*, 239.
14. *Ibid.*, 242.
15. *Ibid.*, 245.
16. *Ibid.*, 241.
17. Macdonald, "No Big Bang: quantum equation predicts that the universe has no beginning".
18. Kumar, *Quantum: Einstein, Bohr and the Great Debate about the Nature of Reality*, 321. Kumar is citing Heisenberg, *Physics and Philosophy*, 117.
19. Hossenfelder, "The problem with quantum measurement".
20. Takho, "Reality has no ultimate building blocks. There may be no fundamental level of reality".
21. Ladyman & Ross, *Every Thing Must Go*, 20. This thesis is flagged up in the title; see especially 148.
22. This is widely quoted. My source is Kirby, "Here's what Neils Bohr says about 'real'".
23. The evocative title of the series of interviews with physicists compiled and edited by P. C. W. Davies and J. R. Brown.
24. It is worth noting that this position is different from the panpsychism that we shall discuss in the next chapter. The consciousness in the atomic ghost is not sealed into individual atoms but belongs to an observer interacting macroscopically with macroscopic equipment.
25. Van Inwagen, *Material Beings*. This startling conclusion is vulnerable for at least two reasons additional to the fact that it is counter-intuitive and counter-experiential, undermining the reality not only of trees, mountains, and buildings but also of human bodies such as those of Peter van Inwagen. Firstly, "persons" are ill-defined. Secondly, as we have seen, it is no longer clear – if it ever was – what counts as an atom.
26. Ladyman & Ross, *Every Thing Must Go*, 20.
27. Cameron, quoted in Takho, "Fundamentality".

28. Ellis & Silk, "Defend the integrity of physics".
29. A similar point is made by Atmanspacher & Graben, "Contextual emergence". And Feynman also pointed out that "Today we cannot see whether Schrodinger's equation contains frogs, musical composers or morality, or whether it does not" ("The flow of wet water"). And in the most widely read passage predicting a Theory of Everything in his *A Brief History of Time*, Stephen Hawking admits that even when that theory had been arrived at, something would be missing. He described that something as "that which would breathe fire into the equations". It seems to me that something more fundamental than fire would be missing: something to breathe stuff, content, variety – that is, actual existence, into the equations.
30. Shrader-Frechette, "Atomism in crisis: an analysis of the current high energy paradigm". The claim is challenged by Weingard, "Do virtual particles exist?".
31. O'Callaghan, "The wonder particle".
32. See, for example, Wu "Everett's Theory of the Universal Wave Function".
33. The journey to, and arguments for, this increasingly popular view are lucidly set out in Pas, "Reality reconstructed". See also Susskind, "The world as a hologram".
34. See Thalos, *Without Hierarchy: The Scale Freedom of the Universe.*
35. See, for example, Wolchover, "A deepening crisis forces physicists to rethink structure of nature's laws".
36. Adlam, "Fundamental?", 5.
37. Wright, "Rival explanations".
38. Searle, *The Mystery of Consciousness*, 8.
39. As argued compellingly in Thalos, *Without Hierarchy.*
40. The thesis is set out in Scheck, *Idealizations in Physics* and beautifully summarized in "The truths in physics are dependent on falsehoods". Another fundamental idealization is one discussed in Tallis, *Freedom*, 36–42: it is the *ceteris paribus* constraint that separates signals from noise, keeping "intrusive" variables at bay by controlling them.
41. Van Inwagen, *Material Beings.*
42. Nature herself does not recognize the categories of signal and noise. She ain't a snob.
43. Harman, "I am also of the opinion that materialism must be destroyed", 774.
44. James, *The Principles of Psychology*, 158–9. This point can be generalized to a critique of the assumption that "parts" and "wholes" – central to emergentist explanations – exist as givens in themselves. They are aspects teased out by conscious subjects, being dependent on the grain of their attention.
45. *Ibid.*, 159.
46. Chalmers, "Strong and weak emergence", 244.
47. Clayton & Davies (eds), *The Re-emergence of Emergence: The Emergentist Hypothesis from Science to Religion*, 13.
48. Hasker, *The Emergent Self*, 190.

CHAPTER 3 THE RECEIVED IDEA OF THE BRAIN AS THE BREWERY OF EXPLICITNESS

1. Chalmers, *The Conscious Mind: In Search of a Fundamental Theory*, 297. Chalmers reaffirmed this distinction a decade later, when he argued that, while consciousness

is a strongly emergent property, "cognitive" and "social" phenomena are only weakly emergent. See Chalmers, "Strong and weak emergence".

2. Seth & Baynes, "Theories of consciousness", 440.
3. Chalmers, *The Character of Consciousness.*
4. Even the first and seemingly least contentious item in Chalmers' list – "the ability to discriminate, categorize, and react to environmental stimuli" – should put us on red alert. Although the notions of "stimuli", "environment" and "categories" can in some sense be used in our description of the behaviour of unconscious organisms such as bacteria, it is reasonable to suspect that these terms acquired their distinctive meaning by backward projection from organisms with experiences, and so ultimately owe their explanatory force to organisms that are conscious in the "hard" sense.
5. See Bourget & Mendelovici, "Phenomenal intentionality".
6. See Tallis, *Freedom*, especially chapter 4.
7. If we extend the idea of intelligence to anything that shapes adaptive behaviour and which consequently need not be associated with phenomenal consciousness, we could also widen it to encompass processes seen inside an individual cell. And this would ensure that the fundamental nature of consciousness is overlooked.
8. Klein, Nguyen & Zhang, "Going out of my head: an evolutionary proposal concerning the 'why' of sentience", 133.
9. See Tallis, "The illusion of illusionism" for a refutation of a particularly desperate attempt to deal with the problem of phenomenal consciousness by claiming that it is an illusion. Of course, the illusion of consciousness is no easier to account for in materialist terms than the reality of consciousness. In fact, it seems more resistant to reduction to neural activity.
10. Crick, *The Astonishing Hypothesis: The Scientific Search for the Soul.*
11. Van Gulick, "Consciousness".
12. An excellent introduction to the philosophically relevant neuroscience of consciousness is Wu, "The neuroscience of consciousness". Even more impressive is Kuhn, "A landscape of consciousness: towards a taxonomy of explanations and implications".
13. Arulshah *et al.*, "Elucidating the visual phenomena in epilepsy: a mini review". Psycho-active drugs that profoundly affect conscious states are associated with major impacts on the electroencephalogram. See, for example, Smausz, Neil & Grigg, "Neural mechanisms underlying psilocybin's therapeutic potential – the need for preclinical in vivo electrophysiology".
14. Penfield & Perot, "The brain's record of auditory and visual experience".
15. See Bickle & Ellis, "Phenomenology and cortical microstimulation".
16. Monti *et al.*, "Willful modulation of brain activity in disorders of consciousness". An even more impressive example of the close correlation between neural activity and conscious experiences has recently been reported by Costandi, "AI reads brain activity to tell what part of a movie you are watching".
17. Most recently in Tallis, *Seeing Ourselves* and *Freedom.*
18. According to Giulio Tononi and others, his integrated information theory can explain this anomaly: the cerebral cortex, unlike the cerebellum, he argues, is characterized by a coexistence of both functional specialization and integration; see Tononi, Massimini & Koch, "Integrated Information theory: from consciousness to its physical substrate". As we shall discuss later in the chapter, it is not at all clear what Tononi's theory

means, but it appears to depend on a "paninformationalism" that requires the brain to do relatively little work in securing the transition from what-is to that-it-is.

19. James, *The Principles of Psychology Volume 1*, 158–9.
20. Since Hume, the causal link seems to explain less and less, whether it is conceived as "oomph" whereby one event brings about another or as glue/cement that makes the universe materially coherent. For a critical discussion of causation as material necessitation, see Tallis, *Freedom*, ch. 3.
21. Warnock quoted in Lockwood, *Mind, Brain and the Quantum*, 299. In some respects, Warnock anticipates the so-called projection theories of consciousness, according to which "we project our neural representations into experiential physical space"; see Pereira Jr, "The projective theory of consciousness: from neuroscience to philosophical psychology".
22. Discussed in Tallis, *Freedom*, ch. 3.
23. See, for example, Mendelovici & Bourget, "Naturalizing intentionality: tracking theories versus phenomenal intentionality theories".
24. Chalmers, *The Conscious Mind*, 297.
25. David Chalmers quoted in Hiley & Pilkkanen, "Can quantum mechanics solve the hard problem of consciousness?".
26. David Chalmers, and K. J. McQueen quoted in Okon & Sebastian, "The subjective-objective collapse model", 24.
27. Quoted by John Horgan in "Why information cannot be the basis of reality".
28. Basil Hiley in Davies & Brown, *The Ghost in the Atom*, 138. This is echoed by Rovelli in *Helgoland*, where he asserts (147) that "quantum physics is the discovery that the physical world is a web of correlations: relative information". And by John Wheeler, for whom, "an electron is nothing more than our information about it" – quoted in Musser, *Putting Ourselves Back in the Equations*, 151.
29. Hiley, in Davies & Brown, 138–9.
30. Ladyman & Ross, *Every Thing Must Go*, 210. It has to be said that Ladyman and Ross seem to embrace paninformationalism rather half-heartedly.
31. The passage is quoted in Robinson, "The limits of information", 22.
32. Weaver, "Recent contributions to the mathematical theory of communication", 265.
33. See Tallis, "The fantasy of conscious machines".
34. Pagnotta, "The use and abuse of information in philosophy", 98.
35. Wellcome Centre for Human Neuroimaging, "The Bayesian brain".
36. Tononi, "Consciousness as integrated information: a provisional manifesto". There is some overlap between integrated information theory (IIT) and global workspace theory (GWT). According to the latter, information becomes conscious when it is widely broadcast in the brain. And this in turn finds echoes in Daniel Dennett's idea of consciousness as "fame in the brain" (Dennett, *Consciousness Explained*, 195), although for Dennett, consciousness is an illusion – one that we are (strangely) conscious of. Jens Zimmermann has pointed out that the notion of information being a fundamental property of physical systems with specific causal properties goes all the way back to Norbert Wiener.
37. Lenharo, "Consciousness theory slammed as 'pseudoscience' – sparking uproar".

38. Tononi & Koch, "Consciousness: here, there, and everywhere?", 11. The tendency to extend the concept of mind is illustrated by this passage from George Musser in his discussion of "the physics of the mind": "If an organism or self-sustaining nonbiological system models the world, it has a kind of mind – rudimentary, perhaps, and not always something we would recognize as a mind, but nonetheless some kind of inner life" (*Putting Ourselves Back in the Equations*, 83). This is further developed when it is suggested that Tononi's "recipe", "introduces concepts that are useful far beyond their original context, providing a way to quantify the complexity of any system, conscious or not" (*ibid.*, 90). And when it is asserted that phi is equated with "the degree of consciousness of a neural network, be it an animal brain, and AI system, or really any network at all" (*ibid.*, 91); "Both predictive coding and IIT imply that minds are everywhere ... Both theories thus embrace a version of panpsychism" (*ibid.*, 94).
39. Seth & Baynes, "Theories of consciousness", 444.
40. Lenharo, "Decades-long bet on consciousness ends – and it's philosophy 1, neuroscience 0".
41. Fleming *et al.*, "The integrated information theory of consciousness as pseudoscience". See also Lenharo, "Consciousness theory slammed as 'pseudoscience' – sparking uproar".
42. Bayne, "On the axiomatic foundations of the integrated information theory of consciousness". It is striking how much explanatory power has been ascribed to IIT. For example, it has been invoked as the cause of quantum collapse; see, for example, Kremnizer & Ranchin, "Integrated information-induced quantum collapse".
43. Edelman & Gally, "Re-entry: a key mechanism for integration of brain function".
44. Early in our discussion of paninformationalism we noted the enthusiasm with which it was embraced by some physicists – notably John Wheeler who is particularly associated with the IT=BIT or IT from BIT vision of the universe. Readers may already be persuaded of the vacuity of this idea but for those who are still unpersuaded, the "itsy-bitsy universe" is dissected in Tallis, "An itsy-bitsy universe?".
45. For an accessible account of the case for panpsychism, see Goff, *Galileo's Error: Foundations for a New Science of Consciousness*.
46. Chalmers, "Idealism and the mind–body problem", 354.
47. Goff, *Galileo's Error*, 52. There is an echo here in the thought that "Life can be known only by life, which provides us, as it were, peepholes into the inwardness of substance" (Sinnott, "Review of Hans Jonas' *The Phenomenon of Life*").
48. See Seager, "Panpsychist infusion".
49. Hiley & Pylkkänen, "Can quantum mechanics solve the hard problem of consciousness?", 433.
50. Albahari, "Panpsychism and the inner-outer gap problem", 39.
51. *Ibid.*, 29. The obstacles to the summing of micro-consciousnesses to, for example, full-blown human subjects are discussed, but not removed by, Goff, Seager & Allen-Hermanson in "Possible solutions to the subject-summing problem".
52. See Goff, Seager & Allen-Hermanson, "Possible solutions to the subject-summing problem". Equally unhelpful is the appeal to integrated information theory to "solve" the combination problem. See, for example, Morch, "Is the integrated information theory of consciousness compatible with Russellian panpsychism?". Morch argues

that only maxima of phi are conscious, so any micro-conscious ingredients will lose their consciousness. There is therefore no conflict between a macro-consciousness and the micro-conscious entities of which they are composed. For this to work as an answer to the subject-summing problem it is necessary to accept both IIT (and the paninformationalism on which it depends) and panpsychism – two pantasies, as I have argued.

Nino Kadić has suggested variations on "monadic panpsychism" according to which "all the individual microsubjects in the brain experience the same thing. This multitude is not apparent within the consciousness of the macro-subject because they are phenomenally or qualitatively identical" (Kadić, "Monadic panpsychism", 38). It is not clear how this magic identity of all the micro-subjects in the macro-object of the brain would be achieved and how it would deliver the rich and evolving variety of the experiences of conscious subjects. Kadić's appeal to "relational properties" between the micro-subjects seems, to say the least, opaque. Equally unconvincing is the suggestion that "entanglement" is the solution to the combination problem. Quantum entanglement of elementary particles hardly captures the unity of the subject – at any one time, never mind over time (just how complex this is will be discussed in Chapter 6). Finally, it is perhaps worth noting that "pancognitivist" (thought) and "panexperientialist" (consciousness) versions of panpsychism are equally unhelpful at addressing the combination and decombination problems.

53. Ganiri & Shani, "What is cosmopsychism?". The paper introduces a special issue devoted to this topic. Cosmopsychism harmonizes with the idea from physics that the universe is a single wave-function and that the universe is a self-training neural network. (See, for example, Alexander *et al.*, "The autodidactic universe"). This universe arrived at its laws and its fine-tuned constants by this means.
54. Goff, *Why? The Purpose of the Universe*, 130.
55. *Ibid.*
56. Protoconsciousness is consciousness without any feeling – presumably consciousness without what we normally associate with consciousness. Tononi who has partly embraced protopanpsychism, has, as we noted earlier, suggested that even a simple system such as a photodiode will be conscious to some degree, if it is not swallowed up within a larger system.
57. There is a striking analogy between the panpsychist's key argument that we access the intrinsic nature of things in virtue of being an example of such things and Schopenhauer's claim in *The World as Will and Idea* that we can get behind the Kantian veil of appearance in virtue of being a thing-in-itself.
58. Rostowski, "Worlds apart", 42.
59. This raises the question as to why the (mis)representation of light energy as colour experiences should be so useful, indeed be more useful than a wired response to electromagnetic radiation which is, after all, colourless.
60. For a brief summary and critique of enactivism, see Tallis, *Freedom*, appendix D, and for a critique of my critique, see Rostowski, "Freedom: an enactive possibility".
61. I owe this point to Soldati, "Representation, intentionality, and consciousness".
62. Seth & Baynes, "Theories of consciousness", 442–3.
63. Levin, "Functionalism".

64. The notion of the organism and the environment as a single coupled system is set out with exemplary clarity by Chemero, "Dynamical explanations and mental representations". It is summarized in Shapiro & Spaulding, "Embodied cognition".
65. For an exemplary discussion of the inverted spectrum thought experiment, see Horgan, "Functionalism, qualia, and the inverted spectrum".
66. Or at least in some cases. Red-green inversion would not be significant so long as there was consistency in the experience of the contrasting pair. There would be no behavioural consequences. This would not be true in the case of pain-pleasure inversion – such that putting one's hand on a hot stove caused pleasure and drinking water when thirsty caused pain – because the behavioural consequences would be disastrous. This, however, underlines how differential experiences only weakly connect with differential behaviour. There are many experiences that do not demand or prescribe particular behaviours defined in terms of their physical properties.
67. Ismael, "Why physics should care about the mind, and how to think about it without worrying about the mind–body problem", 173.
68. Sloman & Chrisley, "Virtual machines and consciousness".
69. Ismael, "Why physics should care about the mind", 175.
70. For a discussion of the functionalist tendency to ascribe "easy problem" consciousness to machines, see Tallis, "The fantasy of conscious machines".
71. Klein, Nguyen & Zhang, "Going out of my head", 4.
72. Shand, "Consciousness: removing the hardness and solving the problem", 1279.
73. *Ibid.*
74. Stan Klein is closer to the mark in Klein, Nguyen & Zhang, "Going out of my head", 2: "[C]onsciousness must, of logical necessity, be directed towards that which it is not – some 'other' that can serve as its target. To understand consciousness, one cannot transform it into an object of consciousness – that is, into a non-sentient entity suitable for conscious apprehension".
75. Chalmers, *Reality +: Virtual Worlds and the Problems of Philosophy*, 279–80.
76. James, *Principles of Psychology Volume 1*, 146. This is echoed in a recent article by Bernardo Kastrup, "Consciousness cannot have evolved". Kastrup argues, among other things, that in a causally closed quantitative world, essentially qualitative consciousness could not have any influence. Unfortunately, his other main argument in this article is that key functions of consciousness – selective attention; discriminating current perceptions from past experiences, making them feel different; motivating behaviour conducive to survival – could be just as well discharged by unconscious mechanisms. Back to the false hard/easy problem distinction.

CHAPTER 4 WHAT-IS AS HOW MUCH: CUTTING MEASUREMENT DOWN TO SIZE

1. Ladyman & Ross, *Every Thing Must Go*, 30.
2. Rovelli, *Helgoland*, 118.
3. Harman, "I am also of the opinion that materialism must be destroyed", 772.
4. Close, *Elusive*, 241.

5. Tegmark & Wheeler, "100 years of the quantum".
6. Interview with John Wheeler in Davies & Brown (eds), *The Ghost in the Atom*, 60.
7. Sean Carroll's tour de force podcast, "Solo: the crisis in physics" should be mandatory listening for anyone talking casually about "the crisis in physics".
8. From a brilliant, brief, lucid summary of the conflict between general relativity and quantum mechanics by Corey S. Powell, "Relativity versus quantum mechanics: the battle for the universe". There are, of course, countless other popular, semi-popular, and scholarly accounts of the problem and attempts to deal with it.
9. Ellis & Silk, "Scientific method: defend the integrity of physics". The fact that string theory allows for a vast number of ground states or vacua – estimated at 10^{500} or perhaps 10^{272000} is not encouraging. (I have these figures from Butterfield's "Essay review of Sabine Hossenfelder, *Lost in Math: How Beauty Leads Physics Astray*", 76.)
10. There is fascinating attempt to disentangle entanglement by ascribing the appearance of it to simple errors such as observation bias by Price & Wharton, "Untangling entanglement". It provides "a magic-free recipe for quantum entanglement".
11. Richard Feynman said this many times in many places. He also (in)famously said "I think I can safely say that nobody really understands quantum mechanics" (*The Character of Physical Laws*, 129).
12. It has been variously attributed to physicist David Mermin (by, among others, David Mermin) and to Richard Feynman (by, among others David Mermin); see Mermin's witty article, "Could Feynman have said this?".
13. Seager, "Strange trails: science to metaphysics", 466.
14. *Ibid.*, 463.
15. John von Neumann to a pupil Felix Smith. I cannot find the original source of this. It is widely quoted and, whenever it is quoted, it is ascribed to him.
16. Seager, "Strange trails", 464.
17. Horgan, "Quantum mechanics gives us power but no answers".
18. The ultimate reification is in the claim by physicist Lawrence Krauss that "'Nothing' is as every bit as physical as 'something'" (*A Universe From Nothing: Why There is Something Rather than Nothing*, xiv.
19. Hossenfelder, *Lost in Math: How Beauty Leads Physics Astray*.
20. Chen, "Realism about the wave function".
21. *Ibid.*
22. Ben-Menahem, "Dummett versus Bell on quantum mechanics", 288.
23. Hance & Hossenfelder, "What does it take to solve the measurement problem?", 1.
24. Eddington, *The Nature of the Physical World*, xv.
25. I have been unable to track down the original source for this well-known anecdote.
26. The contrast between "revisionary" and "descriptive" metaphysics is discussed by P. F. Strawson in the Introduction to *Individuals: An Essay in Descriptive Metaphysics*.
27. See Tallis, "The 'p' word" for a reflection on the fact that philosophy does not seem to make progress in the narrow sense of solving the problems philosophers set themselves.
28. Thomson, "Electrical units of measurement", 73.
29. Lewton, "How can we understand quantum reality if it is impossible to measure?"
30. Sellars, *Empiricism and the Philosophy of Mind*, 83.

31. Yeats, "Under Ben Bulben".
32. Deutsch, *The Fabric of Reality*, 142.
33. In what follows, I will touch on some territory covered in more detail in *Of Time and Lamentation*, especially chapter 3, but I hope it will take the inquiry further – and deeper.
34. Putnam, *Philosophical Papers 1: Mathematics, Matter, and Method*, 73. A striking defence of the realism of scientific theories focuses on the failures rather than the successes of science; see Ben-Menahem, "Realism and the failure of science: reversing the no miracle arguments". If science were not constrained by reality, Ben-Menahem argues, it is difficult to explain why particular theories fail. It is not clear, however, that this is decisive in supporting realistic rather than relativist, conventionalist, or instrumentalist interpretations of the theoretical coherence and practical power of science.
35. Mach, *Knowledge and Error: Sketches on the Psychology of Enquiry*, 171.
36. For a wonderful account of the rich history of measurement and its origins in trade, in social interaction, before it became centred in science, see Vincent, *Beyond Measure: The Hidden History of Measurement*.
37. Acacio de Barros & Montemayor, "Quantum mentality: panpsychism and panintentionalism", 96.
38. See Vincent, *Beyond Measure*.
39. Hossenfelder, *Existential Physics: A Scientist's Guide to Life's Biggest Questions*.
40. Born, *Einstein's Theory of Relativity*, 5.
41. See Vincent, *Beyond Measure*.
42. *Ibid*., 283.
43. Pythagoras, Plato and others – including quite a few contemporary philosophers of mathematics – may, of course, disagree.
44. As Arthur Eddington expressed it, "Intrinsically, Brobdingnag and Lilliput are precisely the same; it needs an intruding Gulliver – an extraneous standard of length – to make them appear different" (quoted in Giovanelli, "But one must not legalize the mentioned sin: phenomenological v dynamical treatment of rods and clocks in Einstein's thought", 1.
45. Einstein, "Autobiographical notes", 59. This is discussed in Giovanelli, "But one must not legalize the mentioned sin".
46. Maudlin, *The Philosophy of Physics: Space and Time*, 106.
47. Rovelli, "Consciousness is irrelevant to quantum mechanics".
48. Rovelli, *Helgoland*, 69.
49. Hossenfelder, *Existential Physics*.
50. On the issue of scales, see Tallis, *Freedom*, ch. 2. The view developed there reflects that developed in Thalos, *Without Hierarchy*.
51. The difficulty of accommodating "The Heisenberg Cut", marking the boundary in fundamental physics between the quantum world that is measured, and the classical world of the measuring device is another expression of the non-natural nature of scales.
52. This is particularly difficult for those who embrace quantum monism – the notion that there is a single universal wave function that describes the quantum state of the

universe, and that the universe as a whole is more fundamental than its parts; see, for example, Calosi, "Quantum monism: an assessment".

53. Bas van Fraassen quoted in Ladyman & Ross, *Every Thing Must Go*, 98.
54. Shapin, *The Scientific Revolution*, 62.
55. Galileo, "The assayer", 238.
56. This is what is behind Plato's part-whole paradox, according to which the same things(s) cannot be both one thing and many things. A thing is in itself neither one nor many. See, for example, *Parmenides* 129 c5–d2. The reason this issue bothered Plato may relate to the high status he ascribed to mathematics as a fundamental reality, though – as Gregor Cantor said – "a set is a Many that allows itself to be thought of as One".
57. See "Noether's theorem", Wikipedia (accessed 27 October 2023).
58. For the story of Einstein "the Parmenidean", see Tallis, *Of Time and Lamentation*, 180.
59. Heisenberg, *Physics and Philosophy: The Revolution in Modern Science*, 32.
60. Maudlin, "Three measurement problems".
61. Bitbol, "The roles ascribed to consciousness in quantum physics: a revelator of dualist (or quasi-dualist) prejudice", 268.
62. The choice of possible outcomes – dead or alive – is somewhat artificial because, from the standpoint of physics or indeed chemistry, there could be a continuum rather than a dichotomy between them. The supposedly dead cat might have a proportion of processes associated with life – for example certain metabolic or enzyme mediated events – continuing, even though it is not a going concern and from the standpoint of the owner it is an ex-cat. The reduction of outcomes to dead vs alive is thus an artefact, rather like that of the classification of all the possible outcomes of tossing a coin to heads vs tails (see Chapter 9). The dramatization may not, however, be a falsification given that any two states of the cat, for example "feeling well" and "feeling less well", are as incompatible with each other as "dead" and "alive".
63. See, for example, Gao (ed.), *Consciousness and Quantum Mechanics.*
64. This is not the place – and I most certainly am not the person – to discuss the contrast between waves and particles. It is worth noting, perhaps, that the contrast is not as well-defined as some discussions of the quantum revolution have suggested. For waves to be plural, to be countable as 1 or more than 1, they would have to have a discrete identity which means that they would have to have quasi-particulate boundaries. The hedge-betting term "wave packets" says it all. And the notion of a particle as a localized region of excitation in a field, and the idea that matter and energy are interchangeable, further highlights the fuzziness of the boundaries between the concepts. The central fuzziness is captured by d'Hooft, "Confusions regarding quantum mechanics": "Physicists realized that all oscillatory motion apparently comes in energy packets that seem to behave as particles, and that the converse also seems to be true: all particles with definite energy must be associated to [sic] waves".
65. See Davies & Brown, *The Ghost in the Atom*, 35: "If a quantum system is, say, in superposition of, say, n quantum states, then, on measurement, the universe will split into n copies. In most cases n is infinite". Whether an actual infinity can exist is another matter. As has been argued since Aristotle, actual infinities are self-contradictory because any actual quantity can always be added to and hence must fall short of being infinite.

66. Some have claimed that so-called "cold spots" in the universe may indicate the presence of parallel universes. Others have argued that we shall never obtain evidence of parallel universes, so the notion is purely speculative and has no scientific credibility; see, for example, Ellis, "Does the multiverse really exist?"
67. Zimmerman Jones, "The many worlds interpretation of quantum physics".
68. Others see this argument as an example of the reverse gambler's fallacy. See, for example, Goff, *Why?*, 30–32. I enter a gambling den and see that one gambler has accumulated a vast quantity of money as a result of an extraordinary run of luck. This run is one out of 10,000,000 possibilities and from this I conclude that there must be 9,999,999 other gamblers in the den who are not so lucky. The fallaciousness of the conclusion is clear, given that all individual runs are equally possible, equally unlikely. Of course, lucky strikes are fewer than disappointing results. Gambling dens are not charitable organizations.
69. Quoted in Kumar, *Quantum*, 262. This is the most comprehensive, comprehensible and illuminating of the many popular books on quantum mechanics that I have read. Strongly recommended to those who, like me, do not have the maths.
70. John A. Wheeler quoted in Barnes, "Retrocausality, Wheeler's delayed choice, and simulation theory reinterpreted via the participatory universe", 1. Wheeler's assertion may seem like a truism if we think of phenomena etymologically as (necessarily observed) appearances.
71. Max Jammer cited in Mermin, "The quantum measurement problem", 13.
72. Quoted in Kumar, *Quantum*, 245.
73. *Ibid.*
74. See Zyga, "Quantum-to-classical transition may be explained by fuzziness of measurement references". The focus on the measurement process – the timing or angle of that which is measured – rather than the precision of the final detection is novel.
75. Bell, "Against measurement". It is lucidly discussed in Baggott, "Thirty years of 'against measurement'".
76. Bell, *Speakable and Unspeakable in Quantum Mechanics*, 239.
77. Baggott, "Thirty years of 'against measurement'", 33.
78. *Ibid.*, 30.
79. Hance & Hossenfelder, "What does it take to solve the measurement problem?", 2.
80. This applies both to the literal, physical borders between the elements and to the Heisenberg Cut between the quantum and classical realms.
81. Close, *Elusive*, 203.
82. If this book, and in particular this chapter, were it not already too long, I would have included a discussion on "operationalism" as a philosophy of science. Its circularity is illustrated by this passage from Percy Bridgman, its founding father: "To find the length of an object, we have to perform certain physical operations ... [T]he concept of length involves nothing more than the set of operations by which length is determined. In general, we mean by any concept nothing more than a set of operations; the concept is synonymous with the corresponding set of operations" (Bridgman, *The Logic of Modern Physics*, 5). The circularity is highlighted when we ask which of the physical movements that I make when I measure the length of a table count as essential to measuring the length of a table.

83. Bell, "Against measurement", 34.
84. Struyve, "Review of Bell & Gao (eds), Quantum Nonlocality and Reality".
85. Bell, "Against measurement", 35.
86. Quoted in Kumar, *Quantum*, 219.
87. Fuchs, "Notwithstanding Bohr, the reasons for Qbism", 19. Caves, Fuchs & Schack, "Subjective probability and quantum uncertainty" develop the view that "probabilities – and thus quantum states – represent an agent's degree of belief, rather than corresponding to objective properties of the quantum system ... Quantum certainty is ... always some agent's certainty", 1.
88. Fuchs, "Notwithstanding Bohr, the reasons for Qbism", 19.
89. Healey, "Quantum-Bayesian and pragmatist views of quantum theory".
90. Chen, "Realism about the wave function".
91. This "ontological" interpretation does not seem to go very far towards reifying the wave function. A field in high dimensional space and a vector in Hilbert space seem rather ethereal.
92. Mermin, "There is no quantum measurement problem", 62.
93. *Ibid.*, 63.
94. De Finetti, *Theory of Probability: A Critical Introductory Treatment*, x.
95. Feynman, "Negative probability".
96. Dirac, "The physical interpretation of quantum mechanics".
97. Hume, "Of the idea of necessary connection" in *A Treatise of Human Nature Volume 1*.
98. Carnap, "Intellectual autobiography", 37. Einstein's emphasizing that "timing" – as opposed to time – is part of nature is necessary because the notion of the "clockwork" universe still lives on in the philosophy of science. It is echoed in George Musser's question "for what is a clock but a little chunk of the universe correlating with the rest?" (Musser, *Putting Ourselves Back in the Equations*, 210). Correlation becomes the basis for a process timing other processes only when it is picked out by a conscious subject and used by that subject to make a measurement.
99. Mermin, "The quantum measurement problem", 62.
100. Mermin, "Making better sense of quantum mechanics", 1. Mermin said that he wrote this paper – which has no practical significance – in the hope of interesting those "who are impractical enough always to have been bothered, at least a bit, by not knowing what they are talking about" (*ibid.*, 3).
101. Seager, "Strange trails", 472.
102. Wheeler, "Bohr, Einstein, and the strange lesson of the quantum".
103. Davies & Brown, *The Ghost in the Atom*, 26.
104. Dupré & Nicholson, "A manifesto for a processual philosophy of biology", 15.
105. Quoted in Kumar, *Quantum*, 262.
106. Harman, "I am also of the opinion that materialism must be destroyed".
107. *MIT Technology Review* editorial team, "A quantum experiment reveals there is no such thing as external reality".
108. Wigner, "Remarks on the mind–body question".
109. Proietti *et al.* "Experimental test of local observer-independence".
110. Kumar, *Quantum*, 277. He preferred to call himself "a mathematical physicist" rather than "a theoretical physicist".

111. A brave defence of the reality of virtual particles – or at least their claim to be equally real as perfectly respectable particles such as quarks – is Jaeger, "Are virtual particles less real?".
112. Krauss, *A Universe From Nothing.*
113. Close, *Elusive*, 59.
114. Hilary Putnam's view of scientific realism has become more complex. This is not the place to discuss it but it is worth reading Bastianelli, "Putnam's no miracles argument: a defence of scientific realism".
115. See Tallis, *Of Time and Lamentation*, ch. 3, especially §3.6.
116. Dorato, "Why are (most) laws of nature mathematical?".
117. This is discussed in more detail in Tallis, *Freedom*, ch. 2.
118. The maths is simple in the sense of boiling down to single line equations, not in the sense of being accessible to a mathematical simpleton such as the author of this book.
119. Isaacson, *Einstein: His Life and Universe.*
120. Russell, *The Analysis of Matter*, 163.
121. Davies & Brown, *The Ghost in the Atom*, 101.
122. The absence of location in mathematized reality is sometimes obscured by presenting locations as the convergence of values of co-ordinates. Those co-ordinates, however, do not amount to a place without a frame of reference that is applied from a non-mathematical outside. The translation of co-ordinate values 2,3,6 into a point in space requires such a frame of reference – imported by a conscious subject.
123. Briceno and Mumford deliver a thorough demolition of Ladyman's version of relationism in "Relations all the way down? Against ontic structural realism", 215–6.
124. As Alfred Korzybski said, "the map is not the territory" (*Science and Sanity: An Introduction to Non-Aristotelian Systems and General Semantics*).
125. Plato, *The Republic*, bk 7, §527b.
126. This is quoted in Hacking, "What mathematics has done to some and only some philosophers", 26. Just how much of a medley is highlighted by the fact that different branches of mathematics utilize different interpretations of the seemingly transparent and fundamental concept of equality. In certain contexts, "the equals sign is used to indicate that each side of an equation represents the same mathematical object". In other contexts, it is thought to mean a "canonical isomorphism" between two sets; see Wilkins, "Mathematicians cannot agree what 'equals' means, and that's a problem". Of course, the heterogeneity of science is even more striking, prompting Paul Feyerabend to comment that "people who say that it is science that determines the nature of reality assume that science speaks with a single voice. They think there is this monster, SCIENCE, and when it speaks it utters and repeats and repeats and repeats and repeats again a single coherent message. Nothing could be further from the truth. Different sciences have vastly different ideologies" (Feyerabend, *The Tyranny of Science*, 55).
127. It seems that I haven't.
128. Hossenfelder, *Lost in Math.*
129. Quoted in Davies & Brown, *The Ghost in the Atom*, 26.
130. Bitbol, "The roles ascribed to consciousness in quantum physics", 26.
131. *Ibid.*, 27.

132. Sellars, *Empiricism and the Philosophy of Mind*, 83.
133. Tegmark, "Is 'the Theory of Everything' merely the ultimate ensemble theory?"
134. *Ibid*., 1.
135. Hawking, *A Brief History of Time*.
136. Close, *Elusive*, 246.
137. Seager, "Strange trails", 471.
138. Discussed by Hossenfelder, *Lost in Math*, 64–5.
139. It is worth noting that in "The crisis in physics", Sean Carroll does concede that the rate of progress is slower than it was 100 years ago, when quantum mechanics was founded, or 50 years ago when the Standard Model of the atom was being completed. The journey was, perhaps, always headed for the brick wall against which we have been banging our heads in this chapter.
140. Vilenkin & Tegmark, "The case for parallel universes". This extraordinary example of fine-tuning prompted cosmologist Bernard Carr to remark that "if you don't want God you had better have a multiverse" (quoted in Wilkinson, "The multiverse conundrum", 38).
141. Davies & Brown, *The Ghost in the Atom*, 131.

CHAPTER 5 AMBODIMENT: THE MARRIAGE BETWEEN (THAT) IT IS AND (THAT) I AM

1. A recent collection of essays, *The Phenomenology of Embodied Subjectivity* edited by Rasmus Thybo Jensen and Dermot Moran shows just how fertile. Much of what follows in this chapter is also influenced by Jean-Paul Sartre's distinction between the body-as-object and the body-as-lived or the-body-for-itself, discussed most extensively in *Being and Nothingness*.
2. See Tallis, *The Kingdom of Infinite Space*, passim but especially chapter 14, and *Seeing Ourselves*, especially chapter 3.
3. See Tallis, "An invitation to navel gazing".
4. Descartes, *Meditations on First Philosophy*, 56.
5. I discuss what follows in greater length in Tallis, *Logos*, ch. 3. The notion that Kant overlooked the body has been challenged: Carpenter, "Kant, the body, and knowledge" has argued that for most of his career, "Kant believed that the body stands as a condition of knowledge". And he cites a passage from Kant's *Universal Natural History* that is particularly poignant in view of Kant's own dementia towards the end of his life: "The spiritual faculties disappear with the vigour of the body ... The agility of thought, the clarity of representation, the vivacity of wit, and the ability to remember lose their strength and grow frigid". Nevertheless, the body seems to be marginalized in the Kant of the *Critiques* who is such a dominating presence in Western philosophy.
6. And if, on the other hand, space and time were the result of the cooperative activity of many minds, it is difficult to see what would justify an individual subject being in a particular parish of spacetime, rather than in some unimaginable locus constructed out of the consciousness of all living subjects.

7. Unpacked at length in Merleau-Ponty, *The Phenomenology of Perception.*
8. This is an important motif in Heidegger's *Being and Time.*
9. Austin, "Other minds", 76.
10. Most fully in chapter 3 of Tallis, *Seeing Ourselves.*
11. Discussed in Cassam, *Self and World.*
12. The relationship between experience and the sense of that which is experienced as an object existing in itself will vary with the perceptual modality. Smells may hang objectless in the air. Tastes may or may not be attached to objects in the mouth: attached when the tasty object is being tasted, detached in an after-taste. Touch seems to grasp an object that has depths beyond the touched surface. This may be because the object is also experienced as having weight and resistance to movement and to compression also made evident in the hands that make it evident. Sounds may or may not be connected with objects. While a connection may be immediately made between a song and the bird producing it, even though the latter is hidden, many sounds do not reveal the objects that are their source. We wonder what that funny noise is and never track down its origin. The situation may be more complex as when, chewing food, I experience touch and taste simultaneously and refer then both to the same object that transcends the surfaces that are touched and the tastes that are experienced.
13. See Patocka, *Body, Community, Language, World.*
14. Cheng, "Sense, Space, and Self".
15. Cassam, *Self and World,* 31, cited in Cheng, "Sense, Space, and Self", 67. Was this, perhaps, hinted at by Kant when he advanced the thesis that "The mere, but empirically determined, consciousness of my body proves the existence of objects in space outside me" (*Critique of Pure Reason,* 245)?
16. Cheng, "Sense, Space, and Self", 67.
17. A more straightforward expression of the intuition that underlies the belief in the existence of parts of the body that are not being sensed – and hence making the body the primordial instance of an object in a weighty sense – is height. My sense of my height has numerous sources; for example, the vision of my feet at a distance; and the sense delivered by bodily experiences and physical exertion of the continuous carnal reality, of an unbroken vertical axis, linking my head and my feet, notwithstanding many parts that are experientially silent. My height is a strange distance that is measured out as, for example, the separation of the bitter taste in my mouth from the snugness of my feet in my socks with the intermediate landmarks of an itch on my neck, the pressure of my elbow on a table and branch lines such as my arms with stations at the tip of my fingers, my wrist, my elbow, and my shoulders. The complexity is compounded by, for example, distances between my fingertips made explicit as when, for example, I am typing this sentence.

    I could develop this patchy phenomenological of the connectedness of my body in many other directions. I have not, for example, discussed the contributions made by parts of the body engaging with other parts, as when I clasp my hands, fold my arms, cross my legs, or use my tongue to extract meat from between my teeth. But I hope I have said enough on this matter.
18. See Tallis, *Logos,* ch. 3.

19. It might be anticipated that the sense that there is more to my body than meets my senses would be reinforced by the intuition that those senses have a substrate, a physical basis, an intuition rooted in the facts that (a) they are located in space; and (b) that we can deploy or re-deploy them by moving them in space – either by transporting our body or by (as in the case of vision) panning round.
20. See Jensen & Moran, "Introduction", vii, where the concept is helpfully surveyed.
21. Discussed in Moran, "Revisiting Sartre's ontology of embodiment in *Being and Nothingness*".
22. Jensen & Moran, "Introduction", vii–viii.
23. Merleau-Ponty's late lectures, discussed by Halák in "Merleau-Ponty on embodied subjectivity from the perspective of subject-object circularity" are of particular relevance to this issue. As Halák expresses it, "for Merleau-Ponty, there is a circular relationship between the objective and subjective dimensions of the body – between the objective and the lived", 26. The two are united in "the body schema" which we discussed earlier.
24. Jensen & Moran, "Introduction", vii.

CHAPTER 6 SELFHOOD

1. Kant, *Anthropology from a Pragmatic Point of View*, 127. Jens Zimmermann (personal communication) has suggested this useful amendment for the translation: "The fact that human beings are enabled to have within their imagination the 'I' raises him above all living beings on earth".
2. In, for example, Tallis, *Seeing Ourselves*, 145–90.
3. Hume, *A Treatise of Human Nature*, 228.
4. A similar point is made by Roderick Chisholm in *Person and Object: A Metaphysical Study*, 37–40. An earlier writer, however, felt that what Hume was up to – or reported himself as being up to – was not fatal for Hume's bundle theory of the mind, although it supported only a half-hearted version; see Pike, "Hume's bundle theory of the self: a limited defence". Galen Strawson also defends Hume, arguing that he does not endorse a "bundle" theory of mind, deny the reality of the subject of experience, or claim that the subject of experience is encountered in experience; see Strawson, "Hume on personal identity".
5. See *Seeing Ourselves*, 146–56, for a further discussion of some of the more prominent contemporary autocides.
6. Dennett, "The self as a centre of narrative gravity", 275.
7. Metzinger, *Being No-One: The Self-Model Theory of Subjectivity*, 1. My discussion of Metzinger is heavily indebted to Harman, "The problem with Metzinger".
8. Metzinger, *Being No-One*, 52.
9. *Ibid.*
10. *Ibid.*, 21.
11. *Ibid.*, 115.
12. "Darwinitis" is a common source of arguments that undermine themselves; see Tallis, *Aping Mankind*, 1–3. This is something to which I shall return in Chapter 10 when I discuss Donald Hoffman's Darwinitic "case against reality".

13. Persson, "Self-doubt: why we are not identical to things of any kind", 391 (emphasis original)
14. Aristotle, *Metaphysics* 12.9 1074b35/6, quoted in Caston, "Aristotle and consciousness".
15. Zahavi, "Consciousness, self-consciousness, selfhood: a reply to some critics", 2.
16. *Ibid.*, 2.
17. Guillot, "I, me, mine: on a confusion concerning the subjective aspect of experience", quoted in Zahavi, "Consciousness, self-consciousness, selfhood", 1.
18. Likewise, my absolute authority does not extend beyond phenomenal experiences to objects they may signify. You may think the object is a stoat and I may think it is a weasel, but our differences won't be settled by more intense introspection. For a slimmed down account of "privileged access" to our own experiences understood as "mental contents" with "objects" of their own, see Heil, "Privileged access".
19. For a persuasive critique of the claim – particularly associated with Elizabeth Anscombe – that "I" is neither referential nor indexical but a feature-placing term like "it" in "it is raining" or "it is snowing" see Garrett, "Anscombe on 'I'". As for siding with the autocides, Metzinger argues that "The first-person pronoun doesn't refer to an object like a football or a bicycle" [You bet, Professor!] "It just points to the speaker of the current sentence" (Taft, "What is the self? An interview with Thomas Metzinger", 1).
20. A view defended in detail in Strawson, *The Subject of Experience*, ch. 3.
21. Kant, *Critique of Pure Reason*, B131–2.
22. Tallis, *Seeing Ourselves*, ch. 5.
23. There are, in fact, four distinct binding problems. They are identified – perhaps a little eccentrically – by Jerome Feldman as: (1) general considerations on co-ordination; (2) visual-feature binding; (3) variable binding; and (4) the subjective unity of perception; Feldman, "The neural binding problem(s)". The most challenging and philosophically interesting is the problem of the unity of perception and, beyond this, the unity of consciousness at any given time that goes beyond perception and encompasses memory, thought, affect, intentions, etc., that is the present focus.
24. Their views are summarized in Brook & Raymont, "Unity of consciousness", §7.2.
25. See Chalmers & McQueen, "Consciousness and the collapse of the wave function".
26. For a further discussion of this, see Tallis, "Remembering memory".
27. There is an ambivalence in memory: it reminds us of what we have lost; and at the same time, suggests that – if it is shared – it is in some important sense not lost.
28. Locke, *An Essay Concerning Human Understanding*, bk 2, ch. 27.
29. Reid, *Essay on the Intellectual Powers of Man*.
30. Grice, "Personal identity".
31. Schacter, Addis & Buckner, "Remembering the past to imagine the future: the prospective brain".
32. Pinto *et al.*, "Split brain: divided perception but undivided consciousness". See also, Downey, "Split brain syndrome and extended perceptual consciousness".
33. Carrère, *The Moustache*.
34. Nietzsche, *The Genealogy of Morals*, 1–3.
35. Donne, "The Good-Morrow".
36. Subjects are positioned in such a way as their left hand is hidden. A life-like rubber hand is attached to them roughly where the real left hand would be. The hidden left

hand and the visible right hand are stroked simultaneously and in the same direction. The subjects feel the stroking in two "hands": the real right hand and the rubber left hand. Indeed, they begin to feel the rubber hand as their own: when they are asked to point to the left hand with their right hand, they point to the rubber hand. This illusion is not experienced if the real and rubber hands are not stroked simultaneously or in the same direction.

37. It is set out most clearly in his *The Human Animal: Personal Identity without Psychology*.
38. Eric Olson quoted in Baker, "Review of *What Are We? A Study in Personal Ontology*", 1120.
39. To lay the (Cartesian) ghost to rest once and for all, we should remind ourselves that a ghost is something that is made explicit – indeed, for those of us who do not believe in them, exists only insofar as it is made explicit – not something that, like a self, makes things explicit.
40. Shakespeare, Sonnet 129.
41. Frye, "When St. Thomas Aquinas likened his work to straw, was that a retraction of what he wrote?".
42. Wittgenstein, *On Certainty*, para. 676. This was the last sentence in his last notebook.
43. Quoted by Thomas Carlyle in "Novalis".
44. This notion is developed in his *Foundations of the Science of Knowledge* where he argues that "the science of knowledge is based not on fact but on action" and "The action in question is expressed in the statement 'I am I'"; Bubner, "Introduction, Fichte", 72.
45. See Tallis, *Freedom*.
46. Parfit tends to use the terms "person" and "personal identity" in a way that overlaps very closely with "self" as it is used in this work.
47. Parfit, *Reasons and Persons*, especially 211–14.
48. See Tallis, *I Am: A Philosophical Inquiry into First-Person Being*.
49. Parfit, *Reasons and Persons*, 309–10.
50. A thought that is unpacked at length in Metzinger, "The ego tunnel".
51. Dennett, *I've Been Thinking*.

CHAPTER 7 AGENCY: EXPLICITNESS IN ACTION

1. I shall cover this only briefly because it is the theme of a recent volume, *Freedom: An Impossible Reality* and of a special edition of the philosophy journal *Human Affairs* devoted to that book: Tartaglia & Leach (eds), Symposium on *Freedom: An Impossible Reality*.
2. Kane, *The Significance of Free Will*, 4.
3. van Inwagen, *An Essay on Free Will*, v.
4. These conclusions do not hold up, as is discussed in Tallis, *Freedom*, 17–20. More recent studies – notably Maoz *et al.*, "Neural precursors of decisions that matter – and ERP study of deliberate and arbitrary choice" – have demonstrated that the brain's anticipation of decisions as signalled by the readiness potential is not seen in decisions

that matter. Sapolsky's *Determined: Life Without Free Will* brilliantly exposes the irrelevance of neurophysiological experiments by Benjamin Libet and John-Dylan Haynes to the question of free will. They focus, he says, only on the last three minutes of the story leading up to choice. However, he looks to the longer-term changes in the brain to demonstrate our dependence on our brain function. He is still, therefore, a neurodeterminist.

5. Mill, "Nature", 152.
6. Hume, *A Treatise on Human Nature*, Part III, §§3–4.
7. See Scott, "Disarming causation in the service of agency: Tallis on Hume".
8. It may be thought that "becausation" is also seen in non-human animals. After all, we may think of a gazelle retreating because it has picked up the scent of a lion. Its flight could be presented as an action triggered by a future possibility. But the possibility does not depend on the animal envisaging it. It is not attached to a future rooted in the singularity of the individual drawing on his or her past to construct. Indeed, it is hardly a choice made by an embodied agent – rather it is the reaction of an organism. How deep the gulf is between human agency and instinctive animal behaviour is open to discussion; see, for example, Steward, *A Metaphysics for Freedom*.
9. van Inwagen, *An Essay on Free Will*, 3.
10. Aristotle, *De Anima*, vol. 3. For a celebration of the hand and its role in making us the unique creatures we are, see Tallis, *The Hand: A Philosophical Inquiry into Human Being*.
11. Hume, *A Treatise of Human Nature*, bk III, pt I, §I.
12. Behind this lurks a ghost: the notion of "moral luck". And behind moral luck there are many other layers of luck; see, for example, Tallis, "Accidental thoughts on luck by an accidental man".

CHAPTER 8 FREE-FLOATING EXPLICITNESS: THINKING ABOUT THINKING

1. Quoted in Rundell, *Super-Infinite: The Transformations of John Donne*, 295.
2. Bayne, *Thought: A Very Short Introduction*, 1.
3. They are lucidly discussed in Wetzel, "Types and tokens".
4. I have strayed into an area of considerable – and justified – controversy. We shall return to this in Section 8.3.
5. See, for example, Hoffman, *The Case Against Reality*, 149.
6. For some thoughts about this thought, see Tallis, *Prague 22*.
7. See Tallis, *Freedom*, ch. 3.
8. The *locus classicus* of the causal theory of reference is Kripke's *Naming and Necessity*.
9. Berk, "Reference and scientific realism", 140.
10. It also deals with the problems associated with the theory of definite descriptions, which claim to explain how names are attached to objects. (For an excellent exposition of this see Michaelson & Reimer, "Reference".) Most importantly, for many of its advocates, notably Hilary Putnam, the causal theory of reference supports scientific realism by seeming to explain how the natural-kind terms used in science refer. As summarized by Berk: "What the theory says about natural-kind terms ... is that they

function more like proper names than is usually supposed. In Kripke's terminology, natural-kind terms, like proper names are 'rigid designators': they have the same extension in any possible world" (Berk, "Reference and scientific realism", 140).

11. Michaelson & Reiner, "Reference", 4.
12. Levine, "Anti-materialist arguments and influential replies", 401.
13. Famously associated with Ferdinand de Saussure and spelled out in his posthumously published lectures *Course in General Linguistics*. His careful arguments about the nature of language unwittingly licensed torrents of nonsense that swept through academic humanities in the final decades of the twentieth century; see Tallis, *Not Saussure*.
14. Wittgenstein, *Philosophical Investigations*, para. 268, 94e.
15. Fernyhough, *The Voices Within: The History and Science of How We Talk to Ourselves*.
16. It is probably not necessary to say that James Joyce's *Ulysses* is the greatest imaginable celebration of the rich, ordered chaos of our inner life.
17. Quoted in Hintikka, "Cogito ergo sum: inference or performative?", 19.
18. See, for example, Bourget & Mendelovici, "Phenomenal intentionality", §6.1.
19. To say this is not to license poststructuralist absurdities such as Jacques Derrida's claim that "there is nothing outside of the text" – that language is solely about the linguistic realm; see Tallis, *Not Saussure*.
20. The question of whether there is phenomenal experience associated with thought, and the different positions philosophers have taken on it, is lucidly discussed in Bayne & Montague's introduction to their collection *Cognitive Phenomenology*, viii–34.
21. This last is one of the so-called "grounding" arguments for cognitive phenomenology, discussed in Bayne & Montague, "Introduction", 27–9.
22. It is perhaps no more surprising than that I ascribe thoughts spoken out loud to myself. In both cases, it is obvious that I am the source: they come from "here" rather than "over there". It is interesting that, according to an article bearing this title, "Thinking out loud to yourself is a technology for thinking" (Ariel).
23. See Hintikka, "Cogito ergo sum"; see also Tallis, "Cogito, ergo sum?".
24. This is reflected in "content externalism" which concerns the content of thought. What is being thought may not be entirely transparent to the thinker, if only because the reference of the words in which we think to ourselves may not be fully evident to us.
25. Intransitive experiences, such as sensations arising spontaneously in my body, disconnected from anything publicly accessible, as we discussed in Chapter 5, may seem more "owned" by me, though my itch seems less of an invitee than a gate crasher.
26. Ryle, *The Concept of Mind*.
27. Or those, at any rate, who do not embrace the idea of "embodied cognition"; see later in this chapter.
28. This has been expressed by Peter Geach: "Even if we accepted the view ... that a judgement is a complex of ideas, we could hardly suppose that in a thought the ideas occur successively, as the words do in a sentence; it seems reasonable to say that unless the whole complex content is grasped all together – unless the ideas ... are all simultaneously present – the thought or judgement just does not exist at all" (Geach, *Mental Acts*, 104).

29. Lockwood, *Mind, Brain, and the Quantum*, 76.
30. James, *The Principles of Psychology* Volume 1, 244.
31. I have discussed this in Tallis, *Freedom*.
32. The title and first line of the poem by Sir Edward Dyer published in 1607.
33. *Hamlet*, Act 2, Scene 2.
34. For a further development of this argument, see Tallis, "Against neural philosophy of mind".
35. See Wikipedia, "Embodied cognition". The literature on this topic is vast and neither the present discussion nor the brief postscript appended to this chapter does it any kind of justice.
36. The reference to "the entire body" highlights the additional problem that comes with liberating conscious experiences from the brain, never mind thoughts. What extra-cerebral aspects of the body are involved in thought? Clearly not bone, muscles, the smorgasbord of tirelessly busy giblets, blood, lymph, and the rest of the corporeal proletariat.
37. Halák, "Embodied higher cognition: insights from Merleau-Ponty's interpretation of motor intentionality", 397.
38. My focus in this chapter has been on self-driven thought. Much of our thinking is, however, prompted and guided by listening to and reading. I am not sure to what extent thought harvested from without is different from spontaneous thought shaped from within. This, however, is more a matter for the psychology rather than the philosophy of thought.
39. The text in the paragraphs that follow is a modified extract from Tallis, "Freedom: An Impossible Reality" (*Human Affairs* 32(4)).
40. Di Paolo, Cuffari & De Jaegher, *Linguistic Bodies: The Continuity between Life and Language*, 6.
41. Cited in Rostowski, "Freedom: an enactive possibilty", 435.
42. Quoted in Kuhn, "A landscape of consciousness", 65.

CHAPTER 9 PURE EXPLICITNESS: POSSIBILITY (AND PROBABILITY)

1. This chapter does not discuss the vital relationship between possibility and human agency. The capacity to envisage possibilities that lies at the heart of free will is discussed in Tallis, *Freedom* and, briefly in Chapter 7 of the present work.
2. Discussed in Tallis, *Of Time and Lamentation*, §8.2.2.
3. Sparkes, "Coin flips don't truly have a 50/50 chance of being head or tails".
4. "Infinite" may be an exaggeration – but not an important one. Even if outcomes have to be differentiated only above the Planck limit, we are still talking about 10^{43} times 10^{45} the numbers of details the battle is judged to have. In other words, quite a few. There is an overlap between the present argument and Henri Bergson's claim that possibilities do not exist prior to actualization. We think they do, he says, only because we retrospectively generate a possibility corresponding, and prior to, something that happens – the retrofitting that I refer to in the main text (see Bergson, "The actual and real"). The view developed in this chapter is that material events are not actualizations

of other natural entities, namely their possibility; rather, they are the actualizations of a happening envisaged by a mind.

5. Another nail in the coffin of paninformationalism (see Chapter 3). The quantification of the bits delivered by any event depends on how it is framed as an anticipated outcome.
6. See *Freedom*, ch. 4 for a more detailed discussion of this point.
7. This points to a possible reconciliation between realist and anti-realist (or instrumentalist) accounts of "natural" laws. The realists are correct to say that the relationship "Pressure (really) is inversely proportional to Volume, if other variables (notably Temperature) are kept constant". The anti-realists would be equally right to say that the law is based on the unnatural separation of pressure, volume, and other circumstances, because they are not separate from each other in unobserved nature. What is more, those variables are subject to evolution as science finds it more useful to look at the world in different ways – for example at different scales.
8. Clark, "Predictive brains, embodied minds, and the puzzle of the evidentiary veil" argues that the prediction does not relate to the future. The brain "guesses the present". Not only is this unclear but "the present" is also tensed.
9. Seth & Baynes, "Theories of consciousness", 440.
10. See Tallis, *The Enduring Legacy of Parmenides.*
11. Nietzsche, *Philosophy in the Tragic Age of the Greeks*, 79.
12. It is not possible to develop an objective or "frequentist" interpretation of probability that acknowledges the role of the conscious subject. Frequencies of outcomes – and hence the probability of any particular outcome – can be meaningfully determined only in the context of observations of certain pre-selected variables in precisely defined systems subject to closely controlled conditions. As Louis Vervoort has expressed it: "Probability is not a property of an object on its own; it is a property of certain systems under repeatable human actions; or, more generally, of certain systems in dynamical evolution in well-defined and massively repeatable conditions, in particular initializing and probing conditions ... In sum, in our view, probability is a property of composed events or composed systems, or of events, and not of events or objects simpliciter" ("A detailed interpretation of probability, and its links with quantum mechanics", 18).

    The central point is this. A seemingly objective, frequentist interpretation of probability, still rests on the choices of the conscious subject, and the customs, practices and standards of the scientific discipline, in which that which is to be observed is defined. Probabilities, that is to say are not properties of objects in the natural world.

CHAPTER 10 WHAT-IS, CONCEIVED AS "REALITY" AND "TRUTH"

1. Described compellingly by Bryan Magee in *Confessions of a Philosopher: A Personal Journey through Philosophy from Plato to Popper*, 8–11.
2. For a secular humanist such as myself, the sense that there is a hidden reality may come with a glow of hope illuminating the space vacated by the departure of the promises of religion: the promises without, perhaps, the threats. Hidden reality,

unlike a God taking a personal interest in the world, would not judge its inhabitants, rewarding and punishing them as He thought appropriate.

3. The interested reader might wish to read three "studies in insincerity" I have published in *Philosophy Now*: "Zhuangzi and that bloody butterfly", "Arguing with a solipsist" and "A long postponed meeting with M. Descartes".
4. Aristotle, *On Generation and Corruption*, 325a, 13–18, quoted in Barnes *The Presocratic Philosophers*, 296.
5. See Tallis, *The Enduring Legacy of Parmenides*.
6. Descartes, *Meditations on First Philosophy*, 25.
7. Quine, "Two dogmas of empiricism", 38.
8. *Ibid.*, 41.
9. Democritus 689B (Diels & Kranz).
10. Galileo, "The assayer", quoted in Drake, *The Discoveries and Opinions of Galileo*, 274.
11. Descartes, *The Principles of Philosophy*, 66–75.
12. Locke, *An Essay Concerning Human Understanding*, bk 2, ch. 8. Contemporary physics has added characteristics such as electric charge and intrinsic spin to the list of primary qualities. See Healey, "Can physics coherently deny the reality of time?" Interestingly, Healey argues that time and change are secondary qualities.
13. Eddington, *The Nature of the Physical World*, xi.
14. Kant, *Prolegomena to Any Future Metaphysics That Will be Able to Come Forward as Science*, 289.
15. Kant, *Critique of Pure Reason*, 41.
16. Breitenbach, "Kant and causal knowledge: causality, mechanism and reflective judgement", 233. Breitenbach attempts to reconcile Kant's principle of causality as a category of understanding in *Critique of Pure Reason* with his regulative conception of mechanical causation in *Critique of Teleological Judgement*. I am not persuaded that this is successful.
17. Oberst, "Two worlds and two aspects: on Kant's distinction between things in themselves and appearances", 53.
18. Allais, "Kant's one world: interpreting 'transcendental idealism'", 657.
19. An interesting question arises in relation to more than one mind. You and I, standing side by side, experience ourselves as being in roughly the same location with the same objects populating it. If, however, we are independently imposing space and time on our experiences, it is not clear why the spatial location ascribed to our two sets of experience should be side by side. This is connected with the ambiguity of the notion of "mind" in Kant – whether it is to be taken to be a universal category that makes things present, or individual minds corresponding to conscious subjects like you and me. I discuss this in Tallis, *Logos*, ch. 3.
20. Nagel, *The View from Nowhere*.
21. Bell, "Against measurement".
22. Ladyman & Ross, *Every Thing Must Go*, 141–5 and *passim*.
23. Rovelli, *The Strange and Beautiful Story of Quantum Physics*, 71.
24. *Ibid.*, 75. This is unpacked a little further in a later passage: "[T]he elements useful for thinking about the world are manifestations of physical systems to each other not absolute properties belonging to each system" (110). "Useful"? "Manifestations"?

Briceno and Mumford deliver a thorough demolition of Ladyman and Ross's version of relationism in "Relations all the way down?". Their central argument is that "a world of pure structure would be no more than a Platonic entity, lacking any resources for concretization" (198). There would not be any resources for individuation.

25. The intuition that there is, at some fundamental level at which what-is is seen for what it is, only one property, is expressed in the trajectory of science towards A Theory of Everything. This homogenizes what-is to the point where it seems to have no intrinsic properties. Hence relational quantum mechanics.
26. Lewis, "Extrinsic properties", 197. In a later paper, co-authored with Rae Langton, Lewis offers a less accommodating definition of intrinsic properties. Building on Jaegwon Kim's suggestion that an intrinsic property is one that an entity could have in a universe in which it was unaccompanied, Langton and Lewis suggest that an intrinsic property is one that is independent of accompaniment or loneliness; see Langton & Lewis, "Defining 'intrinsic'".
27. Mill, *An Examination of the Philosophy of Sir William Hamilton*, cited in Macleod, "John Stuart Mill", 12.
28. Hodes, "The 'causes' of the hard problem: a note", 47.
29. This is a view embraced by many philosophers. See Averill, "Are physical properties dispositions?".
30. This identifies what is true in the argument of the case for panpsychism discussed in Chapter 3. Science tells is what things do rather than what they are. The defence of panpsychism that is built on this relatively firm foundation is, of course, profoundly flawed.
31. Quine in *Word and Object* contrasts "occasion" sentences and "eternal" or "standing" sentences. (In view of this illuminating contrast, he will be forgiven for suggesting that meaning of occasion sentences is "stimulus meaning". The causal theory of meaning is as mistaken as causal theories of perception and of reference.)
32. Quoted in Harman, "The problem with Metzinger", 50.
33. Frith, *Making up the Mind*, 132.
34. Descartes, *Meditations on First Philosophy*, 25.
35. Ryle, *Dilemmas*, 94.
36. See Crane & French, "The problem of perception" for an excellent exposition of this argument.
37. *Ibid.*
38. Seth, *Being You: A New Science of Consciousness*, 5.
39. Magee, *Confessions of a Philosopher*, 458.
40. Frankish, "The mental life of mountains", 38–9.
41. Seth, *Being You*, 76
42. See Tallis, *Of Time and Lamentation*, ch. 5.
43. Seth, *Being You*, 30.
44. Hohwy, "The self-evidencing brain".
45. *Ibid.*, 265.
46. This is how neuroscientist Chris Frith expresses it, "My perception is not of the world, but of my brain's model of the world" (Frith, *Making Up the Mind*). Bertrand Russell said something similar when he stated that "we can witness or observe what goes on

in our heads, and that we cannot witness or observe anything else at all" (Russell, *My Philosophical Development*, 26).

47. Herman von Helmholtz quoted in Zahavi, "Brain, mind, world: predictive coding, neo-Kantianism, and transcendental idealism", 50.
48. Hoffman, *The Case Against Reality*.
49. *Ibid.*, 57.
50. *Ibid.*, 39.
51. Berkeley, *Three Dialogues between Hylas and Philonous*, 152. If things existed only insofar as they were perceived by human or animal minds, it would be difficult to see what constraints there would be on perception – particularly as idealism extends to the body of the perceiving subject and would seem to undermine its standing as a viewpoint.
52. See his lecture, "Perception is a fantasy".
53. Hoffman, *The Case Against Reality*, 107.
54. *Ibid.*, 65.
55. *Ibid.*, 176.
56. *Ibid.*, 120.
57. John Gray quoted in Tallis, *Aping Mankind*, 3.
58. Dennett, *Darwin's Dangerous Idea: Evolution and the Meaning of Life*, 63. The argument that Hoffman's view is self-refuting is also set out in Bagwell, "Debunking interface theory: why Hoffman's skepticism (really) is self-defeating".
59. Dick, "How to build a universe that doesn't fall apart two days later", 261.
60. Austin, *Sense and Sensibilia*, 70. This argument was anticipated by Susan Stebbing in her observation that "we do not know how to use a word that has no sensible alternative".
61. Wittgenstein, *Philosophical Investigations*, 96e, para. 265.
62. Gibson, "The theory of affordances" in *The Ecological Approach to Visual Perception*.
63. James, *Essays in Philosophy, 1876–1910*, 124.
64. Quine, "Two dogmas of empiricism", 41.
65. Gottlob Frege's anticipation of Ramsay is illustrated by this passage: "It is also worthy of notice that the sentence 'I smell the scent of violets' has just the same content as the sentence 'It is true that I smell the scent of violets'. So it seems, then, that nothing is added to the thought by my ascribing to it the property of truth" (Frege, "The thought: a logical inquiry", 293).
66. Ramsay, "The nature of truth".
67. Tarski, "The semantic conception of truth and the foundation of semantics". In fact, it is difficult to be sure what Tarski meant. That which guarantees the truth of the assertion "Paris is the capital of France" lies outside the sentence and any assertion that uses it.
68. As we discussed in Chapter 6, the laws of physical nature were not made explicit through the operation of those laws; nature did not become Nature or "nature" by natural means. See also Tallis, *Freedom*, ch. 2.
69. This, in a sentence, is the case against paninformationalism discussed in Chapter 3.
70. See, for example, Ichikawa, "The analysis of knowledge".
71. See Tallis, *Logos*, ch. 7.
72. Carroll, "Reality as a vector in Hilbert space".

CODA: (IN) CONCLUSION

1. Tallis, *The Explicit Animal.*
2. For example, Tallis, *Not Saussure, In Defence of Realism* and *Theorrhoea and After.*
3. Tallis, *Enemies of Hope: A Critique of Contemporary Pessimism.*
4. Stephenson, "On the widespread use of an inappropriate model of the literary aphorism", 3.
5. Francis Bacon quoted in Stephenson, "On the widespread use of an inappropriate model …", 3.
6. Blake, "Jerusalem (The Holiness of Minute Particulars)".
7. Wittgenstein, *Philosophical Investigations*, §107.
8. Explored in Tallis, *Prague 22.*

References

Acacio de Barros, J. & C. Montemayor. "Quantum Mentality: Panpsychism and Panintentionalism". In S. Gao (ed.), *Consciousness and Quantum Mechanics*. Oxford: Oxford University Press, 2022.

Adams, F. "The Degree of Fine-Tuning in Our Universe – and Others". *Physics Reports* 807 (2019), 1–111.

Adlam, E. "Fundamental?" In A. Aguirre, B. Foster & Z. Merali (eds), *What is Fundamental?* Cham, CH: Springer Nature, 2019.

Albahari, M. "Panpsychism and the Inner-Outer Gap Problem". *The Monist* 105 (2022), 25–42.

Allais, L. "Kant's One World: Interpreting 'Transcendental Idealism'". *British Journal for the History of Philosophy* 12:4 (2004), 655–84.

Alexander, S. *et al.* "The Autodidactic Universe". 29 March 2021. https://doi.org/10.48550/arXiv.2104.03902.

Ariel, N. "Thinking out loud to yourself is a technology for thinking". *Psyche Ideas*, 23 December 2020.

Aristotle, *De Anima* ("On the Soul") Volume 3. Translation by E. M. Edghill. No publisher.

Aristotle, *On Generation and Corruption* 325a 13–18. In J. Barnes (ed.), *The Presocratic Philosophers*. London: Routledge, 1982.

Arulshah, A. *et al.* "Elucidating the Visual Phenomena in Epilepsy: A Mini Review". *Epilepsy Research* 190 (Feb 2023), 107093.

Atmanspacher, H. & P. beim Graben. "Contextual Emergence". *Scholarpedia* 4:3 (2009): 7997.

Austin, J. L. *Sense and Sensibilia*. Reconstructed from manuscript notes by G. Warnock. Oxford: Oxford University Press, 1962.

Austin, J. L. "Other Minds". In J. Urmson & G. Warnock (eds), *Philosophical Papers*, second edition. Oxford: Oxford University Press, 1970.

Averill, E. W. "Are Physical Properties Dispositions?" *Philosophy of Science* 57 (1990), 118–32.

Baggott, J. "Thirty Years of 'Against Measurement'". *Physics World* (Dec 2020), 30–34.

Bagwell, J. N. "Debunking Interface Theory: Why Hoffman's Skepticism (Really) is Self-defeating". *Synthese* 201:25 (2023).

Baker, Lynne Rudder Review of *What Are We? A Study in Personal Ontology* Eric T. Olson *Mind* Volume 117 468 October 2008 1120-1122.

Barthes, R. "The Death of the Author". In *Image, Music, Text*, 143–8. Translated by S. Heath. London: Fontana, 1977.

Barnes, M. "Retrocausality, Wheeler's Delayed Choice, and Simulation Theory Reinterpreted via the Participatory Universe, 'It from Bit', Time Travel and the Everett/Wheeler Hypothesis". ResearchGate Technical Report, December 2016. https://www.vixra.org/abs/1612.0254.

Bastianelli, M. "Putnam's No Miracles Argument: A Defence of Scientific Realism". *European Journal of Pragmatism and American Philosophy* 13:2 (2021).

Bayne, T. *Thought: A Very Short Introduction.* Oxford: Oxford University Press, 2013.

Bayne, T. "On the Axiomatic Foundations of the Integrated Information Theory of Consciousness". *Neuroscience of Consciousness* 1 (2018). doi:10.1093/nc/niy007. ecollection 2018.

Bayne, T. & M. Montague (eds). *Cognitive Phenomenology.* Oxford: Oxford University Press, 2011.

Ben-Menahem, Y. "Dummett versus Bell on Quantum Mechanics". *Studies in the History and Philosophy of Modern Physics* 28:2 (1997), 277–90.

Ben-Menahem, Y. "Realism and the Failure of Science: Reversing the No Miracle Arguments". *Academia Letters*, May 2021, 1–4.

Bell, J. "Against Measurement". In A. Miller (ed.), *Sixty-Two Years of Uncertainty*, 17–32. New York: Plenum, 1990.

Bell, J. *Speakable and Unspeakable in Quantum Mechanics.* Cambridge University Press, 2004.

Bergson, H. "The Possible and the Real" [1930]. https://bergsonian.org/the-possible-and-the-real/.

Berk, E. "Reference and Scientific Realism". *Southwestern Journal of Philosophy* 10:2 (1979), 139–46.

Berkeley, G. *Three Dialogues between Hylas and Philonous.* In *The Works of George Berkeley Bishop of Cloyne, Volume 1.* London: Thomas Nelson, 1949.

Bickle, J. & R. Ellis. "Phenomenology and Cortical Microstimulation". In D. W. Smith & A. Thomasson (eds), *Phenomenology and Philosophy of Mind*, 140–65. Oxford: Clarendon Press, 2005.

Biltot, M. "The Roles Ascribed to Consciousness in Quantum Physics: A Revelator of Dualist (or Quasi-Dualist) Prejudice". In S. Gao (ed.), *Consciousness and Quantum Mechanics.* Oxford University Press, 2022.

Borges, J. L. "The Wall and the Books". In *Labyrinths.* London: Penguin Classics, 2000.

Born, M. *Einstein's Theory of Relativity*. Revised edition. New York: Dover Publications, 1962.

Bourget, D. & A. Mendelovici, "Phenomenal Intentionality". Stanford Encyclopedia of Philosophy (Fall 2019 edition), Edward N. Zalta (ed.), https://plato.stanford.edu/archives/fall2019/entries/phenomenal-intentionality/.

Breitenbach, A. "Kant and Causal Knowledge: Causality, Mechanism and Reflective Judgement". In K. Allen & T. Stoneham (eds), *Causation and Modern Philosophy*, 201–19. London: Routledge, 2011.

Briceno, S. & S. Mumford. "Relations All the Way Down? Against Ontic Structural Realism". In A. Marmodoro & D. Yates (eds), *The Metaphysics of Relations*, 198–217. Oxford: Oxford University Press, 2015.

Bridgman, P. W. *The Logic of Modern Physics*. New York: Macmillan, 1927.

Brook, A. & P. Raymont. "The Unity of Consciousness". Stanford Encyclopedia of Philosophy (Summer 2021 edition), Edward N. Zalta (ed.), https://plato.stanford.edu/archives/sum2021/entries/consciousness-unity/.

Bubner, R. "Introduction to Fichte". In *German Idealist Philosophy*. London: Penguin Classics, 1997.

Butterfield, J. "Essay Review of Sabine Hossenfelder *Lost in Math: How Beauty Leads Physics Astray*". *Physics in Perspective* 21 (2019), 63–86.

Calosi, C. "Quantum Monism: An Assessment". *Philosophical Studies* 175 (2018): 3217–36.

Carlyle, T. "Novalis" [1829]. In *The Works of Thomas Carlyle*. Cambridge: Cambridge University Press, 2010. https://doi.org/10.1017/CBO9780511697296.

Carnap, R. "Intellectual Autobiography". In P. A. Schilpp (ed.), *The Philosophy of Rudolf Carnap*. LaSalle, IL: Open Court, 1963.

Carpenter, A. "Kant, the Body, and Knowledge". *Paideia* (n.d.). https://www.bu.edu/wcp/Papers/TKno/TKnoCarp.htm.

Carrere, E. *The Moustache*. London: Sceptre, 1988.

Carroll, S. M. "Reality as a Vector in Hilbert Space". arXiv, 17 March 2021. https://doi.org/10.48550/arXiv.2103.09780.

Carroll, S. M. "245 Solo: The Crisis in Physics". Mindscape podcast, 31 July 2023. https://www.preposterousuniverse.com/podcast/2023/07/31/245-solo-the-crisis-in-physics/.

Cassam, Q. *Self and World*. Oxford: Oxford University Press, 1999.

Caston, V. "Aristotle and Consciousness". *Mind* 111 (2002), 751–815.

Caves, C., C. Fuchs & R. Schack. "Subjective Probability and Quantum Uncertainty". arXiv, 26 Jan 2007. https://doi.org/10.48550/arXiv.quant-ph/0608190.

Chalmers, D. *The Conscious Mind: In Search of a Fundamental Theory*. Oxford: Oxford University Press, 1997.

Chalmers, D. "Strong and Weak Emergence". In P. Clayton & P. Davies (eds), *The Re-Emergence of Emergence*. Oxford: Oxford University Press, 2006.

Chalmers, D. *The Character of Consciousness*. Oxford: Oxford University Press, 2010.

Chalmers, D. "Idealism and the Mind-Body Problem". In W. Seager (ed.), *Routledge Handbook of Panpsychism*. Abingdon: Routledge, 2020.

Chalmers, D. *Reality+: Virtual Worlds and the Problems of Philosophy*. London: Penguin, 2023.

Chalmers, D. & K. McQueen. "Consciousness and the Collapse of the Wave Function". In S. Gao (ed.), *Consciousness and Quantum Mechanics*. Oxford: Oxford University Press, 2022.

Chemero, A. "Dynamical Explanations and Mental Representations". *Trends in Cognitive Sciences* 5:4 (2001), 141–2.

Chen, E. K. "Realism About the Wave Function". *Philosophy Compass* 2019: 14:e12611.

Cheng, H.-Y. Sense, Space, and Self. PhD thesis, Department of Philosophy, University College London, 2019. https://discovery.ucl.ac.uk/id/eprint/10064788/.

Chisholm, R. *Person and Object: A Metaphysical Study*. London: Routledge, 2004.

Clark, A. "Busting Out: Predictive Brains, Embodied Minds, and the Puzzle of the Evidentiary Veil". *Nous* 51:4 (2016), 727–53.

Clayton, P. & P. Davies. "The Re-Emergence of Emergence: The Emergentist Hypothesis from Science to Religion". Oxford: Oxford University Press, 2008.

Close, F. *Elusive: How Peter Higgs Solved the Mystery of Mass*. London: Allen Lane, 2022.

Costandi, M. "AI reads brain activity to tell what part of a movie you are watching". *New Scientist*, 26 June 2024.

Crane, T. "Cosmic Hermeneutics vs Emergence: The Challenge of the Explanatory Gap". In C. Macdonald & G. Macdonald (eds), *Emergence in Mind*, 22–34. Oxford: Oxford University Press, 2010.

Crane, T. & C. French. "The Problem of Perception". *Stanford Encyclopedia of Philosophy* (Fall 2021 edition), Edward N. Zalta (ed.), https://plato.stanford.edu/archives/fall2021/entries/perception-problem/.

Crick, F. *The Astonishing Hypothesis: The Scientific Search for the Soul*. New York: Scribner, 1994.

Davies, P. & J. Brown (eds), *The Ghost in the Atom*. Cambridge: Cambridge University Press, 1993.

de Finetti, B. *Theory of Probability: A Critical Introductory Treatment*. Chichester: Wiley, 2017.

Dennett, D. *Consciousness Explained*. London: Penguin, 1991.

Dennett, D. "The Self as a Centre of Narrative Gravity". In F. Kessel, P. Cole & D. Johnson (eds), *Self and Consciousness: Multiple Perspectives*. Hillsdale, NJ: Erlbaum, 1992.

Dennett, D. *Darwin's Dangerous Idea: Evolution and the Meaning of Life*. London: Penguin, 1995.

Dennett, D. *I've Been Thinking*. New York: Norton, 2023.

Descartes, R. "Principles of Philosophy". In *The Philosophical Writings of Rene Descartes Volume 1*. Translated and edited by J. Cottingham, R. Stoothoff & R. Murdoch. Cambridge: Cambridge University Press, 1984.

Descartes, R. *Meditations on First Philosophy*. Translated and edited by J. Cottingham. Cambridge: Cambridge University Press, 2013.

Deutsch, D. *The Fabric of Reality*. London: Allen Lane, 1997.

Di Paolo, E., E. Cuffari & H. De Jaegher. *Linguistic Bodies: The Continuity Between Life and Language*. Cambridge, MA: MIT Press, 2018.

Dick, P. K. "How to Build a Universe That Doesn't Fall Apart Two Days Later". In *The Shifting Realities of Philip K. Dick: Selected Literary and Philosophical Writings*. New York: Vintage 1996.

P. Dirac. "The Physical Interpretation of Quantum Mechanics". *Proceedings of the Royal Society A* 180:980 (1942), 1–40.

Dorato, M. "Why Are (Most) Laws of Nature Mathematical?" In J. Faye, P. Needham & M. Urchs (eds), *Nature's Principles*, 55–75. Cham, CH: Springer, 2005.

Downey, A. "Split Brain Syndrome and Extended Perceptual Consciousness". *Phenomenology and the Cognitive Sciences* 17(4), 787–811.

Dreben, B. "Putnam, Quine – and the Facts". *Philosophical Topics* 20:1 (1992), 293–315.

Dupré, J. & D. Nicholson. "A Manifesto for a Processual Philosophy of Biology". In D. Nicholson & J. Dupré (eds), *Everything Flows: Towards a Processual Philosophy of Biology*, 3–47. Oxford: Oxford University Press, 2018.

Eddington, A. *The Nature of the Physical World* Gifford Lectures of 1927. Introduction to the Reith Lecture) An Annotated Edition H.G. Carraway (2014) Cambridge Scholar.

Edelman, G. & J. Gally. "Re-entry: A Key Mechanism for Integration of Brain Function". *Frontiers in Integrative Neuroscience* 7:63 (2013).

Einstein, A. "Autobiographical Notes". In P. A. Schilpp (ed.), *Einstein: Philosopher Scientist, Volume 1*. Lasalle, IL: Open Court, 1949.

Ellis, G. "Does the Multiverse Really Exist?" *Scientific American* 305:2 (2011).

Ellis, G. & J. Silk. "Scientific Method: Defend the Integrity of Physics". *Nature* 516 (2014), 321–3.

Feldman, J. "The Neural Binding Problem(s)". *Cognitive Neurodynamics* 7:1 (2013), 1–11.

Fernyhough, C. *The Voices Within: The History and Science of How We Talk to Ourselves*. London: Wellcome Collection, 2016.

Feyerabend, P. *The Tyranny of Science*. Cambridge: Polity, 2011.

Feynman, R. *The Character of Physical Laws*. London: BBC Publications, 1965.

Feynman, R. "Negative Probability". In B. Hiley & D. Peat (eds), *Quantum Implications: Essays in Honour of David Bohm*, 235–48. London: Methuen, 1987.

Feynman, R. *The Feynman Lectures on Physics* [1963–65]. Edited by R. Feynman, R. Leighton & M. Sands. https://www.feynmanlectures.caltech.edu/.

Fichte, J. G. *The Science of Knowledge*. Edited and translated by P. Heath & J. Lachs. Cambridge: Cambridge University Press, 2008.

Fleming, S. *et al*. "The Integrated Information Theory of Consciousness as Pseudoscience". Posted online 15 September 2023. https://osf.io/preprints/psyarxiv/zsr78_v1.

Frankish, K. "The Mental Life of Mountains". *New Humanist*, Spring 2020, 38–41.

Frege, G. "The Thought: A Logical Inquiry". Translated by A. M. & M. Quinton. In P. F. Strawson (ed.), *Philosophical Logic*. Oxford: Oxford University Press, 1967.

Frith, C. *Making Up the Mind: How the Brain Creates Our Mental World*. Oxford: Blackwell, 2007.

Frye, P. "When St. Thomas Aquinas Likened his Work to Straw, was that a Retraction of What he Wrote?" Catholic Answers Q&A. https://www.catholic.com/qa/when-st-thomas-aquinas-likened-his-work-to-straw-was-that-a-retraction-of-what-he-wrote.

Fuchs, C. "Notwithstanding Bohr, the Reasons for QBism". arXiv, 11 November 2018. arXiv:1705.03483v2 [quant-ph].

Galileo, G. "The Assayer" [1623]. In S. Drake (edited and translated), *Discoveries and Opinions of Galileo*. New York: Doubleday, 1957.

Gallup, G. "Self-Recognition in Primates: A Comparative Approach to the Bidirectional Properties of Consciousness". *American Psychologist* 32 (1977), 329–38.

Ganiri, J. & I. Shani. "What is Cosmopsychism?" *The Monist* 105: 1 (2022), 1–5.

Garrett, B. "Anscombe on 'I'". *Philosophical Quarterly* 47:189 (1997), 507–11.

Geach. P. *Mental Acts*. London: Routledge & Kegan Paul, 1957.

Gibson, J. J. "The Theory of Affordances". In *The Ecological Approach to Visual Perception*. Boston, MA: Houghton, Mifflin, Harcourt, 1979.

Giovanelli, M. "'But One Must Not Legalize the Mentioned Sin': Phenomenological v Dynamical Treatment of Rods and Clocks in Einstein's Thought". *Studies in History and Philosophy of Modern Physics* 48 (2014), 20–44.

Goff, P. *Galileo's Error: Foundations for a New Science of Consciousness*. New York: Pantheon, 2019.

Goff, P. *Why? The Purpose of the Universe*. Oxford: Oxford University Press, 2023.

Goff, P., W. Seager & S. Allen-Hermanson. "Panpsychism". Stanford Encyclopedia of Philosophy (Summer 2022 edition), Edward N. Zalta (ed.), https://plato.stanford.edu/archives/sum2022/entries/panpsychism/.

Godfrey-Smith, P. *Other Minds: The Octopus and the Evolution of Intelligent Life*. London: Collins, 2016.

Gray, J. *Straw Dogs: Thoughts on Humans and Other Animals*. London: Granta, 2002.

Grice, H. P. "Personal Identity". In J. Perry (ed.), *Personal Identity*, 73–95. Berkeley, CA: University of California Press, 2008.

Grosseteste, R. *On Light or the Beginning of Forms*. Translated by C. Reidl. Milwaukee, WI: Marquette University Press, 1942.

Guillot, M. "I, Me, Mine: On a Confusion Concerning the Subjective Aspect of Experience". *Review of Philosophy and Psychology* 8 (2017), 23–53.

Hacking, I. "What Mathematics Has Done to Some and Only Some Philosophers". *Proceedings of the British Academy* 103 (2000), 83–138.

Halák, J. "Merleau-Ponty on Embodied Subjectivity from the Perspective of Subject-Object Circularity". *Acta Universtitatis Carolinae Kinanthropologica* 52:2 (2016), 26–40.

Halák, J. "Embodied Higher Cognition: Insights from Merleau-Ponty's Interpretation of Motor Intentionality". *Phenomenology and Cognitive Sciences* 22 (2023), 369–97.

Hance, J. R. & S. Hossenfelder. "What Does It Take to Solve the Measurement Problem?" arXiv, 28 June 2022. arXiv:2206.10445v2 [quant-ph].

Harman, G. "I Am Also of the Opinion that Materialism Must be Destroyed". *Environment and Planning D: Society and Space* 28 (2010), 772–90.

Harman, G. "The Problem with Metzinger". *Cosmos and History* 7:1 (2011), 7–36.

Hasker, W. *The Emergent Self*. Ithaca, NY: Cornell University Press, 1999.

Hawking, S. *A Brief History of Time: From the Big Bang to Black Holes*. London: Bantam, 1988.

Healey, R. "Can Physics Coherently Deny the Reality of Time?" *Royal Institute of Philosophy Supplements* 50 (2002), 293–316.

Healey, R. "Quantum-Bayesian and Pragmatist Views of Quantum Theory". Stanford Encyclopedia of Philosophy (Winter 2023 edition), Edward N. Zalta & Uri Nodelman (eds.), https://plato.stanford.edu/archives/win2023/entries/quantum-bayesian/.

Heidegger, M. "The Way Back into the Ground of Metaphysics" [1949]. In W. F. Kauffman (ed.), *Existentialism from Dostoyevsky to Sartre*. New York: Methuen, 1958.

Heidegger, M. *Being and Time*. Translated by J. Stambaugh. New York: SUNY Press, 1996.

Heil, J. "Privileged Access". *Mind* 97:386 (1988), 238–51.

Heisenberg, W. *Physics and Philosophy: The Revolution in Modern Science*. London: Allen & Unwin, 1959.

Hiley, B. "Basil Hiley". In P. Davies & J. Brown (eds), *The Ghost in the Atom*, 135–48. Cambridge: Cambridge University Press, 1993.

Hiley, B. & P. Pilkkanen. "Can Quantum Mechanics Solve the Hard Problem of Consciousness?" In S. Gao (ed.), *Consciousness and Quantum Mechanics*. Oxford: Oxford University Press, 2022.

Hintikka, J. "*Cogito Ergo Sum*: Inference or Performative? *Philosophical Review* 71:1 (1962), 3–32.

Hodes, G. P. "The Causes of the Hard Problem: A Note". *NeuroQuantology* 16:9 (2018), 46–9.

Hoffman, D. *The Case Against Reality: How Evolution Hid the Truth From Our Eyes*. London: Penguin, 2020.

Hoffman, D. "Perception as a fantasy" [video]. IAI Player 2024. /video/perception-as-a-fantasy-donald-hoffman.

Hohwy, J. "The Self-Evidencing Brain". *Nous* 50:2 (2016), 259–85.

D'Hooft, G. "Confusions Regarding Quantum Mechanics". Letter to the editor in response to "The Yang-Mills Model". *Inference* 5:3 (2020).

Hopwood, N. "Not Birth, Marriage or Death, But Gastrulation: The Life of a Quotation in Biology". *British Journal for the History of Science* 55:1 (2023).

Horgan, J. "Why Information Cannot be the Basis of Reality". *Scientific American*, blog posting, 7 March 2011.

Horgan, J. "Quantum Mechanics Gives us Power but No Answers". IAI News, 13 April 2023.

Horgan, T. "Functionalism, Qualia, and the Inverted Spectrum". *Philosophy and Phenomenological Research* 44:4 (1984), 453–69.

Hossenfelder, S. 'The Quantum Measurement Problem" Video, 22 October 2019.

Hossenfelder, S. *Lost in Math: How Beauty Leads Physics Astray*. New York: Basic Books, 2018.

Hossenfelder, S. *Existential Physics: A Scientist's Guide to Life's Biggest Questions*. London: Atlantic, 2022.

Hossenfelder, S. "The Problem with Quantum Measurement". YouTube [video]. https://www.youtube.com/watch?v=Be3HlA_9968.

Hume, D. *A Treatise of Human Nature*, Volume 1. Edited by D. Norton & M. Norton. Oxford: Clarendon Press, 2001.

Ichikawa, J. J. & M. Steup. "The Analysis of Knowledge". Stanford Encyclopedia of Philosophy (Fall 2024 edition), Edward N. Zalta & Uri Nodelman (eds), https://plato.stanford.edu/archives/fall2024/entries/knowledge-analysis/.

Inwood, M. "Does the Nothing Noth?" *Royal Institute of Philosophy Supplement* 44 (2010), 271–90.

Isaacson, W. *Einstein: His Life and Universe*. New York: Simon & Schuster, 2007.

Ismael, J. "Why Physics Should Care about the Mind, and How to Think about it Without Worrying about the Mind-Body Problem". In S. Gao (ed.), *Quantum Mechanics and Consciousness*. Oxford: Oxford University Press, 2022.

Jaeger, G. "Are Virtual Particles Less Real?" *Entropy* 21 (2019), 141–58.

James, W. *The Principles of Psychology* Volume 1. New York: Henry Holt, 1890.

James, W. "The Pragmatic Method". In *Essays in Philosophy, 1876–1910*. Cambridge, MA: Harvard University Press, 1978.

Jensen, R. T. & D. Moran. "Editors' Introduction". In *The Phenomenology of Embodied Subjectivity*. Cham, CH: Springer, 2013.

Jonas, H. *The Phenomenon of Life: Towards a Philosophical Biology*. New York: Harper & Row, 1966.

Jones, A. Z. "The Many Worlds Interpretation of Quantum Physics". Thought.Co, 31 January 2018. https://www.thoughtco.com/many-worlds-interpretation-of-quantum-physics-2699358.

Kadic, N. "Monadic Panpsychism". *Synthese* 203:2 (2024), 1–18.

Kane, R. *The Significance of Free Will*. Oxford: Oxford University Press, 1998.

Kant, I. *Critique of Pure Reason*. Translated by N. Kemp Smith. London: Macmillan, 1933.

Kant, I. *Prolegomena to Any Future Metaphysics That Will be Able to Come Forward as Science*. Translated by G. Hatfield. In *Immanuel Kant: Theoretical Philosophy after 1781, Volume 4* edited by H. Allison & P. Heath. Cambridge: Cambridge University Press, 2002.

Kant. I. *Anthropology from a Pragmatic Point of View*. Translated by R. B. Louden. Cambridge: Cambridge University Press, 2014.

Kastrup, B. "Consciousness cannot have evolved". *IAI News*, 5 February 2020.

Kirby, S. "Here's what Neils Bohr says about 'real'". *Unpopular Opinions*, 16 December 2022.

Klein, S., B. Nguyen, B. Zhang. "Going Out of My Head: An Evolutionary Proposal Concerning the 'Why' of Sentience". *Psychology of Consciousness: Theory, Research, and Practice* 12:1 (2025), 131141.

Korzybski, A. *Science and Sanity: An Introduction to Non-Aristotelian Systems and General Semantics*. Brooklyn, NY: Institute of General Semantics, 1933.

Krauss, L. *A Universe from Nothing: Why there is Something Rather than Nothing*. New York: Free Press, 2012.

Kremnizer, K. & A. Ranchin. "Integrated Information-Induced Quantum Collapse". *Foundations of Physics* 45:8 (May 2015), 889–99.

Kripke, S. *Naming and Necessity*. Oxford: Blackwell, 1980.

Kuhn, R. L. "A Landscape of Consciousness: Towards a Taxonomy of Explanations and Implications". *Progress in Biophysics and Molecular Biology* 190 (2024), 28–129.

Kumar, M. *Quantum: Einstein, Bohr and the Great Debate about the Nature of Reality*. London: Icon, 2009.

Ladyman, J. & D. Ross. *Every Thing Must Go: Metaphysics Naturalized*. Oxford: Oxford University Press, 2007.

Laing, R. D. *The Politics of Experience*. London: Routledge & Kegan Paul, 1967.

Langton, R. & D. Lewis. "Defining 'Intrinsic'". *Philosophy and Phenomenological Research* 58:2 (1998), 333–45.

Lenharo, M. "Consciousness theory slammed as 'pseudoscience' – sparking uproar". *Nature News*, 20 September 2023.

Lenharo, M. "Decades-long bet on consciousness ends – and it's philosophy 1, neuroscience 0". *Nature* 619 (2023), 14–15.

Levin, J. "Functionalism". *Stanford Encyclopedia of Philosophy* (Summer 2023 edition), Edward N. Zalta & Uri Nodelman (eds), https://plato.stanford.edu/archives/sum2023/entries/functionalism/.

Levine, J. "Materialism and Qualia: The Explanatory Gap". *Pacific Philosophical Quarterly* 64 (1983), 354–61.

Levine, J. "Anti-Materialist Arguments and Influential Replies". In S. Schneider & M. Velmans (eds), *Blackwell Companion to Consciousness*. Second edition. Oxford: Wiley-Blackwell, 2017.

Lewis, D. "Extrinsic Properties". *Philosophical Studies* 44 (1983), 197–200.

Lewis, D. *Philosophical Papers Volume 2*. New York: Oxford University Press.

Lewis, P. "Against 'Experience'". In S. Gao (ed.), *Consciousness and Quantum Mechanics*, 140–55. Oxford: Oxford University Press, 2022.

Lewton, T. "How can we understand quantum reality if it is impossible to measure?" *New Scientist*, 10 January 2023.

Locke, J. *An Essay Concerning Human Understanding*. Edited by Peter Nidditch. Oxford: Oxford University Press, 1975.

Lockwood, M. *Mind, Brain, and the Quantum: The Compound "I"* . Oxford: Blackwell, 1986.

Macdonald, F. "No Big Bang. Quantum equation predicts that the universe has no beginning". *Science* alert, 10 February 2015.

Mach, E. *Knowledge and Error: Sketches on the Psychology of Enquiry*. Translated by T. J. McCormack & P. Foulkes. Dordrecht: Reidel, 1976.

Magee, B. *Confessions of a Philosopher: A Personal Journey through Philosophy from Plato to Popper*. London: Weidenfeld & Nicolson, 1997.

Macleod, C. "John Stuart Mill". *Stanford Encyclopedia of Philosophy* (Summer 2020 edition), Edward N. Zalta (ed.), https://plato.stanford.edu/archives/sum2020/entries/mill/.

Maoz, U. *et al*. "Neural Precursors of Decisions that Matter: An ERP Study of Deliberate and Arbitrary Choice". *eLife* Research Communication 2019.

Marcel, G. *The Mystery of Being* I .Chicago, IL: Henry Regnery Company, 1960.

Maudlin, T. "Three Measurement Problems". *Topoi* 14 (1995), 7–15.

Maudlin, T. *The Philosophy of Physics: Space and Time*. Princeton, NJ: Princeton University Press, 2012.

Maudlin, T. "The Nature of the Quantum State". In A. Ney & D. Albert (eds), *The Wave Function: Essays on the Metaphysics of Quantum Mechanics*, 126–53. Oxford: Oxford University Press, 2013.

Meillassoux, Q. *After Finitude: An Essay on the Necessity of Contingency*. Translated by R. Brassier. London: Bloomsbury Continuum, 2010.

Mendelovici, A. & D. Bourget. "Naturalizing Intentionality: Tracking Theories versus Phenomenal Intentionality Theories". *Philosophy Compass* 9 (2014), 325–37.

Merleau-Ponty, M. *The Phenomenology of Perception*. Translated by C. Smith. London: Routledge & Kegan Paul, 1963.

Mermin, D. "Could Feynman Have Said This?" *Physics Today* 57:5 (2004), 10–11.

Mermin, D. "The Quantum Measurement Problem". *Physics Today* 75:11 (2022), 13–14.

Mermin, D. "There is No Quantum Measurement Problem". *Physics Today* 75:6 (2022), 62–3.

Mermin, D. "Making better sense of quantum mechanics" *Reports on Progress in Physics* 82:1 (2019), 012002.

Metzinger, T. *Being No-One: The Self-Model Theory of Subjectivity*. Cambridge, MA: MIT Press, 2003.

Metzinger, T. "The Ego Tunnel". TEDxRheinMain 2, February 2011. https://www.youtube.com/watch?v=ZFjY1fAcESs.

Michaelson, E. "Reference". Stanford Encyclopedia of Philosophy (Fall 2024 edition), Edward N. Zalta & Uri Nodelman (eds), https://plato.stanford.edu/archives/fall2024/entries/reference.

Mill, J. S. "Nature". In H. Aiken (ed.), *The Age of Ideology: The Nineteenth-Century Philosophers.* New York: Mentor, 1956.

MIT Technology Review. "The Foundation of Reality: Information or Quantum Mechanics?" *MIT Technology Review*, 18 May 2009.

MIT Technology Review. "A quantum experiment reveals there is no such thing as external reality". *MIT Technology Review*, 25 January 2023.

Monti, M. *et al.* "Willful Modulation of Brain Activity in Disorders of Consciousness". *New England Journal of Medicine* 362 (2010), 579–89.

Moore, A. W. "Immanuel Kant's *Prolegomena to any Future Metaphysics Able to Come Forward as Science*". *Topoi* 33 (2014), 277–83.

Moran, D. "Revisiting Sartre's Ontology of Embodiment in *Being and Nothingness*". In V. Petrov (ed.) *Ontological Landscapes: Recent Thought on Conceptual Interfaces between Science and Philosophy*, 263–93. Frankfurt: Ontos-Verlag, 2010.

Morch, H. H. "Is the Integrated Information Theory of Consciousness Compatible with Russellian Panpsychism?" *Erkenntnis* 84:5 (2018), 1065–85.

Morris, K. (ed.). *Sartre on the Body* Basingstoke: Palgrave Macmillan, 2010.

Musser, G. *Putting Ourselves Back in the Equations: Why Physicists are Studying Consciousness and AI to Unravel the Mysteries of the Universe.* London: Oneworld, 2023.

Nagel, T. *The View from Nowhere.* Oxford: Oxford University Press, 1986.

Nietzsche, F. *On the Genealogy of Morals.* London: Penguin Classics, 2013.

Oberst, M. "Two Worlds *and* Two Aspects: On Kant's Distinction between Things in Themselves and Appearances". *Kantian Review* 20:1 (2015), 53–75.

O'Callaghan, J. "The Wonder Particle". *New Scientist*, 2 December 2023.

O'Connor, T. "Emergent Properties". Stanford Encyclopedia of Philosophy (Winter 2021 edition), Edward N. Zalta (ed.), https://plato.stanford.edu/archives/win2021/entries/properties-emergent/.

Okon, E. & M. A. Sebastian. "The Subjective-Objective Collapse Model. Virtues and Challenges". *Synthese* 200:399 (2022).

Olson, E. *The Human Animal. Personal Identity without Psychology*. Oxford: Oxford University Press, 1997.

Olson. E. "Review of *What Are We? A Study in Personal Ontology* by Lynne Rudder Baker". *Mind* 117:468 (2008), 1120–22.

Pagnotta, M. "The Use and Abuse of Information in Philosophy". *The New Atlantis: A Journal of Technology and Society* 51 (Winter 2017), 93–109.

Parfit, D. *Reasons and Persons*. Oxford: Clarendon Press, 1986.

Parmenides. *On Nature.* In F. Nietzsche, *Philosophy in the Tragic Age of the Greeks.* Translated by M. Cowan. Chicago, IL: Gateway, 1962.

Pas, H. "Reality Reconstructed". *New Scientist*, 8 July 2023.

Patocka, J. *Body, Community, Language, World.* Translated by E. Kohák. Chicago, IL: Open Court, 1998.

Penfield, W. & P. Perot. "The Brain's Record of Auditory and Visual Experience". *Brain* 86:4 (1963), 595–696.

Pereira Jr, A. "The Projective Theory of Consciousness: From Neuroscience to Philosophical Psychology". *Trans/Form/Ação* 41 (2018), 199–232.

Persson, I. "Self-Doubt: Why We Are Not Identical to Things of Any Kind". *Ratio* (New Series) XVII (2004).

Pike, N. "Hume's Bundle Theory of the Self: A Limited Defence". *American Philosophical Quarterly* 4:2 (1967), 159–65.

Pinto, Y. et al. "Split Brain: Divided Perception But Undivided Consciousness". *Brain* 140 (2017), 1231–7.

Plato. *The Republic*. Perseus Digital Library. Plato in 12 Volumes. https://www.perseus.tufts.edu/hopper/text?doc=Perseus:text:1999.01.0168.

Podovic-Callaghan, K. "Quantum 'super behaviour' could create energy seemingly from nothing". *New Scientist*, 29 June 2024.

Powell, C. S. "Relativity versus quantum mechanics: the battle for the universe". *The Guardian*, 4 November 2015.

Price, H. & K. Wharton. "Untangling Entanglement". *Aeon*, 29 June 2023.

Proietti, M. *et al.* "Experimental Test of Local Observer-Independence". arXiv, 4 November 2019. https://doi.org/10.48550/arXiv.1902.05080.

Putnam, H. *Philosophical Papers Volume 1: Mathematics, Matter, and Method*. Cambridge: Cambridge University Press, 1975.

Quine, W. V. "Two Dogmas of Empiricism". *Philosophical Review* 60 (1951), 20–43.

Quine, W. V. *Word and Object*. Cambridge, MA: MIT Press, 1960.

Quine, W. V. *From Stimulus to Science*. Cambridge, MA: Harvard University Press, 1995.

Ramsay, F. "The Nature of Truth". In *F. P. Ramsay: Philosophical Papers* edited by D. H. Mellor. Cambridge: Cambridge University Press, 1990.

Reicher, M. "Nonexistent Objects". Stanford Encyclopedia of Philosophy (Winter 2022 edition), Edward N. Zalta & Uri Nodelman (eds), https://plato.stanford.edu/archives/win2022/entries/nonexistent-objects/.

Reid, T. *Essay on the Intellectual Powers of Man*. Abridged. Cambridge, MA: John Bartlett, 1850.

Robinson, D. "The Limits of Information". *The New Atlantis*, 12 December 2013.

Rostowski, A. "Worlds Apart: The Kantian Roots of the 'Representation Wars'". *Kant Çalışmaları Journal* 2 (September 2023).

Rostowski, A. *Human Affairs* 32:4 (2022), 427–39.

Rovelli, C. *Helgoland: The Strange and Beautiful Story of Quantum Physics*. London: Allen Lane, 2021.

Rovelli, C. "Consciousness is Irrelevant to Quantum Mechanics". Institute of Art and Ideas, 18 July 2022.

Rundell, K. *Super-Infinite: The Transformations of John Donne*. London: Faber, 2022.

Russell, B. *The Analysis of Matter*. London: Kegan Paul, 1927.

Russell, B. *My Philosophical Development*. London: Allen & Unwin, 1959.

Ryle, G. *The Concept of Mind*. London: Hutchinson, 1949.

Ryle, G. *Dilemmas*. Cambridge: Cambridge University Press, 1954.

Sapolsky, R. *Determined: Life Without Free Will*. London: Bodley Head, 2023.

Saran, K. "Recovering Hope in an Age of Despair: A Critique of the Metaphysics of Modernity". Unpublished thesis submitted for Master of Arts, Trinity Western University, January 2024.

Sartre, J. P. *Being and Nothingness: An Essay on Phenomenal Ontology*. Translated by H. Barnes. New York: Washington Square Press, 1984.

de Saussure, F. *Course in General Linguistics*. Translated by W. Baskin. London: Fontana, 1974.

Schacter, D., D. Addis & R. Buckner. "Remembering the Past to Imagine the Future: The Prospective Brain". *Nature Reviews: Neuroscience* 8 (Sep 2007), 657–61.

Scheck, E. "The Truths in Physics are Dependent on Falsehoods". Institute of Art and Ideas, 14 March 2023.

Scheck. E. *Idealizations in Physics: Elements in the Philosophy of Physics*. Cambridge: Cambridge University Press, 2023.

Schopenhauer, A. *The World as Will and Representation*. New York: Dover Publications, 1969.

Scott, D. "Disarming Causation in the Service of Agency: Tallis on Hume". *Human Affairs* 32:4 (2022), 373–88.

Seager, W. "Panpsychist Infusion". In G. Bruntrupp & L. Jashalka (eds), *Panpsychism: Contemporary Readings*. Oxford: Oxford University Press, 2016.

Seager, W. "Strange Trails: Science to Metaphysics". In S. Gao (ed.), *Consciousness and Quantum Mechanics*. Oxford: Oxford University Press, 2022.

Searle, J. *The Construction of Social Reality*. New York: Free Press, 1995.

Searle, J. *The Mystery of Consciousness*. London: Granta, 1997.

Sellars, W. "Empiricism and the Philosophy of Mind". In H. Feigl & M. Scriven (eds), *Minnesota Studies in the Philosophy of Science*, vol. 1, 253–329. Minneapolis, MN: University of Minnesota Press, 1956.

Seth, A. *Being You: A New Science of Consciousness*. London: Faber, 2022.

Seth, A. & T. Bayne. "Theories of Consciousness". *Nature Reviews: Neuroscience*, 23 July 2022, 439–52.

Shand, J. "Consciousness: Removing the Hardness and Solving the Problem". *Revista Portugesa de Filosofia* 77:4 (2021), 1279–96.

Shapin, S. *The Scientific Revolution*. Chicago, IL: University of Chicago Press, 1996.

Shapiro, L. & S. Spaulding. "Embodied Cognition". Stanford Encyclopedia of Philosophy (Fall 2024 edition), Edward N. Zalta & Uri Nodelman (eds), https://plato.stanford.edu/archives/fall2024/entries/embodied-cognition/.

Shrader-Frechette, K. "Atomism in Crisis: An Analysis of the Current High Energy Paradigm". *Philosophy of Science* 44 (1977), 409–40.

Simoes, A. "Dirac's claim and the chemists". *Physics Perspectives* 4 (2002), 253–66.

Sinnott, E. "Review of Hans Jonas' *The Phenomenon of Life: Towards a Philosophical Biology*". *Human Biology* 38:4 (Dec 1966), 445–9.

Sloman, A. & R. Chrisley. "Virtual Machines and Consciousness". *Journal of Consciousness Studies* 10:4/5 (2003), 133–72.

Smausz, R., J. Neil & J. Grigg. "Neural Mechanisms Underlying Psilocybin's Therapeutic Potential: The Need for Preclinical In Vivo Electrophysiology". *J. Psychopharmacology* 36:7 (2022), 781–93.

Soldati, G. "Representation, Intentionality, and Consciousness". In A. Holderegger, R. Gunter & S.-L. Beat (eds), *Hinsforchung und Menschenbild*. Basel: Schwarz, 2008.

Sparkes, M. "Coin flips don't truly have a 50/50 chance of being head or tails". *New Scientist*, 17 October 2023.

Stebbing, L. S. *Philosophy and the Physicists*. Harmondsworth: Penguin, 1944.

Stephenson, R. H. "On the Widespread Use of an Inappropriate Model of the Literary Aphorism". *Modern Language Review* 75:1 (1980), 1–17.

Steward, H. *A Metaphysics for Freedom*. Oxford: Oxford University Press, 2012.

Strawson, G. "Hume on Personal Identity". In P. Russell (ed.), *The Oxford Handbook of Hume*. Oxford: Oxford University Press, 2014.

Strawson, G. *The Subject of Experience*. Oxford: Oxford University Press, 2017.

Strawson, P. F. *Individuals: An Essay in Descriptive Metaphysics*. London: Methuen, 1959.

Struyve, W. 2018. "Review of *Quantum Nonlocality and Reality: 50 Years of Bell's Theorem*". Notre Dame Philosophical Reviews, 2018.

Suddendorf, T. *The Gap: The Science of What Separates Us from Other Animals*. New York: Basic Books, 2013.

Susskind, L. "The World as a Hologram". *Journal of Mathematical Physics* 36:1 (1995), 6377–96.

Taft, M. "What is the Self? An Interview with Thomas Metzinger". Deconstructing Yourself, 10 September 2017. https://deconstructingyourself.com/what-is-the-self-metzinger.html.

Takho, T. "Fundamentality". Stanford Encyclopedia of Philosophy (Winter 2023 edition), Edward N. Zalta & Uri Nodelman (eds), https://plato.stanford.edu/archives/win2023/entries/fundamentality/.

Takho, T. "Reality has no ultimate building blocks. There may be no fundamental level of reality". Institute of Art and Ideas News, 25 April 2023.

Tallis, R. *Not Saussure: A Critique of Post-Saussurean Literary*. Basingstoke: Macmillan Press, 1988; second edition 1995.

Tallis, R. *The Explicit Animal: A Defence of Human Consciousness*. Basingstoke: Macmillan Press, 1990, 1999.

Tallis, R. *Enemies of Hope: A Critique of Contemporary Pessimism*. Basingstoke: Macmillan Press, 1997, 1999.

Tallis, R. *In Defence of Realism*. Second edition. London: Edward Arnold, 1998.

Tallis, R. *Theorrhoea and After*. Basingstoke: Macmillan Press, 1999.

Tallis, R. *The Hand: A Philosophical Inquiry into Human Being*. Edinburgh: Edinburgh University Press, 2003.

Tallis, R. *I Am: A Philosophical Inquiry into First-Person Being*. Edinburgh: Edinburgh University Press, 2004.

Tallis, R. *The Enduring Significance of Parmenides: Unthinkable Thought*. London: Bloomsbury Continuum, 2007.

Tallis, R. *The Kingdom of Infinite Space*. London: Atlantic, 2008.

Tallis, R. "Zhuangzi and that Bloody Butterfly". *Philosophy Now* 76 (Nov/Dec 2009), 46–7.

Tallis, R. *Michelangelo's Finger: An Exploration of Everyday Transcendence*. London: Atlantic, 2011.

Tallis, R. *Aping Mankind: Neuromania, Darwinitis, and the Misrepresentation of Humanity*. Durham: Acumen, 2011.

Tallis, R. "The 'p' Word". *Philosophy Now* (April/May 2015), 52–3.

Tallis, R. *Of Time and Lamentation: Reflections on Transience*. Newcastle upon Tyne: Agenda, 2017.

Tallis, R. *Logos: The Mystery of How We Make Sense of the World*. Newcastle upon Tyne: Agenda, 2018.

Tallis, R. *Seeing Ourselves: Reclaiming Humanity from God and Science*. Newcastle upon Tyne: Agenda, 2020.

Tallis, R. "Against Neural Philosophy of Mind". *Philosophy Now* (April/May 2020), 51–2.

Tallis, R. "Arguing with a Solipsist". Philosophy Now 141 (Dec 2020/Jan 2021), 58–9.

Tallis, R. *Freedom: An Impossible Reality*. Newcastle upon Tyne: Agenda, 2021.

Tallis, R. "The Fantasy of Conscious Machines". *Philosophy Now* 152 (Oct/Nov 2022), 58–9.

Tallis, R. "An Invitation to Navel Gazing". *Philosophy Now* 153 (Dec 2022/Jan 2023), 60–61.

Tallis, R. "Remembering Memory". *Philosophy Now* 156 (Jun/July 2023).

Tallis, R. "An Itsy-Bitsy Universe?" *Philosophy Now* 158 (Oct/Nov 2023), 62–3.

Tallis, R. *Prague 22: A Philosopher on a Tram*. London: Philosophy Now Publications, 2024.

Tallis, R. "Cogito, Ergo Sum?". *Philosophy Now* 160 (Feb/Mar 2024), 60–61.

Tallis, R. "A Long Postponed Meeting with M. Descartes". Philosophy Now 160 (Feb/Mar 2024).

Tallis, R. "The Illusion of Illusionism". *Philosophy Now* 161 (Apr/May 2024), 58–9.

Tallis, R. "Accidental Thoughts on Luck by an Accidental Man". *Philosophy Now* 165 (Dec 2024/Jan 2025), 64–5.

Tang, J. *et al*. "Semantic Reconstruction of Continuous Language from Non-Invasive Brain Recordings". *Nature Neuroscience* 26 (2023), 858–66.

Alfred Tarski "The Semantic Conception of Truth and the Foundation of Semantics" *Philosophy and Phenomenological Research* 1944 4:341-395.

Tartaglia, J. & S. Leach (eds). A Book Symposium on Raymond Tallis's *Freedom: An Impossible Reality. Human Affairs* 32:4 (2022).

Taylor, C. W. W. (ed.), *The Atomists: Leucippus and Democritus Fragments*. A Translation with Text and Commentary. Toronto, ON: University of Toronto Press, 1999.

Tegmark, M. "Is 'the Theory of Everything' Merely the Ultimate Ensemble Theory?" *Annals of Physics* 270:1 (1998), 1–51.

Tegmark, M. & J. A. Wheeler. "100 years of the Quantum". arXiv. quant:ph/0101077v1.

Thagard, P. "When did consciousness begin? Consciousness may have originated with humans, mammals, fish, or bacteria". *Psychology Today*, 11 January 2019.

Thalos, M. *Without Hierarchy: The Scale Freedom of the Universe*. Oxford: Oxford University Press, 2013.

Thomson, W. "Electrical Units of Measurement". In *Popular Lectures and Addresses* Vol. 1, 73–136. London: Macmillan, 1889.

Tonner, P. "Are Animals Poor in the World? A Critique of Heidegger's Anthropocentrism". In R. Boddice (ed.), *Anthropocentrism: An Investigation into the History of an Idea*, 203–21. Leiden: Brill, 2011.

Tononi, G. "Consciousness as Integrated Information: A Provisional Manifesto". *Biological Bulletin* 215 (2008), 216–42.

Tononi, G. & C. Koch. "Consciousness: Here, There, and Everywhere?" *Philosophical Transactions of the Royal Society: Biological Sciences* 370:1668 (2015), 117–34.

Tononi, G., M. Boly & C. Koch. "Integrated Information Theory: From Consciousness to its Physical Substrate". *Nature* 17 (July 2016), 450–61.

Van Gulick, R. "Consciousness". *Stanford Encyclopedia of Philosophy* (Winter 2022 edition), Edward N. Zalta & Uri Nodelman (eds), https://plato.stanford.edu/archives/win2022/entries/consciousness/.

van Inwagen, P. *An Essay on Free Will*. Oxford: Oxford University Press, 1983.

van Inwagen, P. *Material Beings*. Ithaca, NY: Cornell University Press, 1990.

Varela, F. & H. Maturana. *Autopoiesis and Cognition: The Realization of the Living*. Dordrecht: D. Reidel, 1980.

Veervoort, L. "A detailed interpretation of probability, and its link with quantum mechanics". arXiv, 29 November 2010. https://doi.org/10.48550/arXiv.1011.6331.

Vilenkin, A. & M. Tegmark. "The Case for Parallel Universes". *Scientific American*, 19 July 2011.

Vincent, J. *Beyond Measure: The Hidden History of Measurement*. London: Faber, 2022.

Wasserman, R. "Material Constitution". Stanford Encyclopedia of Philosophy (Fall 2021 edition), Edward N. Zalta (ed.). https://plato.stanford.edu/archives/fall2021/entries/material-constitution/.

Watkins, E. "Kant, Sellars, and the Myth of the Given". *Philosophical Forum* 45:3 (2012), 311–36.

Weaver, W. "Recent Contributions to the Mathematical Theory of Communication". In C. Shannon & W. Weaver (eds), *A Mathematical Model of Communication*. Urbana, IL: University of Illinois Press, 1949.

Weingard, R. "Do Virtual Particles Exist?" *Proceedings of the Annual Meeting of the Philosophy of Science Association*. Contributed Papers Volume 1 (1982), 235–42.

Wellcome Centre for Human Neuroimaging. "The Bayesian Brain". https://www.fil.ion.ucl.ac.uk/bayesian-brain/.

Wetzel, L. "Types and Tokens". *Stanford Encyclopedia of Philosophy* (Fall 2018 edition), Edward N. Zalta (ed.). https://plato.stanford.edu/archives/fall2018/entries/types-tokens/.

Wheeler, J. A. "Bohr, Einstein, and the Strange Lesson of the Quantum". In R. Elver (ed.), *Mind in Nature*. Nobel Conference XVII, Gustavus Adolphus College, St Peter, Minnesota, 1–23. San Francisco, CA: Harper & Row, 1982.

Wheeler, J. A. "Law without Law". In J. A. Wheeler & W. Zurek (eds), *Quantum Theory and Measurement*, 182–213. Princeton, NJ: Princeton University Press, 1983.

Wheeler, J. A. "Joh Archibald Wheeler". In P. Davies & J. Brown (eds), *The Ghost in the Atom*. Cambridge: Cambridge University Press, 1993.

Wigner, E. "Remarks on the Mind-Body Question". In I. Good (ed.), *The Scientist Speculates: An Anthology of Partly-Baked Ideas*, 284–302. London: Heineman, 1961.

Wikipedia. "Noether's theorem". Wikipedia. https://en.wikipedia.org/wiki/Noether%27s_theorem.

Wilkins, A. "Mathematicians cannot agree what 'equals' means, and that's a problem". *New Scientist*, 5 June 2024.

Wilkinson, T. "The Multiverse Conundrum". *Philosophy Now* 89 (Mar/Apr 2012).

Wilson, J. *Metaphysical Emergence*. Oxford: Oxford University Press, 2021.

Wittgenstein, L. *Tractatus Logico-Philosophicus*. Translated by D. Pears & B. McGuinness. London: Routledge & Kegan Paul, 1961.

Wittgenstein, L. *Philosophical Investigations*. Translated by G. E. M. Anscombe. Oxford: Blackwell, 1963.

Wittgenstein, L. *On Certainty*. Translated by D. Paul & G. E. M. Anscombe. Oxford: Blackwell, 1974.

Wolchover, N. "A deepening crisis forces physicists to rethink structure of nature"s laws". *Quanta Magazine*, 1 March 2022.

Wright, L. "Rival Explanations". *Mind* 82 (1973), 509–10.

Wu, B. "Everett's Theory of the Universal Wave Function" *The European Physical Journal* Volume 46 (7) 2021.

Wu, W. & J. Morales. "The Neuroscience of Consciousness". Stanford Encyclopedia of Philosophy (Summer 2024 edition), Edward N. Zalta & Uri Nodelman (eds). https://plato.stanford.edu/archives/sum2024/entries/consciousness-neuroscience/.

Zahavi, D. "Consciousness, Self-Consciousness, Selfhood: A Reply to Some Critics". *Review of Philosophy and Psychology* 9 (2018), 703–18.

Zahavi, D. "Brain, Mind, World: Predictive Coding, Neo-Kantianism, and Transcendental Idealism". *Husserl Studies* 34 (2018), 47–61.

Zyga, L. "Quantum-to-classical transition may be explained by fuzziness of measurement references". *Phys.org*, 14 January 2014. https://phys.org/news/2014-01-quantum-to-classical-transition-fuzziness.html.

Index